UNDERSTANDING
THE UNIVERSE
from Quarks to the Cosmos

UNDERSTANDING
THE UNIVERSE
from Quarks to the Cosmos

Don Lincoln

Fermi National Accelerator Laboratory, USA

 World Scientific

NEW JERSEY • LONDON • SINGAPORE • BEIJING • SHANGHAI • HONG KONG • TAIPEI • CHENNAI

Published by

World Scientific Publishing Co. Pte. Ltd.

5 Toh Tuck Link, Singapore 596224

USA office: Suite 202, 1060 Main Street, River Edge, NJ 07661

UK office: 57 Shelton Street, Covent Garden, London WC2H 9HE

Library of Congress Cataloging-in-Publication Data
Lincoln, Don.
　　Understanding the Universe: from quarks to the cosmos / by Don Lincoln.
　　　p. cm.
　　Includes indexes.
　　ISBN 981-238-703-X -- ISBN 981-238-705-6 (pbk)
　　1. Particles (Nuclear physics) -- Popular works. I. Title.

QC793.26.L56 2004
539.7'2--dc22　　　　　　　　　　　　　　　　　　　2004041411

British Library Cataloguing-in-Publication Data
A catalogue record for this book is available from the British Library.

Typeset by Stallion Press

Printed by FuIsland Offset Printing (S) Pte Ltd, Singapore

To

Sharon for giving me life,
Diane for making it worthwhile

&

Tommy, Veronica and David for making it interesting

and to

Marj Corcoran, Robin Tulloch, Charles Gaides and all the others
for directions along the path.

Contents

❖

Foreword

One hot summer day in July of 392 BC, it might have been a Tuesday, the Greek philosopher Democritus of Abdera asserted that everything we see is made of common, fundamental, invisible constituents; things that are so small we don't see them in our everyday experience. Like most great ideas, it wasn't exactly original. Democritus's teacher, Leucippus of Miletus, probably had the same atomistic view of nature. The concept of atomism remained just a *theory* for over two millennia. It wasn't until the 20th century that this exotic idea of "atoms" proved to be correct. The atomistic idea, that there are discernable fundamental building blocks, and understandable rules under which they combine and form everything we see in the universe, is one of the most profound and fertile ideas in science.

The search for the fundamental building blocks of nature did not end with the 20th century discovery of atoms. Atoms are divisible; inside atoms are nuclei and electrons, inside nuclei are neutrons and protons, and inside them are particles known as quarks and gluons. Perhaps quarks are not the ultimate expression of the idea of atomism, and the search for the truly fundamental will continue for another century or so. But they may be! What we do know about quarks and other seemingly fundamental particles provides a remarkably complete picture of how the world works. In fact, not only of how the world works, but of how the entire cosmos works!

The study of nature is traditionally divided into different disciplines: astronomy, biology, chemistry, geology, physics, zoology, etc. But nature itself is a seamless fabric. The great American naturalist John Muir expressed this idea when he said, "When we try to pick out anything by itself, we find it hitched to everything else in the universe." When Don Lincoln and his colleagues at Fermilab in Batavia, Illinois explore the inner space of quarks they are also exploring the outer space of the cosmos. Quarks are hitched to the cosmos. Understanding nature's fundamental particles is part of the grand quest of understanding the universe. Don Lincoln never lets us forget that on this journey from quarks to the cosmos! The spirit of Leucippus of Miletus and Democritus of Abdera is still alive in Don of Batavia.

Don is a physicist at Fermi National Accelerator Laboratory (Fermilab), the home of the Tevatron, the world's most powerful accelerator. Currently Don is a member of one of the two very large colliding beams experiments at Fermilab. Such experiments are dedicated to the study of the nature of fundamental particles when protons and antiprotons collide after being accelerated near the velocity of light. He works at the very frontier of the subject about which he writes.

Don writes with the same passion he has for physics. After years of explaining physics to lay audiences, he knows how to convey the important concepts of modern particle physics to the general public.

There are many books on fundamental particle physics written for the general public. Most do a marvelous job of conveying what we know. Don Lincoln does more than tell us *what* we know; he tells us *how* we know it, and even more importantly, *why* we want to know it!

Understanding the Universe is also a saga of the people involved in the development of the science of particle physics. Don tells the story about how an important experiment was conceived over a lunch of egg rolls at New York's Shanghai Café on January 4th, 1957. He also describes life inside the 500-person collaboration of physicists of his present experiment. Great discoveries are not made by complex detectors, machinery, and computers, but by even more complex people. If you ever wondered what compels scientists to work for years on the world's most complicated experiments, read on!

Rocky Kolb
Chicago, Illinois

❖

Preface (And so ad infinitum)

The most incomprehensible thing about the universe is that it's comprehensible at all …

— Albert Einstein

The study of science is one of the most interesting endeavors ever undertaken by mankind and, in my opinion, physics is the most interesting science. The other sciences each have their fascinating questions, but none are so deeply fundamental. Even the question of the origins of life, one of the great unanswered mysteries, is likely to be answered by research in the field of organic chemistry, using knowledge which is already largely understood. And chemistry, an immense and profitable field of study, is ultimately concerned with endless and complicated combinations of atoms. The details of how atoms combine are rather tricky, but in principle they can be calculated from the well-known ideas of quantum mechanics. While chemists rightfully claim the study of the interactions of atoms as their domain, it was physicists who clarified the nature of atoms themselves. Although the boundaries between different fields of scientific endeavor were somewhat more blurred in earlier eras, physicists first discovered that atoms were not truly elemental, but rather contained smaller particles within them. Also, physicists first showed that the atom could in some ways be treated as a solar system, with tiny electrons orbiting a dense and heavy nucleus. The realization that this simple model could not possibly be the entire story led inexorably to the deeply mysterious realm of quantum mechanics. While the nucleus of the atom was first considered to be fundamental, physicists were surprised to find that the nucleus contained protons and neutrons and, in turn, that protons and neutrons themselves contained even smaller particles called quarks. Thus the question of exactly what constitutes the smallest constituent of matter, a journey that began over 2500 years ago, is still an active field of scientific effort. While it is true that our understanding is far more sophisticated than it was, there are still indications that the story is not complete.

Even within the field of physics, there are different types of efforts. Research into solid state physics and acoustics has solved the simple questions and is now attacking more difficult and complex problems. However, there remain physicists who are interested in the deepest and most fundamental questions possible. There are many questions left, for example: What is the ultimate nature of reality? Are there smallest particles or, as one looks at smaller and smaller size scales, does space itself become quantized and the smallest constituents of matter can be more properly viewed as vibrations of space (the so-called superstring hypothesis)? What forces are needed to understand the world? Are there many forces or few? While particle physicists can hope to study these questions, the approach that they follow requires an ever-increasing concentration of energy into an ever-decreasing volume. This incredible concentration of energy has not been generally present in the universe since the first fractions of a second after the Big Bang. Thus, the study of particle physics provides guidance to another deeply fundamental question, the creation and ultimate fate of the universe itself.

The current state of knowledge cannot yet answer these questions, however progress has been made in these directions. We now know of several particles that have thus far successfully resisted all attempts to find structure within them. The particles called quarks make up the protons and neutrons that, in turn, make up the atom's nucleus. Leptons are not found in the nucleus of the atom, but the most common lepton, the electron, orbits the nucleus at a (relatively) great distance. We know of four forces: gravity, which keeps the heavens in order and is currently (although hopefully not forever) outside the realm of particle physics experimentation; the electromagnetic force, which governs the behavior of electrons around atomic nuclei and forms the basis of all chemistry; the weak force, which keeps the Sun burning and is partly responsible for the Earth's volcanism and plate tectonics; and the strong force, which keeps quarks inside protons and neutrons and even holds the protons and neutrons together to form atomic nuclei. Without any of these forces, the universe would simply not exist in anything like its current form. While we now know of four forces, in the past there were thought to be more. In the late 1600s, Isaac Newton devised the theory of universal gravitation, which explained that the force governing the motion of the heavens and our weight here on Earth were really the same things, something not at all obvious. In the 1860s, James Clerk Maxwell showed that electricity and magnetism, initially thought to be different, were intimately related. In the 1960s, the electromagnetic and weak forces were actually shown to be different facets of a single electro-weak force. This history of unifying seemingly different forces has proven to be very fruitful and naturally we wonder if it is possible that the remaining four (really three) forces could be shown to be different faces of a more fundamental force.

All of creation, i.e. all of the things that you can see when you look about you, from the extremely tiny to the edge of the universe, can be explained as endless combinations of two kinds of quarks, an electron and a neutrino (a particle which we haven't yet discussed). These four particles we call a generation. Modern experiments have shown that there exist at least two additional generations (and probably only two), each containing four similar particles, but with each subsequent generation having a greater mass and with the heavier generation decaying rapidly into the familiar particles of the first generation. Of course, this raises yet even more questions. Why are there generations? More specifically, why are there three generations? Why are the unstable generations heavier, given that otherwise the generations seem nearly identical?

Each of the four forces can be explained as an exchange of a particular kind of particle, one kind for each force. These particles will eventually be discussed in detail, but their names are the photon, the gluon, the W and Z particles and (maybe) the graviton. Each of these particles are bosons, which have a particular type of quantum mechanical behavior. In contrast, the quarks and leptons are fermions, with completely different behavior. Why the force-carrying particles should be bosons, while the matter particles are fermions, is not understood. A theory, called supersymmetry, tries to make the situation more symmetric and postulates additional fermion particles that are related to the bosonic force carriers and other bosonic particles that are related to the mass-carrying fermions. Currently there exists no unambiguous experimental evidence for this idea, but the idea is theoretically so interesting that the search for supersymmetry is a field of intense study.

While many questions remain, the fact is that modern physics can explain (with the assistance of all of the offshoot sciences) most of creation, from the universe to galaxies, stars, planets, people, amoebae, molecules, atoms and finally quarks and leptons. From a size of 10^{-18} meters, through 44 orders of magnitude to the 10^{26} meter size of the visible universe, from objects that are motionless, to ones that are moving 300,000,000 meters per second (186,000 miles per second), from temperatures ranging from absolute zero to 3×10^{15}°C, matter under all of these conditions is pretty well understood. And this, as my Dad would say, impresses the hell out of me.

The fact that particle physics is intimately linked with cosmology is also a deeply fascinating concept and field of study. Recent studies have shown that there may exist in the universe dark matter ... matter which adds to the gravitational behavior of the universe, but is intrinsically invisible. The idea of dark energy is a similar answer to the same question. One way in which particle physics can contribute to this debate is to look for particles which are highly massive, but also stable (i.e. don't decay) and which do not interact very much with ordinary matter (physics-ese for invisible). While it seems a bit of a reach to say that particle physics is related to cosmology, you must recall that *nuclear* physics, which is particle physics' lower-energy cousin, has made critical contributions to the physics of star formation, supernovae, black holes and neutron stars. The fascinating cosmological questions of extra dimensions, black holes, the warping of space and the unfathomably hot conditions of the Big Bang itself are all questions to which particle physics can make important contributions.

The interlinking of the fields of particle physics and cosmology to the interesting questions they address is given in the figure below. The answer to the questions of unification (the deepest nature of reality), hidden dimensions (the structure of space itself) and cosmology (the beginning and end of the universe), will require input from many

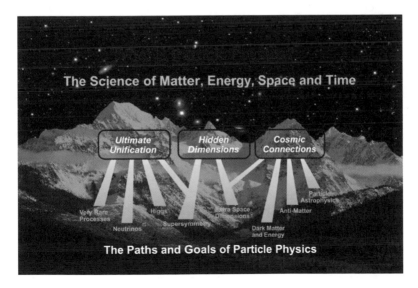

Figure The intricate interconnections between the physics of the very small and the very large. (Figure courtesy of Fermilab.)

fields. The particle physics discussed in this book will only provide a part of the answer; but a crucial part and one richly deserving study.

Naturally, not everyone can be a scientist and devote their lives to understanding all of the physics needed to explain this vast range of knowledge. That would be too large a quest even for professional scientists. However, I have been lucky. For over twenty years, I have been able to study physics in a serious manner and I was a casual student for over ten years before that. While I cannot pretend to know everything, I have finally gained enough knowledge to be able to help push back the frontiers of knowledge just a little bit. As a researcher at Fermi National Accelerator Laboratory (Fermilab), currently the highest energy particle physics laboratory in the world, I have the privilege of working with truly gifted scientists, each of whom is driven by the same goal: to better understand the world at the deepest and most fundamental level. It's all great fun.

About once a month, I am asked to speak with a group of science enthusiasts about the sorts of physics being done by modern particle physics researchers. Each and every time, I find some fraction of the audience who is deeply interested in the same questions that researchers are. While their training is not such that they can contribute directly, they want to *know*. So I talk to them and they understand. Physics really isn't so hard. An interested layman can understand the physics research that my colleagues and I do. They just need to have it explained to them clearly and in a language that is respectful of what they know. They're usually very smart people. They're just not experts.

So that's where this book comes in. There are many books on particle physics, written for the layman. Most of the people with whom I speak have read many of them. They want to know more. There are also books, often written by theoretical physicists, which discuss speculative theories. And while speculation is fun (and frequently is how science is advanced), what we *know* is interesting enough to fill a book by itself. As an experimental physicist, I have attempted to write a book so that, at the end, the reader will have a good grasp on what we know, so that they can read the theoretically speculative books with a more critical eye. I'm not picking on theorists, after all some of my best friends have actually ridden on the same bus as a theorist. (I'm kidding, of course. Most theorists I know are very bright and insightful people.) But I would like to present the material so that not just the ideas and results are explained, but also so that a flavor of the experimental techniques comes through … the "How do you do it?" question is explained.

This book is designed to stand on its own. You don't have to read other books first. In the end you should understand quite a bit of fundamental particle physics and, unlike many books of this sort, have a pretty good idea of how we measure the things that we do and further have a good "speculation" detector. Speculative physics is fun, so towards the end of the book, I will introduce some of the unproven ideas that we are currently investigating. Gordon Kane (a theorist, but a good guy even so) in his own book *The Particle Garden*, coined the phrase "Research in Progress" (RIP) to distinguish between what is known and what isn't known, but is being investigated. I like this phrase and, in the best scientific tradition, will incorporate this good idea into this book.

Another reason that I am writing this book now is that the Fermilab accelerator is just starting again, after an upgrade that took over five years. The primary goal (although by no means the only one) of two experiments, including one on which I have been working for about ten years, is to search for the Higgs boson. This particle has not been observed (RIP!), but if it exists will have something to say about why the various known particles have the masses that they do. While the Higgs particle may not exist, something similar to it must, or our understanding of particle physics is deeply flawed. So we're looking and, because it's so interesting, I devote a chapter to the topic.

This is not a history book; it's a book on *physics*. Nonetheless, the first chapter briefly discusses the long interest that mankind has had in understanding the nature of nature, from the ancient Greeks until the beginning of the 20th century. The second chapter begins with the discovery of the electrons, x-rays and radioactivity (really the beginning of modern particle physics) and proceeds through 1960, detailing the many particle discoveries of the modern physics era. It was in the 1960s that physicists really got a handle on what was going on. Chapter 3 discusses the elementary particles (quarks and leptons) which could neatly explain the hundreds of particles discovered in the preceding sixty years. Chapter 4 discusses the forces, without which the universe would be an uninteresting place. Chapter 5 concentrates on the Higgs boson, which is needed to explain why the various particles discussed in Chapter 3 have such disparate masses and the search for (and hopefully discovery of) will consume the efforts of so many of my immediate colleagues. Chapter 6 concentrates on the experimental techniques needed to make discoveries in modern accelerator-based particle physics experiments. This sort of information is often given at best in a skimpy fashion in these types of books, but my experimentalist's nature won't allow that. In Chapter 7, I outline mysteries that are yielding up their secrets to my colleagues as I write. From neutrino oscillations to the question of why there appears to be more matter than antimatter in the universe are two really interesting nuts that are beginning to crack. Chapter 8 is where I finally indulge my more speculative nature. Modern experiments also look for hints of "new physics" i.e. stuff which we might suspect, but have little reason to expect. Supersymmetry, superstrings, extra dimensions and technicolor are just a few of the wild ideas that theorists have that just might be true. We'll cover many of these ideas here. In Chapter 9, I will spend some time discussing modern cosmology. Cosmology and particle physics are cousin fields and they are trying to address some similar questions. The linkages between the fields are deep and interesting and, by this point in the book, the reader will be ready to tackle these tricky issues. The book ends with several appendices that give really interesting information that is not strictly crucial to understanding particle physics, but which the adventurous reader will appreciate.

The title of this preface comes from a bit of verse by Augustus de Morgan (1806–1871) (who in turn was stealing from Jonathan Swift) from his book *A Budget of Paradoxes*. He was commenting on the recurring patterns one sees as one goes from larger to smaller size scales. On a big enough scale, galaxies can be treated as structure-less, but as one looks at them with a finer scale, one sees that they are made of solar systems, which in turn are made of planets and suns. The pattern of nominally structure-less objects eventually revealing a rich substructure has continued for as long as we have looked.

> Great fleas have little fleas,
> upon their back to bite 'em,
> little fleas have lesser fleas,
> *and so ad infinitum...*

He goes on to even more clearly underscore his point:

> And the great fleas themselves, in turn,
> have greater fleas to go on;
> While these again have greater still,
> And greater still, and so on.

I hope that you have as much fun reading this book as I had writing it. Science is a passion. Indulge it. Always study. Always learn. Always question. To do otherwise is to die a little inside.

Don Lincoln
Fermilab
October 24, 2003

❖

Acknowledgements

In a text of this magnitude, there is always a series of people who have helped. I'd like to thank the following people for reading the manuscript and improving it in so many ways. Diane Lincoln was the first reader and suffered through many an incarnation. Her comments were very useful and she also suggested adding a section that most of the following readers said was the best part of the book.

Linda Allewalt, Bruce Callen, Greg Jacobs, Barry Panas, Jane Pelletier, Marie & Roy Vandermeer, Mike Weber, Connie Wells and Greg Williams all read the manuscript from a "test reader" point of view. Linda especially noted a number of points missing in the original text. These points are now included. Since many of these people are master educators, their suggestions all went a long way towards improving the clarity of the book.

Monika Lynker, Tim Tait, Bogdan Dobrescu, Steve Holmes and Doug Tucker all read the manuscript from an "expert" point of view. All made useful comments on better ways to present the material. Tim was especially helpful in making a number of particularly insightful suggestions.

With their generous help, both the physics and readability of the text have much improved. Any remaining errors or rough edges are solely the responsibility of Fred Titcomb. Actually, Fred doesn't even know this book is being written and I've seen him rarely these past twenty years, but I've known him since kindergarten and routinely blamed him for things when we were kids. While it's true that any remaining errors are my fault, I don't see any reason to stop that tradition now.

I am grateful to Rocky Kolb for contributing a foreword for this book. Rocky is a theoretical cosmologist with a real gift for science communication. His inclusion in this book is in some sense a metaphor for the book's entire premise…the close interplay between the fields of cosmology and particle physics; experimental and theoretical.

In addition, there were several people who were instrumental in helping me acquire the figures or the rights to use the figures. I'd like to thank Jack Mateski, who provided the blueprints for Figure 6.22 and Doug Tucker who made a special version of the Las Campanas data for me. Dan Claes, a colleague of mine on DØ, graciously contributed a number of hand drawn images for several figures. It seems quite unfair that a person could have both considerable scientific and artistic gifts. I'd also like to thank the public affairs and visual media departments at Fermilab, CERN, DESY, Brookhaven National Laboratory and The Institute for Cosmic-Ray Research at the University of Tokyo for their kind permission to use their figures throughout the text. I am also grateful to NASA for granting permission to use the Hubble Deep Space image that forms the basis of the book cover. I should also like to thank the editorial, production and marketing staffs of World Scientific, especially Dr. K.K. Phua, Stanley Liu, Stanford Chong, Aileen Goh, and Kim Tan, for their part in making this book a reality.

Finally, I'd like to mention Cyndi Beck. It's a long story.

chapter 1

Early History

Whatever nature has in store for mankind, unpleasant as it may be, men must accept, for ignorance is never better than knowledge.

— Enrico Fermi

Billions of years ago, in a place far from where you are sitting right now, the universe began. An enormous and incomprehensible explosion scattered the matter that constitutes everything that you have ever seen across the vast distances that make up the universe in which we live. It would not be correct to call the temperatures hellish in that time following the Big Bang…it was far hotter than that. The temperature at that time was so hot that matter, as we generally understand it, could not exist. The swirling maelstrom consisted of pure energy with subatomic particles briefly winking into existence before merging back into the energy sea. On quick inspection, that universe was as different from the one in which we live as one can imagine. Basically, everywhere you looked, the universe was the same. This basic uniformity was only modified by tiny quantum fluctuations that are thought to eventually have seeded the beginnings of galaxy formation.

Fast forward to the present, ten to fifteen billion years after the beginning. In the intervening years, the universe has cooled and stars and galaxies have formed. Some of those stars are surrounded by planets. And on an unremarkable planet, around an unremarkable star, a remarkable thing occurred. Life formed. After billions of years of change, a fairly undistinguished primate evolved. This primate had an upright stance, opposable thumbs and a large and complex brain. And with that brain came a deep and insatiable curiosity about the world. Like other organisms, mankind needed to understand those things that would enhance its survival — things like where there was water and what foods were safe. But, unlike any other organism (as far as we know), mankind was curious for curiosity's sake. Why are things the way they are? What is the meaning of it all? How did we get here?

Early creation beliefs differed from the idea of the Big Bang, which modern science holds to be the best explanation thus far offered. One people held that a giant bird named Nyx laid an egg. When the egg hatched, the top half of the shell became the heavens, while the bottom became the earth. Another people believed that a man of the Sky People pushed his wife out of the sky and she fell to Earth, which was only water at the time. Little Toad swam to the bottom of the ocean and brought up mud that the sea animals smeared on the back of Big Turtle, which became the first land and on which the woman lived. Yet a third group asserted that the universe was created in six days. A common theme of all of these creation ideas is the fact that we as a species have a need to understand the pressing question: "From where did we come?"

While the modern understanding of the origins of the universe fulfills a need similar to that of its predecessors, it is unique in a very important way. It can be tested. It can, in principle, be proven wrong. In carefully controlled experiments, the conditions of the early universe, just fractions of a second after the Big Bang, can be routinely recreated. This book tries to describe the results of those experiments in ways that are accessible to all.

First Musings

The path to this understanding has not been very straight or particularly easy. While much of the understanding of the universe has come from astronomy, the story of that particular journey is one for another time. An important and complementary approach has come from trying to understand the nature of matter. Taken on the face of it, this is an extraordinary task. When you look around, you see a rich and diverse world. You see rocks and plants and people. You see mountains, clouds and rivers. None of these things seem to have much in common, yet early man tried to make sense of it all. While it is impossible to know, I suspect that an important observation for early man was the different aspects of water. As you know, water can exist in three different forms: ice, water and steam. Here was incontrovertible proof that vastly different objects: ice (hard and solid), water (fluid and wet) and steam (gaseous and hot); were all one and the same. The amount of heat introduced to water could drastically change the material's properties and this was a crucial observation (and probably the most important idea to keep in your mind as you read this book). Seemingly dissimilar things can be the same. This is a theme to which we will often return.

The observation that a particular material can take many forms leads naturally to what is the nature, the very essence, of matter. The ancient Greeks were very interested in the nature of reality and offered many thoughts on the subject. While they preferred the use of pure reason to our more modern experimental approach, this did not mean that they were blind. Like Buddha, they noticed that the world is in constant flux and that change seems to be the normal state of things. Snow comes and melts, the Sun rises and sets, babies are born loud and wet and old people die and fade into dry dust. Nothing seems to be permanent. While Buddha took this observation in one direction and asserted that nothing physical is real, the early Greeks believed that there must be *something* that is permanent (after all, they reasoned, we always see *something*). The question that they wanted answered was "What is permanent and unchanging among this apparent turmoil and chaos?"

One train of thought was the idea of opposing extremes. The thing that was real was the essence of opposites: pure hot and cold, wet and dry, male and female. Water was mostly wet, while ice had a much higher dry component. Different philosophers chose different things as the "true" opposing extremes, but many believed in the basic, underlying concept. Empedocles took the idea and modified it somewhat. He believed that the things we observe could be made from a suitable mix of four elements: *air, fire, water* and *earth*. His elements were pure; what we see is a mix, for instance, the fire that we observe is a mixture of *fire* and *air*. Steam is a mix of *fire, water* and *air*. This theory, while elegant, is wrong, although it did influence scientific thinking for thousands of years. Empedocles also realized that force was needed to mix these various elements. After some thought, he suggested that the universe could be explained by his four elements and the opposing forces of *harmony* and *conflict* (or *love* and *strife*). Compare the clouds on a beautiful summer day to a violent thunderstorm and you see *air* and *water* mixing under two extremes of his opposing forces.

Another early philosopher, Parmenides, was also an esoteric thinker. He did not worry as much about *what* were the fundamental elements, but more on the nature of their permanence. He believed that things could not be destroyed, which was in direct conflict with observation. Things do change; water evaporates (maybe disappears or is destroyed), vegetables rot, etc. However, he might have offered in counterpoint a wall surrounding an enemy citadel. After the city is captured and the wall pulled down by the conquerors, the wall, while destroyed, still exists in the form of a pile of rubble. The essence of the wall was the stones that went into it. The wall and rubble were just two forms of rock piles.

This prescient insight set the stage for the work of Democritus, who is traditionally mentioned in these sorts of books as the first to offer something resembling a modern theory of matter. Democritus was born circa 400 B.C., in Abdera in Thrace. He too was interested in determining the unchanging structure of matter. One day during a prolonged fast, someone walked by Democritus with a loaf of bread. Long before he saw the bread, he knew it was there from the smell. He was struck by that fact and wondered how this could work (apparently fasting made him dizzy too). He decided that some small bread particles had to travel through the air to his nose. As he couldn't see the bread particles, they had to be very small (or invisible). This thought led him to wonder about the nature of these small particles. To further his thinking, he considered a piece of cheese (he seemed to have a thing with food, perhaps because of all of those fasts). Suppose you had a sharp knife and continuously cut a piece of cheese. Eventually you would come to the smallest piece of cheese possible, which the knife could no longer cut. This smallest piece he called atomos (for uncuttable), which we have changed into the modern word "atom."

If atoms exist, then one is naturally led to trying to understand more about them. Are all atoms the same? If not, how many kinds are there and what are their properties? Since he saw that different materials had different properties, he reasoned that there had to be different types of atoms. Something like oil might contain smooth atoms. Something like lemon juice, which is tart on the tongue and hurts when it gets into a cut, would contain spiny atoms. Metal, which is very stiff, might contain atoms reminiscent of Velcro, with little hooks and loops that connected adjacent atoms together. And so on.

The concept of atoms raised another issue. It concerned the question of what is between the atoms. Earlier, some philosophers had asserted that matter always touched matter. They used as an example the fish. Fish swim through water. As they propel themselves forward, the water parts in front of them and closes behind them. Never is there a void that contains neither water nor fish. Thus, matter is always in contact with matter.

The idea of atoms somehow belies this assertion. If there exists a smallest constituent of matter, this implies that it is somehow separate from its neighbor. The stuff that separates the atoms can be one of two things. It can be matter, but a special kind of matter, just used for separating other matter. But since matter is composed of atoms, then this material must also contain atoms and the question arises of just what separates them. So this hypothesis doesn't really solve anything. An alternative hypothesis is that the atoms are separated by empty space, not filled with anything. This space is called the void.

The idea of nothingness is difficult to comprehend, especially if you're an early Greek philosopher. While today we are comfortable with the idea of the vacuum of outer space or in a thermos bottle, the Greeks had no such experience. Try as they might, they could find no place where they could point and say, "There is nothing." So the void idea wasn't very popular. Democritus finally reasoned that the atoms must be separated by an empty space, because one could cut a piece of cheese. There had to be a space between the cheese atoms for the knife-edge to penetrate. This argument is interesting, but ultimately not completely compelling.

The ideas of the Greeks came into being during the Golden Age of Greece, circa 500 B.C. This time was exceptional in that it allowed (and even encouraged) people (mostly rich, slave-owning men, it's true) to think about the cosmos, the nature of reality and the very deep and interesting questions that still cause modern man trouble. For the next 2000 years, there was not the right mix of circumstances to encourage such a lofty debate. The Roman era was marked by a concern for law, military accomplishments and great feats of engineering. The Dark Ages, dominated by the Catholic Church and small kingdoms, was more concerned with matters spiritual than scientific and even learned men of that time deferred to the Greeks on these topics. Even the lesser-known Golden Age of Islam, notable for its remarkable accomplishments in arts, architecture, cartography, mathematics and astronomy, did not add appreciably to mankind's knowledge of the nature of reality. (A mathematical smart-aleck might say that it added zero.)

Before we switch our discussion to the next era in which substantial progress was made in these weighty matters, some discussion of the merits of the Greeks' early ideas is warranted. Books of this type often make much of the success of some of the Greeks in guessing the nature of reality. Some guesses were right, while most were wrong. This "canonization" is dangerous, partially because it confuses non-critical readers, but even more so because writers of books on the subject of New Age spirituality usurp this type of writing. These writers steal the language of science for an entirely different agenda. Using crystals to "channel" makes sense because scientists can use crystals to tune radio circuits. Auras are real because scientists really speak of energy fields. Eastern mysticism uses a language that sounds similar to the non-discerning reader to that of quantum mechanics. Somehow it seems enough to see that the ancients had many ideas. Some of these ideas look much like the results of modern science. It's clearly, they would assert, just a matter of time until other ancient beliefs are proven to be true too.

Of course the logic of this argument fails. Most speculative ideas are wrong (even ours ... or mine!) The ancient Greeks, specifically the Pythagoreans, believed in reincarnation. While the experimental evidence on this topic is poor, it remains inconclusive. But the fact that the Greeks predicted something resembling atoms has no bearing on, for instance, the reincarnation debate.

I think that the really interesting thing about the Greeks' accomplishments is not that a Greek postulated that there was a smallest, uncuttable component of matter, separated by a void; after all, that model of the atom was wrong, at least in detail. The truly astounding thing was that people were interested in the nature of reality at a size of scale that was inaccessible to them. The fact is that their atoms were so small that they would never be able to resolve the question. Reason is a wonderful skill. It can go a long way towards helping us understand the world. But it is experiment that settles such debates. A primitive tribesman, living in the Amazon jungle, could no more predict ice than fly. Thus it is perhaps not at all surprising that the generations following the Greeks made little progress on the topic. The Greeks had used reason to suggest several plausible hypotheses. Choosing among these competing ideas would await experimental data and that was a long time coming.

The next resurgence of thought on the nature of matter occurred in the years surrounding the beginning of the Italian Renaissance. During this time, alchemists were driven to find the Philosopher's Stone, an object that would transmute base metals (such as lead) into gold. What they did was to mix various substances together. There was little understanding, but a great experimental attitude. Along the way, dyes were discovered, as were different explosives and foul-smelling substances. While the theory of what governed the various mixings (what we call chemistry) was not yet available, the alchemists were able to catalog the various reactions. Centuries of experimentation provided the data that more modern chemists would need for their brilliant insights into the nature of matter. There were many deeply insightful scientists in the intervening centuries, but we shall concentrate on three of the greats: Antoine Lavoisier (1743–1794), John Dalton (1766–1844) and Dmitri Mendeleev (1834–1907).

Better Living through Chemistry

Lavoisier is most known in introductory chemistry classes because of his clarification of the theory of combustion. Prior to Lavoisier, chemists believed that combustion involved a substance known as phlogiston. He showed that combustion was really the combination of materials with oxygen. However, in the context of our interest, the ultimate constituents of matter, he actually should be known for other things. One of his accomplishments was notable only long after the fact. He completely revamped the chemical naming convention. Prior to Lavoisier, the names of the various substances manufactured by the alchemists were colorful, but not informative. Orpiment was a particular example. What Lavoisier did was rename the substances in such a way that the name reflected the materials involved in the reaction. For instance, if one combined arsenic and sulfur, the result was arsenic sulfide, rather than the more mysterious orpiment. While Lavoisier was more concerned with the fact that arsenic and sulfur were combined to make the final product, we now know that the final product contains atoms of arsenic and sulfur. Just the more organized naming somehow helped scientists to think atomically.

Another important discovery by Lavoisier concerned water. Recall that the ancients treated water as an element (recall *fire, air, earth* and *water?*). Lavoisier reacted two materials (hydrogen and oxygen gas) and the result was a clear liquid. This experiment is repeated in high-school chemistry labs today. Hydrogen and oxygen are first isolated (another Lavoisier effort) and then recombined using a flame. After a "pip" (a little explosion), the same clear liquid is observed. This liquid is water. So first Lavoisier proved that water was truly not elemental. An even greater observation was the fact that in order to get the two gases to react fully, they had to be combined in a weight ratio of one to eight (hydrogen gas to oxygen gas). No other ratio would use up all of both reactants, which somehow suggested pieces of hydrogen and oxygen were coming together in fixed combinations. Lavoisier also reversed the process, separating hydrogen and oxygen from water and also observed that the resultant gases had the same ratio by weight: eight parts oxygen to one part of hydrogen. While Lavoisier was not focused on the atomic nature of matter, his meticulous experimental technique provided evidence that lesser scientists could easily see as consistent with the atomic nature of matter. Lavoisier's brilliance was tragically extinguished on the guillotine in 1794 as part of the blood purge that was France's Reign of Terror.

John Dalton was an amateur chemist who expanded on Lavoisier's earlier observations. Although Lavoisier did not focus on the theory of atoms, Dalton did. While some historians of science have suggested that Dalton has received an undue amount of atomic glory, he is generally credited with the first articulation of a modern atomic theory. Democritus postulated that the basic difference between different kinds of atoms was shape, but for Dalton the distinguishing factor was weight. He based his thesis on the observation that the products of a chemical reaction always had the same weight as the materials that were reacted. Like Lavoisier's earlier observations of the mixing ratio of oxygen and hydrogen, Dalton mixed many different chemicals together, weighing both the reactants and the products. For instance, when mixing hydrogen and sulfur together, he found that by weight one needed to mix one part of hydrogen to sixteen parts of sulfur to make hydrogen sulfide. Mixing carbon and oxygen together proves to be a bit trickier, because one can mix them in the ratio of twelve to sixteen or twelve to thirty-two. But this can be understood if there exist atoms of oxygen and carbon. If the ratio of weights is 12:16 (twelve to sixteen), then this can be explained by the formation of carbon monoxide, which consists of one atom of carbon and one atom of oxygen. If, in addition, it was possible to combine one atom of carbon with two atoms of oxygen, now to make carbon dioxide, then one could see that the ratio of weights would be 12:32. The mathematically astute reader will note that the ratio 12:16 is identical to 3:4 and 12:32 is identical to 3:8. Thus the reason that I specifically chose a ratio of twelve to sixteen was due to additional knowledge. In the years since Dalton, scientists have performed many experiments and shown that hydrogen is the lightest element and thus its mass has been assigned to be one. This technique is moderately confusing until one thinks about more familiar units. A one-pound object is a base unit. A five-pound object weighs five times as much as the base unit. In chemistry, the base unit is the hydrogen atom and Dalton and his contemporaries were able to show that a unit of carbon weighed twelve times more than a unit of hydrogen. So carbon is said to have a mass of twelve.

Dalton is credited with making the bold assertion that certain materials were elements (for example hydrogen, oxygen, nitrogen and carbon) and that each element had a smallest particle called an atom. The different elements had different masses and these were measured. The modern model of chemical atoms was born.

In the years following Dalton's assertion, many chemical experiments were done. Chemists were able to isolate many different elements and, in doing so, they noticed some peculiar facts. Some chemicals, while of significantly differing masses, reacted in very similar ways. For instance, lithium, potassium and sodium are all similarly reactive metals. Hydrogen, fluorine and chlorine are all highly reactive gases, while argon and neon are both highly non-reactive gases.

These observations were not understood and they posed a puzzle. How was it that chemically similar materials could have such disparate masses? The next hero of our tale, Dmitri Mendeleev, was extremely interested in this question. What he did was to organize the elements by mass and properties. He wrote on a card the name of the element, its mass (determined by the experiments of Dalton and his contemporaries) and its properties. He then ordered the known elements by mass and started laying the cards down from left to right. However, when he reached sodium, which was chemically similar to lithium, he put the sodium card under lithium and continued laying down the cards again towards the right, now taking care to group chemically similar elements in columns. Mendeleev's real genius was that he didn't require that he know of all possible elements. It was more important that the columns be chemically similar. One consequence of this choice was that there were holes in his table. This "failure" was the source of considerable derision directed at Mendeleev's organizing scheme. Undaunted, Mendeleev asserted that his principle made sense and also he made the bold statement that new elements would be discovered to fill the holes. Two of the missing elements were in the slots under aluminum and silicon. Mendeleev decided to call these as-yet undiscovered elements eka-aluminum and eka-silicon. (Note that "eka" is Sanskrit for "one." When I was a young student and told of this tale, I was informed that "eka" meant "under," a myth which I believed for over twenty years until I started writing this book.) In the late 1860s, this assertion was a clear challenge to other chemists to search for these elements. Failure to find them would discredit Mendeleev's model.

In 1875, a new element, gallium, was discovered that was clearly consistent with being eka-aluminum. Also, in 1886, germanium was discovered and shown to be eka-silicon. Mendeleev was vindicated. This is not to say that his table, now called the Periodic Table of the elements and displayed in every chemistry classroom in the nation, was understood. It wasn't. But the repeating structure clearly pointed to some kind of underlying physical principle. Discovery of what this underlying principle *was* would take another sixty years or so. We will return to this lesson later in the book.

Mendeleev died in 1907, without receiving the Nobel Prize even though he lived beyond its inception, a tragedy in my mind. Like Lavoisier's rationalization of chemical names, the mere fact that Mendeleev was able to organize the elements in a clear and repeating pattern gave other scientists guidance for future research. By the time of Mendeleev, atoms were firmly established, although interesting questions remained. The studies of these questions have led to the science that is the topic in this book.

With the chemical knowledge of about 1890, chemists were pretty certain that they had finally discovered the atoms originally postulated by Democritus, nearly 2400 years before. Elements existed and each was associated with a unique smallest particle called an atom. Each atom was indestructible and all atoms of a particular element were identical. All of the various types of matter we can see can be explained as endless combinations of these fundamental particles called atoms. Given the scientific knowledge of the time, this was a brilliant achievement.

May the Force Be with You

The existence of atoms did not answer all questions. Thus far, we have not addressed what keeps the atoms together. Something bound the atoms together to make molecules and molecules to make gases, liquids and solids. The obvious question then becomes: "What is the nature of this force?"

Asking the question of the nature of the inter-atomic force opens an even larger question. What sorts of forces are there? We know of gravity of course, and static electricity and magnetism. Are these forces related or completely different phenomena? If they are related, how can one reconcile this with the obvious differences between the forces? Just what is going on?

Leaving aside the question of inter-atomic forces (we will return to this a few more times in this book), let's discuss other forces, starting with gravity. As stated earlier in this chapter, this is not a book about astronomy, so we pick up our story when the scientific community had accepted that there were several planets and that their motion could be best explained as orbiting a central point, specifically the Sun. Since Aristotle had claimed that the natural state of matter was that all objects eventually slow and stop moving, many outlandish theories had been proposed for why the planets continue to move (including the idea that the planets' motion was caused by angels beating their wings). But the real understanding of the motion of planets would await Sir Isaac Newton.

Isaac Newton (1642–1727) was one of the greatest scientists who ever lived, arguably the greatest. In addition to having brilliant insights into optics and other fields, Newton postulated that objects in motion tended to continue moving until acted upon by an outside force. He combined this observation with the contention that the same gravity that keeps us firmly ensconced on Earth is responsible for providing the force that keeps the planets in their paths. Oh, and by the way, to solve the problems generated by his theories, he was forced to invent calculus. When these ideas were combined, he was able to describe the orbits of the planets with great precision. His theory also agreed with the observation that a person's weight did not appear to depend on elevation. Newton's work on gravity was characteristically brilliant, but in addition to his scientific success, one should stress a specific insight. Newton was able to show that different phenomena, a person's weight and the motion of the heavens, could be explained by a single unifying principle. We say that the theory of gravity *unified* the phenomena of weight and planetary motion. The idea that a single physics theory can unite what appeared previously to be unrelated phenomena is one to which we will return fairly frequently in subsequent chapters.

Newton's theory of gravity stood unchallenged for centuries until an equally brilliant man, Albert Einstein, recast the theory of gravity as the bending and warping of the structure of space itself. While Einstein's General Theory of Relativity is not terribly relevant to the topic of this book (until a much later chapter), one point of great interest concerns the melding of the concept of force and the structure of space. This concept remains unclear and thus continues to be a topic of active research. The concept that the very structure of space and time can be related in a very fundamental way to energy and forces is so interesting that all physicists (and anyone else who has considered the topic) eagerly await the illuminating idea that sheds light on this fascinating question.

An important failure of Einstein's idea is the fact that it is currently completely incompatible with that other great theory: quantum mechanics. Since the original publication of the theory of general relativity in 1916 and the subsequent development of quantum mechanics in the 1920s and 1930s, physicists have tried to merge general relativity and quantum mechanics to no avail (quantum mechanics and *special* relativity could be reconciled much more easily). As we shall see in Chapter 4, other forces have been successfully shown to be consistent with quantum mechanics and we will discuss some of the modern attempts to include gravity in Chapter 8.

While gravity is perhaps the most apparent force, there exists another set of forces that are readily observed in daily life. These two forces are magnetism and static electricity. Most of us have played with magnets and found that while one end of a magnet attracts the end of another magnet, if one magnet is flipped (but not the other), the magnets then repel. Similarly, one can comb one's hair on a dry winter's day and use the comb to pick up small pieces of tissue paper. Alternatively, one learns about static electricity when socks stick to sweaters in the clothes dryer.

During the 1800s, scientists were fascinated with both the forces of electricity and magnetism and spent a lot of time unraveling their properties. Earlier, the electric force was shown to become weaker as the distance between the two things that attracted (or repelled) became larger. Scientists even quantified this effect by showing that the force lessened as the square of the distance (physics-ese for saying that if two objects felt a particular force at a particular distance, when the distance doubled, the force was $1/2 \times 1/2 = 1/4$ the original force; similarly if the distance was tripled, the force was reduced to $1/3 \times 1/3 = 1/9$ that of the original force). Other experiments showed that there appeared to be two kinds of electricity. These two types were called positive $(+)$ and negative $(-)$. It was found that while a positive charge repelled a positive one and a negative charge repelled a negative one, a positive charge attracted a negative charge. In order to quantify the amount of electricity, the term "charge" was coined. The unit of charge is a Coulomb (which is sort of like a pound or foot, i.e. a pound of weight, a foot of length and a coulomb of charge) and you could have an amount of positive or negative charge.

It was further shown that if the correct sequence of metals and felt were stacked in a pile, then wetted with the proper liquids, electricity would move through the wire connected to the layers. (This is what Americans call a battery, but it explains why it is called a "pile" in many European languages.) These studies were originally accomplished using recently-severed frogs' legs (which kind of makes you wonder about some of the early scientists...). These experiments showed that electricity was somehow related to life, as electricity could make the legs of dead frogs twitch. It was this observation that provided the inspiration of Mary Shelly's *Frankenstein*.

Magnetism was most useful to the ancients in the form of a compass. The north end of a compass points roughly north, irrespective of where on the globe one is sitting. This was not understood until it was shown that a magnet could deflect a compass needle. This demonstrated that the entire Earth was a giant magnet and further that the compass needle itself was a magnet.

In 1820, Hans Christian Oersted made a truly remarkable discovery. When a current-carrying wire (physics-ese for a wire through which electrical charge was flowing) was placed near a compass, the compass' needle deflected. When the current in the wire was stopped, the compass again pointed in its natural direction (i.e. the direction determined by the Earth). This discovery seemed to suggest that while electricity didn't cause magnetism, *moving* electricity did. Further experimentation showed that a current-carrying wire was surrounded by magnetism.

Reasoning that if current caused magnetism, then perhaps magnetism caused current, scientists took a permanent magnet (like the ones that hold your kid's art to the refrigerator) near a wire that was hooked to a current measuring device. They measured exactly zero current. So no luck. However, when they moved the magnet, they saw current in the wire (actually they used many loops of wire fashioned into a coil, but that's not completely critical although it does make the experiment much easier). Since the strength of the force caused by a magnet is related to the distance from the magnet, when the magnet is moved the strength of magnetism seen by the wire changed. Thus it was shown that it wasn't that a magnet caused a current, but when a magnet's strength *changes* that causes a current.

While the reader is most familiar with the forces caused by static electricity and magnetism (e.g. when you pick something up with a magnet), it is necessary to introduce a new concept here, the concept of fields. The concept of a field is most easily introduced by using the familiar phenomenon of gravity. When you stand in a particular place, you feel a downward force from gravity (see Figure 1.1). If you then move from that spot to another spot, you now feel gravity at the new spot (profound isn't it?) But what happened at the original spot? Is gravity still operating there? You think it somehow should still be there, but how do you know? You could walk back to the original spot, but that doesn't address the question, as you want to know about the

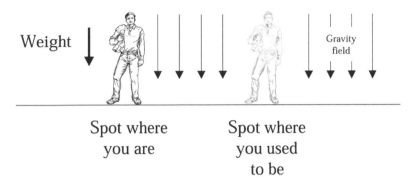

Figure 1.1 Even though gravity manifests itself to the observer as a force, the gravity field exists everywhere. The idea of a field is important and extends to all of the known forces. (Drawing courtesy of Dan Claes.)

spot when you're not there. You could put something else there (say a cat) to see if it felt a force and stayed there (although a cat, knowing that you wanted it to stay there, would probably wander off, so perhaps that's a bad example). The fact is, that while you must put an object there for something to feel a force, we're pretty confident that even when there's nothing there, gravity is still present. We say that there is a gravity field everywhere and always pointing downward.

By analogy, there also exist electric and magnetic fields. While it takes at least two charges or two magnets to feel a force, a single electric charge has an electric field surrounding it and similarly a single magnet is surrounded by a magnetic field. During the period of 1861–1865, James Clerk Maxwell took the experimental observations mentioned above and combined them with the concept of fields, added a dash of tricky math and was able to show that the concepts of electricity and magnetism were not, in fact, isolated phenomena but rather two facets of a single unifying phenomenon, now called electromagnetism. This remarkable feat was made even more amazing by an observation, made somewhat later, that the new electromagnetic theory also explained light...an unforeseen accomplishment. This achievement certainly rivaled and perhaps surpassed Newton's earlier unifying theory of gravity.

While Maxwell's theory of electromagnetism was wildly successful, the question of the ultimate nature of electricity was not yet resolved. The original idea of electric current was believed to be similar to fluid flow. An electrical liquid would flow through wires much in the same way that water flows through pipes. While we now know that charge comes in discrete chunks (like marbles), this was not obvious at the time. In fact, the discovery of the discrete nature (often called the quantized nature) of charge marks the beginning of modern particle physics and, as such, this story will be told in the next chapter. Interestingly enough, the idea of atomism (i.e. the idea that there might be a smallest imaginable piece of charge) was not so obvious, even as the great atom debate described earlier was being pursued.

The question of what kept atoms together, alluded to earlier, was not resolved by the successful theories of Newton and Maxwell. However experiments had been done that did provide some useful guidance and these are worth discussing. Early in the saga of the electrical experiments, physicists had been able to distinguish between two types of materials; those called conductors (like metals) and those called insulators (like rubber or wood). Conductors allowed electricity to flow and insulators stopped any sort of current flow. The interesting experiment occurred when a wire, which allowed electricity to flow, was cut and insulated so that on the ends of the wire only bare metal remained. These bare ends were placed into a jar containing water. When this was done (with the circuit set up so that current would flow if the wire ends touched), bubbles of gas formed on each wire. When the gas was captured and identified, it turned out that oxygen was being formed on one wire and hydrogen on the other. Twice as much hydrogen was formed as oxygen (by volume), in agreement with Lavoisier's earlier measurements. But unlike Lavoisier's experiments, which used heat to break water apart, these experiments used electricity. Further, the amount of gas produced was proportional to the amount of charge that flowed through the water. Measuring the volume of gas is easy, but you might ask how you measure the total charge. Since current is the amount of charge per unit time that flows through a circuit (similar to the gallons of water per minute that comes out of a water faucet), one could measure the electrical current and total time and thus determine the total charge.

Other studies showed equally interesting effects. Suppose one of the wires described above was attached to something (say a cup or plate made of an inexpensive metal) and the object was placed in a vat containing a mixture of water and silver nitrate. If the other wire were attached to a piece of silver, also immersed in the liquid, and current allowed to flow through the liquid, the cup or plate would begin to be covered by a thin coating of silver. This had great commercial value, as a person could now own a dinner service that looked like solid silver without paying the expense incurred if the entire plate or cup were actually solid silver. This phenomenon is called electroplating.

Both electroplating and the breakup of water into oxygen and hydrogen led scientists to the inescapable conclusion that somehow electricity is related to the force that holds atoms together. Further, the facts that (a) chemicals (such as water) were shown to be composed of atoms of elements and (b) the amount of that chemical that was disassociated into its constituent elements is proportional to the total charge passing through the liquid suggest that perhaps electricity might come in chunks of "electricity atoms" as well. The case for this assertion was strong, but not iron clad and the final proof of this idea would await 1897, when J.J. Thomson discovered the electron and started the modern era of particle physics.

The twilight years of the 19th century found many physicists and chemists inordinately pleased with themselves. The long struggle to understand the chemical elements seemed to be complete, although one might wonder why the Periodic Table had the structure it did. Physicists were perhaps even cockier. By using the juggernaut theories of Newton and Maxwell, they could explain almost every phenomenon that they observed. The motion both of objects in the heavens and here on Earth were explained, as was the subtle interplay of electricity and magnetism and light. There was still the question of what was the nature of the medium through which light propagated (the so-called aether), as well as the mildly disturbing fact that contemporary theories predicted that hot objects would radiate more short-wavelength radiation than was observed (the ultraviolet catastrophe). There was also the nagging question that Lord Rayleigh raised when he calculated that the Sun should have used up its fuel in about 30,000 years if it burned chemically; a realization that was quite troubling in light of the fact that even then the Sun was known to be much older than that. No less a luminary than Lord Kelvin was deeply concerned with the first two questions (the question of the medium that transmitted light and why the radiation of hot objects was incorrectly predicted). In his famous Baltimore lecture "Nineteenth Century Clouds over the Dynamical Theory of Heat and Light" at the Royal Institution of Great Britain, Lord Kelvin could not help but comment on these striking failures. He said in the July 1901 issue of the *Philosophical Magazine*

> The beauty and clearness of the dynamical theory, which asserts heat and light to be modes of motion, is at present obscured by two clouds. I. The first came into existence with the undulatory theory of light, and was dealt with by Fresnel and Dr. Thomas Young; it involved the question, How could the Earth move through an elastic solid, such as essentially is the luminiferous ether? II. The second is the Maxwell-Boltzmann doctrine regarding the partition of energy... I am afraid we must still regard Cloud No. I. as very dense... What would appear to be wanted is some escape from the destructive simplicity of the general conclusion. The simplest way of arriving at this desired result is to deny the conclusion; and so, in the beginning of the twentieth century, to lose sight of a cloud which has obscured the brilliance of the molecular theory of heat and light during the last quarter of the nineteenth century.

One might paraphrase his paper (which is rather dense) as stating that the prevailing theory of the nature of light, as well as the understanding of how energy was shared among atoms in matter, was clearly not completely understood. Nonetheless, with the exception of these two small clouds on the horizon, the world was well explained. Kelvin was more insightful than he knew, for these "small clouds" were soon unleashing the violent thunderstorms of relativity and quantum mechanics.

Our discussion of the nature of matter, forces and electromagnetism brings us to the final days of the 19th century. This journey has been truly rapid and by no means is it intended to be a thorough treatment. The interested reader is invited to peruse the bibliography for this chapter where many delightful books are listed that discuss this history in far greater detail. These early achievements set the stage for the deluge of discoveries that was soon to come.

chapter 2

The Path to Knowledge
(History of Particle Physics)

The most exciting phrase to hear in science, the one that heralds new discoveries, is not 'Eureka!' (I found it), but 'That's funny ...'

— Isaac Asimov

The close of the 19th century was marked by an unnatural confidence in scientists' understanding of the nature of nature. John Trowbridge, head of Harvard's physics department was discouraging young scientifically inclined students who were interested in physics. Everything was understood, at least in principle, he told them. Advances in physics would not be made like the astounding discoveries in electricity and magnetism that had marked the 19th century, but rather by making ever more precise measurements. Physics, to quote Albert Michelson, was "to be looked for in the sixth place of decimals." It is a marvelous irony that Michelson was one of the architects of *the* seminal experiment that presaged Einstein's Theory of Special Relativity and signaled the death knell of classical physics.

Long before these egregious errors in judgment, physicists had observed phenomena that, when properly interpreted, would lead the way to the quantum revolution and to today's modern world-view of particle physics. There are two phenomena that provided critical guideposts along the way. The first set of phenomena are still familiar today: phosphorescence and fluorescence. Phosphorescence occurs when a material, placed in the light, would continue to emit light after the light source was removed. Today's modern "Glow in the Dark" paint provides an excellent example of phosphorescence. Fluorescence, on the other hand, is somewhat different. A fluorescent material only emits light when being illuminated by another light, although the emitted fluorescent light could be of a substantially different color. Today's black light posters provide familiar examples of fluorescence. While neither phenomena was understood, each was present and played an important role in some early moments of epiphany, during which certain lucky scientists knew that they had discovered something truly new.

Cathode Rays

Another interesting phenomenon came from an ongoing interest in electricity and how it worked. Early in the history of the investigation of electricity, scientists were fascinated by sparks. In the 1740s, William Watson, a friend of Benjamin Franklin said "It was a most delightful spectacle, when the room was darkened, to see the electricity in its passage." One question that was asked addressed the effect of the composition and the pressure of the gas that separated the two sides of the spark. When you shuffle your feet and touch a doorknob, the spark jumps through air. By the first half of the 19th century, scientists were able to generate pure samples of many different types of gas ... oxygen, hydrogen and nitrogen to name a few. Studies were undertaken whereby one could use the pure gases and explore just how much electricity was required to make a spark in each kind of gas.

One could do this by having a glass blower make a flask with two openings for gas flow. In addition, two plates, each end connected to what was in effect a very powerful battery, would be the surfaces between which the sparks would be generated. Scientists would then blow a gas (say hydrogen) into the flask until the air had been completely displaced. The problem was that it took a great deal of time (and amount of hydrogen gas) to completely displace the air. Clearly what was needed was a method to first completely remove the air and then bleed in a measured amount of the gas of interest.

Scientists saw that the spark would eventually be replaced by a glowing, purplish snake, similar to the "Eye of the Storm" globes that can be purchased today that look like captive lightning storms. In 1855, Heinrich Geissler invented the mercury vacuum pump that allowed experimenters the ability to easily remove the air from a glass flask. In about 1875, William Crookes built a tube (later called the Crookes tube) to carefully measure the voltage needed to get a spark between the plates. However, prior to a full-fledged spark, he found that as he increased the voltage, he could see an electrical current in his circuit. Since the plates weren't touching, the electricity had to flow through the gas or vacuum, if one pumped long enough. Since gas and vacuum was considered to be an insulator rather than a conductor, it was only with accurate instruments that he could measure this small current flow. In addition, with a suitable choice of gas and pressure, he could see the flow of electricity through the gas, as the gas emitted light (although the exact source of the light was not immediately apparent). Crookes investigated the flow of electricity and determined that electricity (perhaps) was flowing from the plate connected to the negative side of the battery (this plate was called the cathode) towards the plate connected to the positive side of the battery (called the anode). Subsequently, in 1876, the German physicist Eugen Goldstein, a contemporary of Crookes whose most active research period was earlier, had named this flow "cathode rays." Crookes tube (shown in Figure 2.1a) was modified by his contemporaries to better inspect their properties by putting a small hole in the plate connected

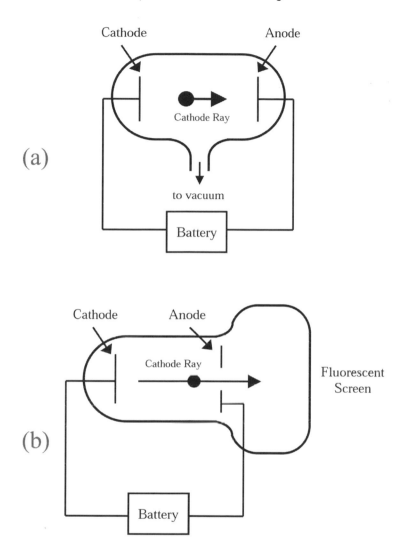

Figure 2.1 Diagrams of variants of Crookes' tubes.

to the positive side of the battery. This hole allowed the cathode rays to pass through and hit the far end of the glass vessel.

With this improvement on Crookes' design, the study of cathode rays could begin in earnest. It was found that the cathode rays traveled in straight lines and could cause, by their impact on the end of the glass vessel, great heat. Crookes knew that earlier studies had proven that charged particles would move in a circle in the presence of a magnetic field. When cathode rays also were shown to be deflected in the presence of a magnetic field, Crookes concluded that cathode rays were a form of electricity. Using the improved version of the Crookes tube, one could coat the end of the glass vessel with a phosphorescent material like zinc sulfide and observe that cathode rays caused the zinc sulfide to glow. The astute reader will recognize in these early experiments the origin of their computer monitor or television, also called a CRT or cathode ray tube.

Crookes believed that he had discovered a fourth state of matter, which he called "radiant matter." But Crookes' theoretical or explanatory abilities did not match his experimental skills, which were considerable, so his explanation of cathode rays proved to be incorrect. Luckily Crookes lived until 1919 and was able to see some of the extraordinary spin-offs of the Crookes tube. His work in many areas of scientific investigation was impressive and for this work he was knighted in 1897 and in 1910 he received the Order of Merit. In addition to his invention of the device that was to become the television and computer monitor, we are familiar with another one of Crookes' inventions, the radiometer. The radiometer is that glass device, shaped like a clear light bulb, which contains within it four vanes with alternate sides painted white or black. When placed near a light source, the vanes spin.

Later in life, Crookes investigated radioactivity, the discovery of which we will discuss presently. He found that "p-particles," which were the particles ejected from radioactive materials (like uranium and radium), when made to impinge on zinc sulfide, would result in a small burst of light, with each impact. This technique was very important and we will see it again when the nature of the atom was ascertained. Crookes' discoveries, while not directly related to particle physics, indirectly set the stage for the dazzling discoveries at the opening of the 20th century. Crookes tubes are not available for purchase anymore, for reasons that will soon become apparent.

The study of cathode rays continued for many years after their initial discovery. While the study of the luminous properties of the rays consumed the efforts of many investigators, in 1892 a young assistant of Heinrich Hertz tried a remarkable thing. Philipp Lenard managed to coat the end of a Crookes tube with a very thin layer of aluminum. Much to everyone's surprise, the cathode rays could penetrate the aluminum. Here was evidence that cathode rays, the luminous properties of which most investigators were most interested, could penetrate an opaque surface...a solid wall of metal. Very mysterious. While Lenard's later staunch support of the Nazi Party caused embarrassment for his scientific colleagues, there is no denying that this result was an important piece of the puzzle.

X-Rays

Several years later, things got extremely interesting. In the last decades of the 19th century, Wilhelm Konrad Roentgen was a rather ordinary physics professor. His work was noteworthy for his meticulous attention to detail rather than for its extraordinary insights. By 1895, Roentgen had been transferred four times within the German university system and was unlikely to go much higher. It's probably useful to note that the German university system (especially of that era) is different from the American one. At each university, each subject had only one, or at most just a few, professors. What would be viewed in the American system as other junior professors, in the German system, were assistants to "the" professor. The senior professor ran his laboratory or institute with considerable authority over his underlings. Some of my older German colleagues relate stories in which they were required to mow the lawn and wash the car of their thesis professor. Roentgen did not have the élan necessary to rise to the upper levels of the German university system and lead a major physics institute.

Nonetheless, Roentgen's work was perfectly respectable and in 1895 he was a professor of physics and rector at the University of

Wurzburg. On the evening of November 8th, Roentgen was working alone in his laboratory, trying to understand the effect observed by Lenard; the question of how cathode rays could penetrate an opaque wall of metal. Quite by happenstance, about six feet from the end of his cathode ray tube was a screen that had been coated by the salt barium platinum cyanide. While the night was dark and the lighting poor (recall that electric lighting wasn't as good as it is today), there was enough light to cause problems with his investigation of cathode rays which, as we recall, were dim glows in glass tubes full of gas. In order to reduce the problem, Roentgen covered his tube with a black, opaque box. As luck would have it, his tube happened to be pointed at the screen and the screen was in a dark corner of the lab. When Roentgen paused from his studies and rubbed his weary eyes, he noticed a peculiar thing … the screen was glowing!!! When he turned off the power to his tube, the glow disappeared.

Roentgen responded as any responsible physicist would. He said "Huh," or however they say that in German. He moved the screen closer and farther away and saw little effect except that the glowing spot got a little bigger and smaller, the behavior looking much like moving a piece of paper towards and away from a flashlight. He turned the screen around. It still glowed. He pointed the tube away and the glow disappeared. He swapped the tube for a couple of variants of the Crookes tube. The glow persisted. He then started putting various things between the tube and the screen; things like paper, pens, books, etc. He found that most things did not make the glowing stop … they appeared to be transparent to whatever was causing the glow. In fact, he needed to put relatively thick metal objects between the screen and the tube to cut off the glow. But the really fascinating effect he observed was when he put his hand between the tube and the screen. It turned out that much of his hand was transparent too, but some parts stopped the glow. When he looked at the dark spots carefully, he realized that *he was looking at his own bones!!*

Roentgen had worked with cathode rays for some time and he knew that they did not have the correct properties to be the cause of the phenomenon he was observing. Cathode rays would be stopped much more easily. Roentgen then deduced that he was observing something entirely new. He called his discovery "x-rays," a name which persists to this day.

Roentgen, we recall, was not a flashy guy; rather he was extremely careful. He didn't quickly contact the press or use his invention to construct a commercial product. Instead, he tried to determine what he could of the properties of x-rays. He even worked out how to capture his images on film. Finally, on January 1st, 1896, he sent copies of his report to many of the premier laboratories across Europe and included as "advertisement" photos of the bones of his own hand. One of the recipients was Henri Poincaré, who forwarded the letter to the French Academy of Science on January 20, 1896. This paper took Europe by storm and was immediately reprinted in *Science, Nature* and other noteworthy journals. Other scientists quickly reproduced his work. Roentgen received hundreds of letters and telegrams of congratulations from around the world. The speed of the spread of the news can be seen by a story on February 9th in *The New York Times*, in which was stated: "The Wizard of New Jersey (Thomas Edison) will try to photograph the skeleton of a human head next week." Edison failed, but clearly the excitement was worldwide. The dangers of x-rays to cause harm in living tissue became apparent fairly quickly, as evidenced by successful lawsuits against doctors (some things never change). Relatively modern safety precautions were undertaken rather early on.

The medical community was quick to appreciate the incredible utility of the discovery. Within three weeks of the announcement of the discovery in Paris, little Eddie McCarthy of Dartmouth, New Hampshire, had his broken arm set after the doctors first viewed the break with an x-ray. Within a year, over a thousand papers were published on the phenomenon. Of course, not all people received the news of the discovery with equal enthusiasm. A London newspaper wrote "On the revolting indecency of this there is no need to dwell, and it calls for legislative restrictions of the severest kind." Quick to exploit some of the public's concern, a leading clothing manufacturer advertised "x-ray proof" women's clothing.

Roentgen gave only one talk on the phenomenon that he had discovered. About a month after he sent out his letters, he spoke to the Physical-Medical Society at the University of Wurzburg, where he received tumultuous applause. Roentgen published only two more papers on x-rays before he rather inexplicably moved on to other fields of investigation. His professional life improved in 1900, when he moved to Munich, where he became the director of the Institute of Experimental Physics. In 1901, Roentgen received the very first Nobel Prize, setting a very high standard for subsequent recipients. Because of Roentgen's retiring nature, he slipped out of Stockholm so as to avoid having to give the public lecture required of a recipient of the prize. By a bit of legal legerdemain, he was able to substitute his earlier lecture at Wurzburg for the required talk. (Current recipients must give a lecture to the Swedish Academy of Sciences.)

Unfortunately for Roentgen, he did not profit financially from his discovery. During World War I, funding for his institute was cut: "There's a war on, you know," and he died in poverty at the age of 73, during the inflationary Weimar period between the two world wars.

Radioactivity

The excitement engendered by Roentgen's discovery of x-rays was felt by all active practitioners of the time. But for one particular person, it provided the inspiration necessary for future investigation that lead in an unexpected direction. A member of the French Academy of Sciences, Antoine Henri Becquerel, was present at many of the meetings that followed the announcement of the discovery of x-rays. Becquerel was a French physicist, as were both his father and his grandfather. Both he and his father had studied phosphorescence and, after seeing many of his colleagues use various versions of Crookes tubes to darken photographic plates, Becquerel had a hunch. Perhaps phosphorescence caused x-rays. Because of the family interest in the phenomenon, he had a large collection of various rocks and woods that would glow in the dark after a suitable period in the Sun to "charge up." Although Becquerel had not been particularly active in his research for some time, his hunch invigorated him. Upon his return to his lab, he took a number of phosphorescent materials, placed them in sunlight to activate them, and then placed them in a dark room, lying on top of a photographic plate. The idea was that phosphorescence caused x-rays and thus would darken the plate. After many experiments with no positive result, things looked bleak. He then decided to try "crystalline lamellas of the double sulfate of potassium and uranium" ($K(UO)SO_4 + H_2O$, for the chemically minded). We shall call this a uranium salt. This particular substance has the property that it glowed when illuminated by ultraviolet light. Becquerel placed the uranium salt on top of an opaque holder which held a photographic plate. He put the plate in the Sun for about 5 hours. Being careful, Becquerel simultaneously placed a second identical photographic plate nearby, this one with no uranium salts. This "control," as it is called, would establish the effect of having a shielded photographic plate sitting in the sunlight for that long.

When he developed the plate not in the presence of the uranium salts, he found the plate unchanged. However, when he developed the one with the uranium salts placed upon it, the photographic plate was exposed with the outline of the uranium salts clearly visible. He had his evidence that phosphorescence caused x-rays. Or did he? Being a careful experimenter, Becquerel tried a number of other experiments. He "charged up" the uranium salts, not by direct sunlight, but by sunlight that had first been reflected by a mirror and passed through a prism. The uranium salts still fogged the photographic plate. He then put a thin plate of copper in the shape of a cross between the photographic plate and the uranium salts, in order to see if the copper could block the "x-rays." He found that while he could see the image of the cross on the photographic plate, the plate was still exposed, even under the cross, indicating that the copper blocked only part of the unknown rays. He repeated the experiments with a cross made of thinner copper and saw that the thinner copper blocked the rays even less.

Becquerel then had some really good luck that appeared to be bad at the time. He prepared some photographic plates on February 26, 1896, with the intent of placing them in the Sun. As luck would have it that day, it was mostly cloudy and remained so for the rest of the week, so he only had a short amount of time of sunlight. He placed the whole contraption in a drawer to await better days. Because it was in the dark, the uranium salts would not phosphoresce and so he expected the plates to be essentially unfogged. However, there had been a little sunlight, so he expected a little fogging and, rather than confusing subsequent experiments by reusing these plates, he decided to develop the plates on March 1. Much to his surprise, he found that the photographic plates were extremely fogged. Most peculiar.

Because Becquerel knew that the phosphorescence of these particular uranium salts persisted for about 1/100 of a second, it appeared that the visible phosphorescence wasn't the cause. In his first paper to the French Academy, he noted the similar behavior to Roentgen's x-rays and hypothesized that perhaps the sunlight activated the x-rays for much longer than it activated the visible phosphorescence.

In late May, Becquerel published another paper, further discussing his discovery. From March 3 to May 3, he had kept some uranium salts always in the dark, thus giving time for the invisible phosphorescence to die out. He found that the uranium salts' ability to fog a photographic plate was undiminished in that time. Further, during that time he continued to experiment with other salts of uranium, including some that did not exhibit any phosphorescent behavior. He came to realize that uranium was the key factor, not phosphorescence, and thus speculated that a disk of pure uranium metal would fog the plates even more, a fact which he then demonstrated. By the end of the summer, he began to believe that he had discovered something different … an "invisible phosphorescence" associated with uranium. By the end of the year, he had shown that the rays he had discovered, while superficially similar to those discovered by Roentgen, had many different properties when investigated in detail. While the word had not yet been coined, radioactivity had been discovered.

In order to better understand radioactivity, the contributions of other people must be considered. The first of these is still a household name, even a hundred years later. Marya Sklodowska was born on November 7, 1867, into a family of teachers. Being a teacher was no more lucrative in the nineteenth century than it is today. Marya was an exceptionally bright young woman and she harbored a desire to become a scientist, something nearly unthinkable at the time. Lacking enough money to attend the university, Marya was forced to become a tutor for the children of wealthier people. She had a very strong sense of family duty and so she lived a very frugal life, sending all of her excess money to her sister Bronia, who was studying medicine in Paris. The idea was that once Bronia had established herself in her medical profession, she would return the favor.

So in 1891, Marya arrived in Paris. In those days, it was possible to get a degree without the whole "number of credits" thing that is necessary today and, after some study, she easily passed her physics degree and continued on to study mathematics. In 1895, she married Pierre Curie, a young scientist known for his work on crystallography and magnetism. In 1897, Marya Sklodowska, now known by her married name of Marie Curie, decided to attempt her physics doctorate. As we have seen, this period was extremely exciting, with the recent discoveries of Roentgen and Becquerel. Despite Becquerel's recent work, not too many people were working to understand "Becquerel Rays," in favor of the more easily manipulated x-rays (and their progenitors, cathode rays). For Marie, this was ideal, as she wouldn't have to plow through a huge literature search and could get directly to work on her own experiments. As you might imagine, if there have been many papers written on a topic, this implies that many experiments have been done and all of the easy results discovered. This makes future discoveries even more difficult. The fact that Becquerel Rays were less known improved her chances for making an interesting discovery in a timelier manner.

In order to understand the Becquerel Rays, she undertook an incredibly impressive program. She looked at all forms of uranium; solid, powdered, wet, dry, or in compound form. Using both Pierre's and her considerable chemical skills, she calculated the amount of uranium present in the various compounds and compared the results to an equivalent amount of solid uranium. In all cases, she found that the only thing that mattered was the quantity of uranium. She tested all other elements known at the time and discovered in April of 1898 that not only uranium, but also thorium could fog a plate. Of course, the question remained whether or not the rays from thorium and uranium were the same.

One improvement made by Curie in the study of radioactivity was the use of a new instrument. As anyone who has looked at a piece of exposed film can attest, it's difficult to assign a number to just how much the film is exposed. Dark is dark and light is light. Becquerel had used an electroscope to show that his rays would make air conductive. In about 1886, Pierre Curie and his older brother Jacque had invented a more sensitive electrometer that could measure very small electrical currents in air. The idea is simple. One takes two plates and connects them to a battery. The plates are separated by air. The radioactive substance is placed near the plates and makes the air slightly conductive. The electroscope measures the small electrical currents that flow in the air. A more radioactive substance will make the air more conductive and more current will be registered. Curie's laboratory was a terrible place to conduct the experiments, being basically a damp potato cellar. The dampness in the air makes the air conductive, in direct competition with the effect she was trying to measure, but she was still able to make accurate and reproducible measurements. Thorium's radioactivity had been discovered. Marie Curie coined the term "radioactivity" after the Latin word for "ray."

It was about this time that two things happened. First, Pierre realized that Marie was investigating something truly innovative and he abandoned his own work on crystals to join her. Secondly, the question of what actually caused the radioactivity began to be seriously

considered. Both Becquerel and Curie had shown that both pure uranium and uranium salts were radioactive, suggesting that perhaps it
was the uranium itself that was radioactive, rather than the chemical
bonds of the salts connected to uranium. Studies of the chemistry of
thorium suggested the same thing.

Of course, this raised a really interesting question. At the time, the
atoms of elements were thought to be fundamental. Each atom was
pointlike and contained no internal structure. Its basic properties were
its mass and its chemical behavior. Now two elements had been shown
to have a unique behavior. The question that was in the forefront of
everyone's mind was "What the heck is it about thorium and uranium
that makes it radioactive?" Curie soon compounded the confusion by
first an inference and then a discovery. She realized that two common
uranium ores, pitchblende and chalcolite, were even more radioactive
than uranium itself. Marie came to believe that the reason that these
minerals were so radioactive was because they contained other elements, not yet discovered, that were even more radioactive. With her
usual determination, she set out to isolate the two new elements. After
very demanding and tedious work, she isolated two different samples,
each highly radioactive. The first sample was mostly barium and the
second mostly bismuth. Since neither barium nor bismuth were
radioactive, she believed that each sample contained an admixture of
the dominant, radioactively inert, element and a trace amount of a
chemically similar, highly radioactive element. This was Mendeleev's
table (discussed in Chapter 1) all over again. In June 1898, the Curies
published a paper, in which they announced the discovery of a new element, called "polonium," after the country of Marie's birth. Polonium
was chemically similar to bismuth, except for the fact that it was
radioactive. In December of that year, they announced the discovery
of radium, the radioactive barium analog. Both of the new elements
were very different chemically, but both were radioactive. Two new
elements had been added to the radioactive pantheon.

With the observation of these new elements, the next step was to
extract pure samples of each. She was given a ton of pitchblende to

process and, after three years, she managed to extract one-tenth of a gram (a gram is about the mass of a paperclip) of radium chloride. She was never able to isolate polonium, the reason for which we now know is that it decays in about three months. So as she was isolating it, it was decaying even more rapidly.

In 1903, Pierre Curie and Henri Becquerel were nominated for the Nobel Prize for the discovery and characterization of radioactivity. A member of the nominating committee, Magnus Goesta Mittag-Leffler, a Swedish mathematician and an early advocate for women scientists, wrote to Pierre to notify him of this injustice. In his reply, Pierre argued most eloquently that a Nobel Prize for the study of radioactivity that did not include Marie would be most unfair. He wrote

> If it is true that one is seriously thinking about me [for the Nobel Prize], I very much wish to be considered together with Madame Curie with respect to our research on radioactive bodies.

In December of 1903, the same year that Marie received her doctorate, the Curies and Becquerel were awarded the Nobel Prize in Physics for radioactivity. The chemistry nominating committee insisted that the physics citation did not mention the discovery of radium, as they wished to consider Marie Curie for a Nobel Prize in Chemistry too. She received that honor in 1911.

Becquerel's and the Curies' legacy to mankind cannot be overstated. The Curies further bequeathed to the world their daughter Irène who, with her husband Frédéric Joliot, working in the same laboratory as her mother, discovered artificially induced radioactivity and thus received their own Nobel Prize in 1935.

Of course, while Becquerel had shown that radioactivity was different from x-rays, the question still remained "What makes up radioactivity?" Was it the emission of a particle or some sort of wave phenomenon? In June of 1903, a New Zealand-born scientist was in Paris and attended the celebration honoring Marie's being awarded her doctorate, the first woman in France to receive such an honor. This scientist was Ernest Rutherford, who had been working on radioactivity himself.

In 1899, Rutherford had discovered two distinctly different types of radioactivity emanating from uranium. This was done by allowing the radiation to pass through a magnetic and electric field and watching the deviation of the rays. There clearly was a negatively-charged component, as demonstrated by its strong bending in the magnetic field. This type of radiation was called beta radiation. In addition, there appeared to be another type of radiation, called alpha radiation, which appeared to not have its path deflected by a magnetic field. More careful study showed that there was a small deflection, indicative of a positively-charged particle. The small deflection was ultimately explained when it became apparent that the particle carrying alpha radiation was extremely heavy. In 1900, Paul Villard, of the École Normale Superiéure in Paris, France, discovered the existence of gamma rays by finding a component of radiation that was not affected by electrical and magnetic fields. Careful work showed that these rays penetrated matter in a manner different from x-rays, so they were considered to be yet another phenomenon. The situation was becoming murkier.

In 1900, Ernest Rutherford made a truly extraordinary discovery. He noticed that the radioactivity of thorium decreased over time. This suggested that radioactivity could go away. Such an observation was very curious, as uranium radiation appeared to be constant. Working in Montreal, Canada with chemist Frederick Soddy, Rutherford theorized a critical mechanism to explain the decrease of radiation by thorium. They believed that the process of radioactivity was nothing less than the transmutation of one element into another. Atoms were thought to be immutable, yet if their idea were true, the days of the unchanging atomic element were over!!! Even perhaps more ironic, the goal set out by those earlier Renaissance alchemists and "proven" to be impossible by 19th century chemists, the techniques of the transmutation of elements... of base metals into gold... had been achieved. The consternation felt by scientists of the time was summarized in the August 1903 issue of *Scientific American*

> Just what shall be done with the newly discovered radioactive substances is a problem that perplexes every thinking physicist. They refuse to fit into our established and harmonious chemical system; they even threaten to undermine the venerable atomic theory, which we have accepted unquestioned for well-nigh a century. The elements, once conceived to be simple forms of primordial matter, are boldly proclaimed to be minute astronomical systems of whirling units of matter. This seems more like scientific moonshine than sober thought; and yet the new doctrines are accepted by Sir Oliver Lodge and by Lord Kelvin himself.

While the immutability of elements, the very cornerstone of 19th century physics and chemistry, was now in doubt, the proliferation of the various rays was another problem. There were x-rays, gamma rays, alpha rays and beta rays. In addition, there were cathode rays and canal rays (an interesting, but ultimately minor, consequence of cathode rays). Through the investigation of Rutherford, Becquerel and others, x-rays and gamma rays were shown to be immune to the influence of electric and magnetic fields. Alpha rays and canal rays were shown to be positively charged particles of a fairly large mass. Finally, beta and cathode radiation behaved like negatively charged particles. Further, after investigating their ability to penetrate matter, cathode rays and beta rays looked suspiciously similar. Somebody needed to make sense out of the chaos. In order to fully understand the nature of things, we need to return to 1897 and enter the life of Joseph John (always called J.J.) Thomson.

The Discovery of the Electron

J.J. Thomson was the director of the Cavendish Laboratory at the University of Cambridge and one of the most respected scientists in Great Britain. He was passionately interested in the nature of cathode rays and towards the study of this question he brought his great experimental skill.

J.J. was born on December 18, 1856, the son of a book publisher and a housewife. His early years were filled with the usual childhood experiences, although he showed an early aptitude for things technical. At the age of 14, he was sent to study at Owens College (now Manchester University). His parents had selected an engineering firm at which J.J. would eventually apprentice. In the meantime, Thomson would study engineering and await an opening for an engineering apprentice in the chosen firm.

When Thomson turned 16, his father died. Suddenly the fee that his family had negotiated with the engineering firm in order for them to take him on was an impenetrable barrier. Saddened, his mother informed him that his plans to be an engineer were no longer possible and instead J.J. moved to Trinity College in Cambridge to study some more engineering. After he arrived, he found that his real interest was mathematics. In 1876, students were strictly ranked and this rank had considerable weight when decisions for future career opportunities were made. As you might imagine, such a system encouraged intense competition between the students and Thomson was ranked second in his class, behind Joseph Lamor, who eventually became a noted theoretical physicist.

J.J. began to work in the Cavendish Laboratory in 1880 under the then-director Lord Rayleigh. Upon Rayleigh's retirement in 1884, much to many people's surprise, Thomson was appointed as the Director of the Laboratory. Given that the first two directors of Cavendish were the legendary Maxwell and Rayleigh, the appointment of such a relatively unknown scientist was not met with universal acclaim. One of the tutors in the college said that things had come to a pretty pass when boys were made Professors. Glazebrook, a demonstrator at the Laboratory, wrote to Thomson "Forgive me if I have done wrong in not writing to you before to wish you happiness and success as Professor. The news of your election was too great a surprise to me to permit me to do so."

Irrespective of these misgivings, the appointment of Thomson as the director of Cavendish proved to be an enlightened choice. Under his guidance, extremely interesting experiments were performed in the fields of electricity and the nature of the atom. Many of his protégés proved their worth, with a number making critical discoveries and attaining high posts throughout Europe. Seven people who began their careers at Cavendish under Thomson were eventually awarded the Nobel Prize and 27 were elected as Fellows of the Royal Society. While Thomson was not particularly skillful in the laboratory techniques, he had a gift for figuring out what the experimental results meant. H.F. Newall, an assistant to a young professor Thomson, wrote "J.J. was very awkward with his fingers and I found it necessary to not encourage him to handle the instruments! But he was very helpful in talking over the ways in which things ought to go."

Among Thomson's many students was young Rose Paget, one of the first women permitted to study advanced physics at Cambridge. Joining Cavendish in 1889, Rose performed experiments on the vibrations of soap bubbles. She and J.J. were married on January 22, 1890 and they had two children. The eldest was George Paget Thomson, who followed in his father's physicist footsteps and was eventually awarded the Nobel Prize in Physics in 1937. Their daughter, Joan Paget Thomson, was devoted to her father and accompanied him in his frequent travels.

When young J.J. Thomson took over the directorship of Cavendish Laboratory, he embarked on an experimental program into the nature of electricity. He was especially interested in understanding cathode rays, about which he wrote in 1893, "There is no other branch of physics which awards us so promising an opportunity of penetrating the secret of electricity."

We recall that cathode rays were formed when two electrodes were placed in a glass tube and the bulk of the air removed by a vacuum pump. When a high voltage was placed between the electrodes, the remaining air would conduct, if the air pressure was appropriate. As the air conducted, it would glow like a writhing, purplish snake. The cause of the fluorescence of the gas was called cathode rays, which flowed from the negative electrode (the cathode) to the positive one (the anode). Where the cathode rays hit the glass envelope, the glass itself would fluoresce. With the creation of Crookes tubes (discussed earlier) and other similar designs, the study of cathode rays began in earnest. They were affected by magnets, but not by electric fields. They caused the glass tube to get hot. They also caused x-rays and could penetrate a thin layer of metal. Just what were these cathode rays? Certainly, there were many theories.

In general, British physicists held that cathode rays were particles. "Particles of what?" was the remaining question. Jean Perrin had shown that cathode rays carried electricity, as they could "charge up" an electroscope. One theory of cathode rays held that they were atoms coming from the cathode that had picked up a negative charge. If so, changing the metal of the cathode would change the nature of cathode rays, as the different elements were known to have differing masses. A contrary theory held that cathode rays themselves were not charged, but rather they caused charge to flow. It's like a river containing water and fish. They both move in the same direction and are related, but they aren't the same.

German physicists held a different view. While it was known that a magnet could deflect cathode rays (as one would expect if they carried an electrical charge), Heinrich Hertz knew that an electric field should have a similar effect. He placed two plates separated by the space through which the cathode rays moved and put a strong electrical field between the plates. If cathode rays were electrical in nature, they should be deflected by the electric field. The result of the experiment was that Hertz' electric field had no effect on the direction of cathode rays. This experiment seemed to provide conclusive proof that cathode rays were not fundamentally electrical in nature. Hertz' student, Philipp Lenard, had placed a thin aluminum foil in the path of cathode rays. The rays, we recall from earlier discussion, penetrated the foil. This seemed to suggest that cathode rays were a vibration, with the model being that the rays caused the foil to vibrate, which in turn caused the space beyond to vibrate, thus allowing the cathode rays to penetrate the metal. It was like talking at a drumhead.

The drumhead vibrates and the sound penetrates to the other side. The question now became "What was vibrating?" The most popular, although not only, explanation was that the aether was the vibrating material. The aether was supposed to be the material that conducted the vibrations of light. Thus perhaps cathode rays were a form of light? But light is not affected by a magnetic field and cathode rays were. Sheesh. … it's no wonder that the nature of cathode rays went unresolved for so long. J.J. Thomson, writing in 1897, said "The most diverse opinions are held as to these rays … it would seem at first sight that it ought not to be difficult to discriminate between views so different, yet experience shows that this is not the case."

In 1897, Emil Wiechert made a puzzling measurement. He was not able to determine the charge, nor the mass of cathode rays, but he was able to measure the ratio of the mass to the charge; the so-called m/e ratio (because e is now used to denote the charge of a cathode ray particle, while m denotes its mass.) The same ratio had been measured for the various atomic elements and the element with the smallest ratio (hydrogen) had a ratio over 1000 times greater than that measured by Wiechert for cathode rays. Taken literally, this could mean that if cathode rays had a mass equal to that of hydrogen, they had an electrical charge one thousand times greater. Alternatively, if cathode rays had the same electrical charge as hydrogen, then their mass must be one thousand times smaller. Because the result was a ratio, either explanation could be true or, for that matter, any number of other combinations.

Into this confusing fray stepped J.J. Thomson and his group of able assistants. To shed light on the topic, he performed three meticulous experiments and changed the world. J.J.'s first experiment was a variation on Jean Perrin's 1895 experiment that indicated that cathode rays were negatively charged particles. The counterargument to this explanation of Perrin's experiment was that perhaps the negatively charged particles and cathode rays were merely going in the same direction, but weren't really related. Perrin had simply put an electricity-measuring device in the way of cathode rays and observed the presence of electricity. J.J. added an external magnet, which deflected the cathode rays.

He found that the electricity always followed the cathode rays. Only if the cathode rays were hitting the electricity-measuring device did it register electricity. While this experiment did not completely rule out the "cathode rays and electricity go in the same direction, but are different" hypothesis, it provided very suggestive evidence that cathode rays and negatively charged particles were one and the same.

Of course, there was Hertz' result that an electric field did not deflect cathode rays. This was in direct conflict with the idea that cathode rays were negatively charged particles. J.J.'s great experience in experimenting with electricity passing through gases led him to a deeply insightful hypothesis. While an electric field will deflect a charged particle, this is true only if the charged particle is not shielded from the electric field (say by a copper tube or mesh). J.J. knew that cathode rays made the gas conducting and thought that perhaps the electrically-charged residual gas would shield the cathode rays from the influence of an external electrical field. Thomson thus went to great pains to completely remove all gases from his tube and built an apparatus like that shown in Figure 2.2.

Cathode rays were made in the traditional way and made to pass through a region with an electric field. They passed on to a screen coated with a fluorescent material to which a ruler had been affixed, so as to be able to measure deviation. When no electric field was applied, the cathode rays went in a straight line and caused a bright spot at the center of the screen. However, when an electric field was applied, *he saw the spot move!!!* J.J.'s explanation for why Hertz had failed to observe a deflection proved to be correct. More importantly, a crucial response to the objection that cathode rays could not be electrically charged particles had been tendered.

Thomson's third experiment would not have been performed, had experiment #2 not been successful. He wanted to measure two things; the first was the speed of cathode rays, as if they were a form of light, they would have to travel at the speed of light. The second thing he wanted to do was check Wiechert's earlier measurement of the mass to charge ratio of cathode rays.

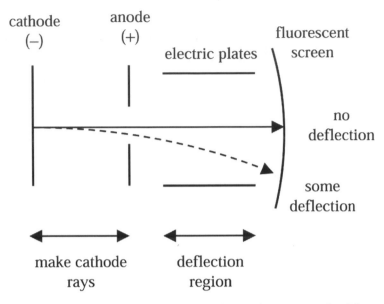

Figure 2.2 Essential aspects of how a modern television works. The region on the left creates cathode rays (accelerates electrons in modern language), while the region on the right deflects them left and right.

In order to measure the speed of cathode rays, he used two bits of physics knowledge. The first was the fact that the strength of the magnetic force on an object is proportional to that object's speed. The second was the fact that since electric fields and magnetic fields can deflect a moving, charged object, if one arranges them so they deflect in opposite directions, the strength of each can be changed until they exactly cancel one another. When this is achieved, the velocity of the particle can be ascertained by a simple calculation. (Note: Modern laboratory demonstrations use this technique. From Thomson's original paper, it appears that his approach, while similar, was somewhat less elegant.) When this experiment was performed, cathode rays were shown to have a velocity much smaller than that of light. Thus cathode rays could not simply be a light phenomenon.

Finally, once the speed of the particles was known, the third experiment could easily determine the ratio of mass to charge of the cathode rays (although to see how obvious this is would require a short excursion into mathematics). Thomson's first result supported Wiechert's findings. However, cognizant of the idea that perhaps cathode rays were atoms of elements from the cathode or gas that had picked up an electric charge, he decided to repeat the experiment with various gases and using different metals to construct his electrodes. In all cases, he found the mass to charge ratio was the same. It sure looked like the material making up the electrodes or surrounding gas didn't matter. In all fairness, one could argue that heavier atoms could perhaps pick up more charge (recall that all he measured was a ratio), but he felt that this was not the case.

Thomson then sat and thought. What could explain the myriad of properties measured by him and others? He noted that his mass over charge ratio was approximately one thousand times smaller than that measured for any known element. Because this measurement was independent of the materials used to make the measurement, he concluded that it was likely that he was seeing something completely new. He wrote in his seminal paper of 1897

> From these determinations we see that the value of m/e is independent of the nature of the gas, and that its value 10^{-7} is very small compared with the value 10^{-4}, which is the smallest value of this quantity previously known, and which is the value for the hydrogen atom in electrolysis.
>
> Thus, for the carriers of electricity in the cathode rays, m/e is very small compared with its value in electrolysis. The smallness of m/e may be due to the smallness of m or the largeness of e, or a combination of the two.

Thomson then goes on to discuss some of Lenard's results, which suggested that the most likely scenario was that the correct view was that the mass of cathode rays was very small, but he does not go so far as to declare this work definitive. Presumably, he felt it necessary to measure the charge or mass of cathode rays directly. This experiment was performed two years later and will be discussed presently.

However, in 1897 there were still mysteries. Nonetheless, J.J. Thomson felt that he had assembled enough information to announce his results to his colleagues. On Friday, April 30, 1897, J.J. lectured his colleagues and some of the "better" people of London, who had gathered to hear what was new in the world of science. Speaking to an audience in the great lecture hall of the Royal Institute of Great Britain, J.J. made an extraordinary announcement. He had discovered a particle that was a component of atoms. All learned people knew that atoms were the smallest particle of an element, with no internal structure. Thomson was telling them that this wasn't true. Atoms had structure. He wrote in his subsequent paper

> The explanation which seems to me to account in the most simple and straightforward manner for the facts is founded on a view of the constitution of the chemical elements which has been favourably entertained by many chemists. This view is that the atoms of the different chemical elements are different aggregations of atoms of the same kind. In the form in which this hypothesis was enunciated by Prout, the atoms of the different elements were hydrogen atoms; in this precise form the hypothesis is not tenable, but if we substitute for hydrogen some unknown primordial substance X, there is nothing known which is inconsistent with this hypothesis, which is one that has been recently supported by Sir Norman Lockyer for reasons derived from the study of the stellar spectra.
>
> If, in the very intense electric field in the neighbourhood of the cathode, the molecules of the gas are dissociated and are split up, not into the ordinary chemical atoms, but these primordial atoms, which we shall for brevity call corpuscles; and if these corpuscles are charged with electricity and projected from the cathode by the electric field, they would behave exactly like the cathode rays.

Thus for many historians of science, this lecture heralded the age of modern particle physics.

Thomson called his discovery "corpuscles," although in this he was quickly out of step with his colleagues. While most were very skeptical of his assertions (a distinguished member of his audience later told Thomson that he thought that Thomson was pulling their leg), evidence quickly grew. Physicists began to call the new particles "electrons," a term coined by G. Johnstone Stoney in 1891, in an entirely different context. Stoney used the term electron to describe the smallest unit of charge found in an experiment that passed current through chemicals. Thomson did not use the term "electron" for more than 20 years.

While Thomson had measured the mass to charge ratio, he could say little about either independently. Using entirely different techniques, two years later Thomson showed that the charge of one of his "corpuscles" was about the same as that carried by a hydrogen ion. Consequently, one was led to the inescapable conclusion that the mass of an electron was very small (modern measurements give it a mass of 1/1886 that of a hydrogen atom). The electron became a very light component of all atoms, each carrying the same charge as an ion. For his brilliant understanding of the data, as well as a few delicate experiments of his own, Thomson was awarded the Nobel Prize in Physics in 1906, for "researches into the discharge of electricity in gases." Knighthood followed in 1908 and the Order of Merit in 1912. Thomson lived until 1940, long enough to see many of the extraordinary consequences of his discovery.

The Nature of the Atom

With the realization that atomic atoms contain more primordial atoms (or at least electrons) within them, scientists realized very quickly that an entirely new field of inquiry had opened up ... that of the nature of the atomic atoms and the understanding of the constituents contained therein. In addition, Rutherford and Soddy's work at McGill University in Montreal had suggested that at least radioactive atoms could transmute among themselves, subject to strict rules, which they elucidated. The simplest explanation of this transmutation is that the as-yet undiscovered constituents of the atomic atoms were rearranging in some not yet understood way. Clearly the rules underlying the transmutation of the elements were mysterious and needed further investigation. It's a rare time that a completely new field of investigation, indeed a completely new paradigm, unfolds. Casting aside the quiet confidence (i.e. smugness) of the late 19th century, physicists traded it for the excitement associated only with new vistas to be explored. There was knowledge to be gained, mysteries to unravel. Life was good.

One of the most pressing questions was the nature of the atomic atom (which, in keeping with common usage, we will simply call an atom) itself. Atoms of the various elements were known to have different masses and were electrically neutral. The only known constituent of the atom, the electron, was known to have a negative charge equal to the smallest charge allowed for an ion (or for an atom) and a mass very much smaller than the atom. Since atoms were known to be electrically neutral, the question of just what was positively charged within the atom, as well as what caused the atomic mass to be so much larger than that of an electron, were considered to be most pressing. Of similar interest was the question of how atoms interacted... basically of how atomic physics could explain chemistry.

The first questions were tackled before the more complicated chemical ones. One interesting model of the atom was put forth by Japanese physicist Hantaro Nagaoka. He suggested that perhaps an atom looked like a little copy of the planet Saturn. A positively charged center was surrounded by a ring of electrons that orbited it. A problem with this model was immediately noted. When a charged particle like an electron moves in a circle, it radiates electromagnetic waves like a little radio transmitter. In radiating, the electron would lose energy and spiral down into the center of the atom. So Nagaoka's atom was neglected. We'll return to this later.

The English physicist, Lord Kelvin, put forth the first model of the atom that received significant attention. This model suggested that the positively charged material within an atom was a semi-liquid substance with small and hard electrons distributed throughout it, like raisins in a cake. Being British, Kelvin drew an analogy with plums in a pudding and thus the model became known as the "Plum Pudding" model.

J.J. Thomson liked the idea so much that in 1904 he calculated some of the possible motions of electrons in the "pudding." Many popular accounts of the Plum Pudding model incorrectly ascribe the original idea to Thomson, but it was Kelvin that should get the credit.

The Plum Pudding model, while inspired, had no evidence showing that it was correct. Before it would be universally accepted, experimental proof was needed. A rough New Zealand physicist, Ernest Rutherford, performed the definitive experiment on the subject.

Ernest Rutherford was born on August 20, 1871, just outside Nelson, New Zealand, the son of a Scottish émigré and an English schoolteacher. Both of Rutherford's parents highly prized education and they made sure that all 12 of their children attended school. Ernest distinguished himself early on for both his mathematical talent and boundless scientific curiosity. Being rather poor, Rutherford's only hope for higher education was to win a scholarship, which he did after his second attempt. Following his brother George to Nelson College, Ernest did quite well academically and played rugby during his final year. (I don't know what it is about physicists and rugby, but when I was in graduate school, most of the rugby team members were physics graduate students.) Rutherford topped his class in every subject in his senior year and won one of ten scholarships awarded in a national competition, although again he had to take the test twice. With this scholarship, he was able to attend what is now the University of Canterbury. In 1892, Rutherford was awarded a B.A. and won the only Senior Scholarship awarded that year in mathematics, which allowed him an additional year at the university, during which he received a M.A. in mathematics and physics. It was during this year that he derived a method for measuring time differences of as little as one hundred-thousandth of a second. In 1894, he completed a Bachelors of Science degree in geology and chemistry and, in 1895, he was awarded a prestigious research scholarship. This scholarship allowed him the opportunity for further study. Cambridge University in England had recently allowed for "foreign students" (i.e. people who had not received an undergraduate degree from Cambridge) to attend for advanced study.

Arriving in 1895, young Rutherford began working for the renowned J.J. Thomson, for whom he devised a method for detecting electromagnetic waves for distances exceeding several hundred meters. At the time, he envisioned the technology as a method for ships to detect lighthouses in exceptionally heavy fog. Following Rutherford's success, Thomson invited him to study the electrical conduction of gases, leaving commercialization of wireless technology (i.e. radio) to Guglielmo Marconi.

Rutherford was resident at Cavendish during the excitement surrounding Thomson's announcement of the discovery of the electron. For his work, Rutherford was awarded a B.A. Research Degree and, when the MacDonald Chair opened at McGill University in Montreal, Canada, Rutherford took it. Being a protégé of Thomson helped, of course. Upon hearing the news, Rutherford wrote his fiancée "I am expected to do a lot of work and to form a research school in order to knock the shine out of the Yankees!" As we will see, they picked the right guy for the task.

It was at McGill that Rutherford finally attained the financial stability needed to marry his long-time fiancée, Mary Georgina Newton. They were married in Christchurch, New Zealand in 1900 and had their only child, Eileen, in 1901.

McGill also proved to be professionally productive for Rutherford. In our discussion on radioactivity, we noted that Rutherford discovered beta radiation, which seemed to be identical to cathode rays. By this time, beta radiation was known to be the spontaneous emission of electrons by certain radioactive elements. In the same paper, Rutherford announced the existence of another type of radiation, termed alpha rays. Following an approach similar to that of his mentor Thomson, by 1903 Rutherford was able to show that alpha rays had a mass to charge ratio consistent with being a doubly-ionized helium atom (i.e. an object with twice the charge and four times the mass of a hydrogen atom).

Around the same time, Rutherford began working with a chemist, Frederick Soddy, trying to understand the nature of radioactivity better. Together, they were able to show that when a pure sample of a particular element was allowed to decay radioactively, what was left was a mixture of chemically-dissimilar elements. Together, they deduced that atoms, thought to be fundamental and immutable smallest examples of particular elements, were not so stable. One element could transmute into another, and the two of them worked out examples of "transmutation chains." The field of nuclear chemistry had begun and for this work, Rutherford was awarded the 1908 Nobel Prize, not in physics, but rather chemistry, for "Investigations into the disintegration of the elements and the chemistry of radioactive substances." Soddy's Nobel Prize came later (in 1921) for "his contributions to our knowledge of the chemistry of radioactive substances, and his investigations into the origin and nature of isotopes."

Discoveries worthy of the Nobel Prize do open doors and in 1907, Rutherford was lured back to England to become the Langworthy Professor of Physics at the University of Manchester. It was at this time that he entered into a debate with Antoine Becquerel on how alpha particles reacted when they were ejected from a radioactive substance. Becquerel had performed an experiment that led him to believe that alpha particles accelerated after they were emitted. This behavior would be bizarre. Rutherford had conducted a similar experiment and determined that alpha particles actually slow down as they travel through the air. Both men disputed the other's results and it was lucky that they were civilized men. Rather than pistols at dawn, they simply redid their experiments. It turns out that Rutherford was right.

Little disagreements like this are part and parcel of the life of a physicist at the frontier of knowledge. Usually it's of no consequence. However, in this case, this minor dispute led to a new way of thinking. Rutherford kept coming back to his own experiment, in which it turned out to be very difficult to measure the path of alpha rays (and such a measurement was necessary to perform the experiment.) While Rutherford was an "idea" and not so much a "detail" guy (he is reported to have said "There is always someone, somewhere, without ideas of his own that will measure that accurately"), he could do careful experiments, when the circumstances warranted. However, try as he might, he couldn't get a very accurate measurement of the path of alpha particles through the air. The alpha particles seemed to be bouncing all over the place. Finally, he decided that measuring exactly how alpha particles scattered was a necessary experiment. Rutherford conceived of an experiment and assigned the problem to a research assistant, Hans Geiger, who was working with an undergraduate student, Ernest Marsden. The experiment was as follows. A thin gold foil separated a radioactive source and a screen coated with zinc sulfide. We noted earlier that Crookes had noticed that a radioactive material would cause zinc sulfide to scintillate (that is, to give off light). Thus, the idea was that Geiger or Marsden would sit in a totally darkened room for 15 minutes or more, in order to let his eyes adjust. Then the source would emit lots of alpha rays that would pass through the gold foil and hit the zinc sulfide screen. The experimenters would note where the blink of light occurred and mark it down. An alpha particle that did not scatter would pass straight through the foil without deviation. A small scatter would manifest itself as a small deviation, with increasingly violent scatters resulting in ever-larger deviations (see Figure 2.3).

By determining the probability for the various degrees of scattering, one could hope to understand the scattering mechanism. Gold foil was chosen as it is dense and thus the scattering material is concentrated at a specific point, in contrast to the air, which extends over a considerable distance. Because there's a lot of mass in the foil, scattering in the air as the particle travels from the source to the screen is of less importance. Designing the apparatus in this way makes the analysis and explanation correspondingly easier.

Depending on the strength of the radioactive source, one must sit in the dark for a long time, straining one's eyes to see the barely perceptible flashes. Many hours are needed to gather enough data and so Geiger and Marsden sat in the dark for a long time. (You see why Geiger was so motivated to invent his radioactivity-detecting Geiger counter…) After considerable data-taking and analysis, they presented their results to Rutherford. After listening carefully, he

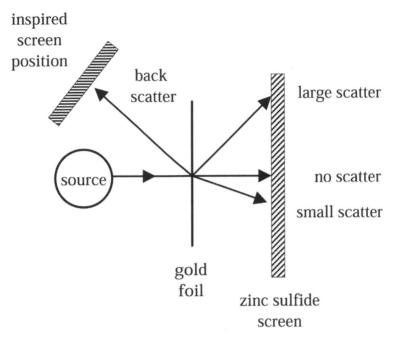

Figure 2.3 Rutherford's experiment. A source of alpha particles is directed at a gold foil, with the intent of understanding how alpha particles are scattered by gold atoms. From the scattering pattern, the nature of the nuclear atom became apparent. The most important information was the unexpected back-scattering.

suggested that they look to see if any of the alpha particles scattered backwards. This suggestion proved to be more inspired than Rutherford could have hoped.

Before we continue, let's think about what we expect. The state of the art model of the atom was Kelvin and Thomson's Plum Pudding; a sort of goopy fluid with a positive electrical charge, with electrons embedded throughout. The alpha particle was known to be a helium nucleus (relatively massive), moving at great speed. Such heavy projectiles should blast through the goopy pudding part of the atom, with only a minimal amount of scattering. Also, most alpha particles will be deviated at least a little bit, as they all have to pass through the entire thickness of the atom. This model is illustrated in Figure 2.4a.

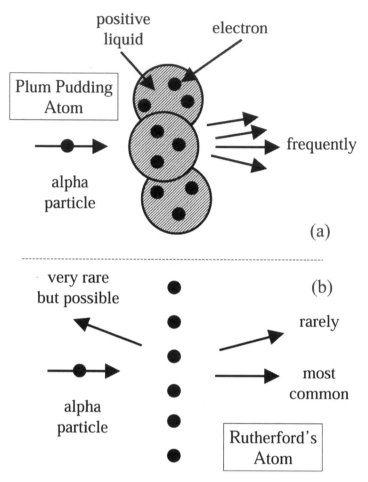

Figure 2.4 (a) Thomson's idea of the atom, the so-called "Plum Pudding" model, in which small and hard negatively-charged electrons exist inside a goopy positively charged fluid. (b) Rutherford's atom, consisting of a compact, positively charged nucleus surrounded by a dispersed cloud of small, negatively charged electrons.

Geiger and Marsden performed the experiment and, against all expectation, they found that about one alpha particle in 8,000 was reflected backwards. Most bizarre. Such a behavior is inexplicable using the Plum Pudding model, in which only fairly low violence scatters are allowed. Later, Rutherford made the much-quoted remark "It was truly remarkable. It was as if we had fired a 15" shell at a piece of tissue paper and it had bounced back."

Rutherford knew that Geiger and Marsden's data was not consistent with the Plum Pudding Model, but what other model could better describe the data? He returned to the administrative chores that go along with being a senior professor, but the conundrum of alpha particle scattering was never far away. Finally, after about 18 months, he had it. He told his colleagues "I know what the atom looks like." Rutherford explained his idea. The atom had to consist of a dense core of charge (at the time, whether the flavor of the charge at the core was positive or negative was not resolved), surrounded by mostly empty space. That way, most alpha particles miss the center of the atom, being deflected only a little. But, every so often the alpha particle hits the core of the atom straight on and then, like a bullet hitting a stone wall, the alpha particle can ricochet backwards.

In February of 1911, Rutherford reported his hypothesis to the Manchester Literary and Philosophical Society, followed by a paper in April. He correctly deduced the basic properties of the atom. An atom consisted of a small and massive nucleus, about 10^{-14} meters in diameter, surrounded by a thin cloud of electrons orbiting the nucleus. The size of the cloud was about 10^{-10} meters, fully 10,000 times greater. To give some perspective, if the nucleus of the atom were the size of a marble, the electrons would swirl around at a distance approximately the size of a football stadium, with the nucleus at the 50-yard line. Thus, one sees that an atom consists of mostly empty space.

Of course, Rutherford's atom has the same fatal property as the model proposed earlier by Nagaoka. Maxwell's theory of electromagnetism could easily prove the model wouldn't work. In order to make the electron move in a circle, it needed to be accelerated. Accelerated charges radiate energy in the form of light. As the electron loses energy, it slows down, thereby traveling in a smaller orbit. The net effect is that the electron would experience a "spiral of death" into the nucleus of the atom. The whole process would take far less than a second. This was rightfully considered by many physicists to be a fatal flaw. Rutherford was aware of this problem and stated in his paper "...the question of the stability of the atom proposed need not be considered at this stage." When Rutherford wrote to other respected physicists throughout Europe, the reception to his idea was at best lukewarm and at worst dismissive. Rutherford seemed a bit taken aback by how others received his brainchild and stopped pushing the idea quite so hard. Instead he turned to writing a book *Radioactive Substances and Their Radiation*. We will return to this technical difficulty presently.

The Nature of the Nucleus

However, let's ignore this problem for the moment. If he were right, what would the nucleus look like? Physicists and chemists thought that they knew the mass and charge of the various elements, although there remained some debate on this topic. The numbers for the first few elements are given in Table 2.1 (taking the mass and charge of the hydrogen atom to be the base unit). Thus beryllium has four times the charge and eight times the mass of that of hydrogen.

As early as 1815, an English chemist named William Prout put forth an idea that was ignored for nearly 100 years. He thought that perhaps all atoms could be made of more and more hydrogen atoms. At the time, mass was the best known property and thus Prout would say that a beryllium atom would consist of eight hydrogen atoms. Of course, we see that this can't be right, as it would also have an

Table 2.1 The mass and charge of the first four elements (taking hydrogen to be the base unit).

Element	Mass	Charge
Hydrogen	1	1
Helium .	4	2
Lithium	6	3
Beryllium	8	4

electrical charge of 8, twice as much as was measured. So no luck for Prout. Of course, the idea changes if one thinks in terms of heavy positive particles and light negative particles. If the positive and negative particles have the same electric charge (but of opposite sign), one could construct a consistent theory. The mass of the atom comes from the heavy positive particles concentrated at the center of the atom. Enough negative electrons sit in the nucleus to cancel out some of the electric charge (remember that $(+1) + (-1) = 0$). Then the remaining electrons swirl around the nucleus of the atom at a great distance, completing the atom. Taking helium as an example, one would need four hydrogen nuclei and four electrons. Two of the electrons remain in the nucleus, while the others orbit. A similar configuration would make up the other atoms. Figure 2.5 shows an example.

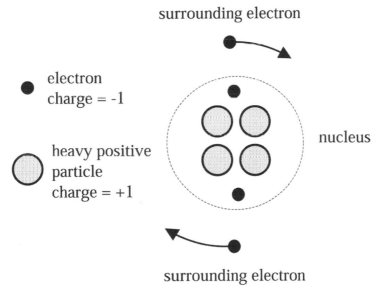

Figure 2.5 Early model of the atom, consisting only of electrons and protons. In the nucleus of the atom a few electrons cancel the charge of the protons, giving the nucleus additional mass without additional net charge. Without the small and low mass electrons, the nucleus would contain only protons and have the wrong amount of charge for a given mass. This model is now discredited.

While such a model is attractive, of course we need confirmation. This experiment took place in 1918–1919 (as Rutherford did war-related work from 1914–1918, rather than pure science). In April 1919, Rutherford published a paper that showed that the nucleus of an atom contained heavy, positively charged particles. His experiment consisted of taking alpha particles and passing them through a flask of hydrogen gas. He knew how far alpha particles would penetrate a gas and he saw that there were particles that penetrated much farther than that. Rutherford realized that the heavy alpha particles were hitting the hydrogen nuclei and accelerating them. This observation wasn't really much of a surprise.

The interesting thing happened when he repeated the experiment and let the alpha particles pass through air. He knew that air consisted of nitrogen, oxygen and carbon dioxide molecules and he could calculate how fast they would move if they were hit by an alpha particle. He saw evidence in line with his predictions. But he also saw that there was a particle that could travel a great distance through the air, just like the hydrogen nucleus. Since he knew that there was little hydrogen in air, it seemed unlikely that he could explain these penetrating particles as simply cases of alpha particles hitting hydrogen. So he did the obvious thing. He generated chemically pure samples of oxygen, nitrogen and carbon dioxide gas. When he repeated the experiment, he saw no deeply penetrating particles with oxygen and carbon dioxide, but he did see them with nitrogen. In his paper, he reasoned that nitrogen had a peculiar nuclear structure and that perhaps it could be thought of as a tightly bound core with a few more loosely bound hydrogen nuclei. Since they were more loosely bound, they could be easily knocked off by the alpha particles. Oxygen and carbon, on the other hand, had only a tightly held nuclear core, with no loosely bound hydrogen nuclei contained within them. Thus, Rutherford had shown that at least a nitrogen nucleus contained within it the more primordial hydrogen nucleus. Rutherford called the hydrogen nucleus a proton after the Greek word "protos" meaning "first" and thus the proton was shown to be a constituent of atomic nuclei.

Further refinements led Rutherford to wonder about the peculiar combination of a proton and an electron in the nucleus of an atom. If such a combination were possible, this particle would be electrically neutral. Such a particle is highly desirable, because it would move freely through matter. Since both atomic nuclei and alpha particles have an electric charge, they interact at great distances through their respective electric fields. A neutral particle would be invisible to the nucleus of an atom and thus be able to get very close. This neutral particle would be an ideal probe of the nucleus of atoms. In Rutherford's 1920 paper on the subject, he explicitly thanked his assistant, James Chadwick. We'll meet Mr. Chadwick again.

Rutherford returned to Cambridge's Cavendish Laboratory in 1919, this time as Director, taking over for his one time mentor J.J. Thomson. His path had come full circle. In addition to Rutherford's many experiments of his own, he proved to be an extraordinary mentor. James Chadwick was a young student of Rutherford, as was Niels Bohr, one of the early architects of Quantum Mechanics. John Cockroft and Ernest Walton were driven by Rutherford to develop the first real particle accelerator, which opened up an entirely new way to do particle physics experiments. All of these young protégés of Rutherford eventually joined him in that most exclusive of clubs, being a recipient of the Nobel Prize. Even Robert Oppenheimer, the so-called "Father of the Atomic Bomb" for his role in the American effort to build the atom bomb, worked for some time under Rutherford's watchful eye.

For his work, Rutherford received the 1908 Nobel Prize in Chemistry and 21 honorary degrees. In 1931, he was awarded a baronetcy and titled himself Baron Rutherford of Nelson, after a town near where he was born. Ever the good son, he wired his mother "Now Lord Rutherford. More your honour than mine. Ernest." The baronetcy was not entirely a happy time. Eight days before he received the honor, his only daughter died due to complications following childbirth with her fourth child.

On October 19, 1937, Ernest Rutherford died, following an operation to repair a minor hernia that occurred while he was cutting down trees on his property. Rutherford's ashes were interred in the nave of Westminster Abbey, just west of Sir Isaac Newton's tomb and near those of Lord Kelvin. Rutherford played a crucial role, both directly and indirectly, in understanding that the atom was not as fundamental as had been thought. Without Rutherford's crucial insights, the upheaval that was Quantum Mechanics that so rocked the worldview of physicists the world over might have been delayed. That a so superbly trained classical physicist could play such a pivotal role in replacing the physics that he had learned shows a great openness of thinking and provides a great example to which young scientists can aspire.

From the efforts of Rutherford and Thomson, we have built up a model of the atom quite different from that supposed in the closing years of the 19th century. A dense nucleus of the atom, consisting of protons and perhaps pairs of protons and electrons, was surrounded by a loose aggregate of electrons swirling at relatively great distances. The proton and electron had been observed, but the neutron, as Rutherford had termed the closely bound state of the electron and the proton, had not. Rutherford did not discover the neutron, but as his legacy includes a number of talented researchers, perhaps he should get some of the credit.

James Chadwick was a student at Manchester University, graduating in 1911. After graduation, Chadwick stayed at the laboratory as Rutherford's research assistant. In 1914, Chadwick traveled to Berlin to work with Hans Geiger, another Rutherford protégé. Trapped by the outbreak of World War I, Chadwick's English citizenship entitled him to accommodations courtesy of the German government as a civilian prisoner of war. Chadwick was treated reasonably well (although he did suffer from malnutrition) and allowed to pursue academic curiosity by reading and chatting with other scientists, but experiments were forbidden. With the end of the war in 1918, Chadwick returned to Manchester. As you recall, this was the period of time when Rutherford was working on the discovery of the proton. In 1919, Rutherford was appointed to the directorship of Cavendish Laboratory and Chadwick followed him to Cambridge. Rutherford oversaw Chadwick's Ph.D. in 1921 and, upon obtaining his doctorate, Chadwick was appointed as the assistant director of Cavendish.

Chadwick was interested in Rutherford's neutron and he looked for it both in 1923 and in 1928, without success. In 1930, experimental results on the European continent piqued his curiosity and he watched with great interest. German physicists Walther Bothe and Herbert Becke had noticed that when they shot alpha particles at a block of beryllium, electrically neutral radiation was produced that could penetrate 20 centimeters (8 inches) of lead. They assumed that this radiation was high-energy gamma rays (i.e. photons).

Irène Joliot-Curie (daughter of Marie and Pierre Curie) and her husband Frédéric Joliot-Curie put a block of paraffin wax in front of the neutral rays. ("Why wax?" you say. I don't know … I asked myself the same question. We now know that paraffin is a good idea because of its large hydrogen content, but what gave them the original idea?) They noticed that protons were leaving the wax. They concurred with Bothe and Becke's evaluation, suggesting that the gamma ray photons were knocking protons out of the wax.

Chadwick disagreed. He did the arithmetic and showed that such an explanation violated the law of the conservation of energy. Instead, he proposed that the neutral radiation was the missing neutron. He set out to test his hypothesis. He repeated Bothe and Becke's experiment, but this time he made the neutral particle hit a hydrogen gas target. When the neutral particles hit the hydrogen, protons flew out.

Because Chadwick could not directly see the neutral particle, he determined its mass by measuring the energy of the proton leaving the hydrogen and worked backwards. He found that the mass of the neutral particle was about 1.006 times greater than that of a proton. (How's that for precision?) The neutron had been found.

Of course, the question of the nature of the neutron (i.e. was it a mix of a proton and an electron or was it an entirely different particle?) was not resolved. In his paper "The Possible Existence of a Neutron," submitted in 1932, Chadwick wrote

> … we must nevertheless suppose that the neutron is a common constituent of atomic nuclei. We may then proceed to build up nuclei

out of α-particles, neutrons and protons, and we are able to avoid the presence of uncombined electrons in a nucleus. ...

...It has so far been assumed that the neutron is a complex particle consisting of a proton and an electron. This is the simplest assumption and is supported by the evidence that the mass of the neutron is about 1.006, just a little less than the sum of the masses of a proton and an electron. Such a neutron would appear to be the first step in the combination of the elementary particles towards the formation of a nucleus. It is obvious that this neutron may help us to visualize the building up of more complex structures, but the discussion of these matters will not be pursued further for such speculations, though not idle, are not at the moment very fruitful. It is, of course, possible to suppose that the neutron may be an elementary particle. This view has little to recommend it at present, except the possibility of explaining the statistics of such nuclei as N^{14}.

Chadwick's experimental results were quickly accepted and when Werner Heisenberg showed that the neutron could not possibly be a combination of a proton and an electron, scientists were accepting of the existence of the neutron as an elementary particle in its own right. For his discovery of the neutron, Chadwick was awarded the Nobel Prize in Physics in 1935, followed by knighthood in 1945.

Quantum Mechanics: An Intermission

The model of the atom in 1932 was as follows. The nucleus, consisting exclusively of protons and neutrons, was surrounded by a cloud consisting exclusively of electrons. The mass and charge of the respective elements of the atom are listed in Table 2.2, using the proton as the basis for comparison. The atom had begun to look like the eventual logo of the Atomic Energy Commission. There was one especially interesting and desirable consequence of this model. Rather than nearly a hundred different atoms, one for each element, now all atoms could be explained as endless combinations of three particles, or possibly four, if one included the alpha particle. This is a clear simplification

Table 2.2 Relative charge and mass of the three elementary particles that make up an atom. All units are given relative to the proton.

Particle	Mass	Charge
Proton	1	+1
Neutron	1.006	0
Electron	1/1886	−1

and suggests that our understanding of the nature of the universe had much improved. There was only one problem. Everyone knew that the whole thing was entirely hogwash. Electrons simply couldn't orbit the nucleus as described. Something was badly wrong.

The resolution of this conundrum is the story of Quantum Mechanics. In 1900, Max Planck had postulated that energy came in discrete chunks, rather than a continuous spectrum of possibilities. In 1913, a protégé of Rutherford, Niels Bohr, adopted Planck's ideas and combined them with Rutherford's original model to get around the objections put forth to both Nagaoka and Rutherford's model. During the 1920s, things really heated up, with the legendary founders of Quantum Mechanics being especially productive. Pauli, Heisenberg, Schroedinger, Dirac and Born all played prominent roles. While the story of the beginnings of Quantum Mechanics is fascinating reading, it is really outside the scope of this book. Once the atom became a conglomeration of other (and even more fundamental, particles) the frontiers of particle physics had moved on. The interesting story of quantum mechanics can be read in the references given in the suggested reading for Appendix D.

If the acquisition of knowledge can be represented as an endless staircase that we are meant to climb, the appreciation that elements are made up of atoms that are smallest examples of each element is but a step. The deeper understanding that each of these atoms aren't fundamental at all, but rather contain within them protons, neutrons and electrons, arranged in intricate and complex ways, is the next step. Most casual students of science stop their ascent here. Rutherford's model is very nice and explains much of the world we see around us, but not all. And you, gentle reader, by continuing to read, will be taking additional steps along that long staircase, rising to ever more interesting heights. Some of your less enlightened friends and colleagues might not understand your need to know but, to borrow from Thoreau, if you walk to a different beat than your contemporaries, perhaps it is because you hear a different drummer. And besides, the ongoing story of particle physics is a fascinating one and, with each step, we can come closer to understanding the universe at the deepest and most fundamental level. The need for something more than Rutherford's model was clear very early on. Even as Rutherford and Bohr made their initial attempts at explaining the "planetary system" model of the atom and even before Chadwick had unambiguously determined the existence of the neutron, the very beginning rumblings of the first frenzied years of particle physics was being heard. While the idea of protons and electrons had been kicked around by chemists for years, something entirely new was becoming apparent.

A mystery of the 1900s was the nature of light. Heated gases had been shown to emit light of particular colors. Each element emitted a different set of colors; in fact, each set can be thought of as a "fingerprint" of the respective element. Amazingly, the element helium was discovered in October 1868, by Sir Joseph Lockyer by analyzing light from the Sun. The Sun contained helium. It wasn't until 1895 that Sir William Ramsay detected helium here on Earth in a uranium-bearing mineral, cleveite. Because Ramsay did not have a good spectroscope, he sent samples to both Lockyer and Crookes, of the Crookes tube fame. They confirmed his discovery.

As discussed in Chapter 1, Maxwell had shown that light and electromagnetism were two facets of the same underlying phenomenon. With Rutherford's atom, it was understood that the light was being emitted by the electrons surrounding the nucleus. Since each atom had a different number of electrons in different configurations, this might explain the "fingerprint" of each element. As usual, there was a problem. While Maxwell's theory could explain how the electrons could emit light, it could not explain how each atom could emit only specific colors of light. Related to earlier criticisms of Nagaoka's and Rutherford's atom, the electrons would emit a continuous spectrum of colors as they spiraled down into the nucleus, never to return. Niels Bohr, who simply added a requirement, saved this model of the atom in 1913. He said that the electrons were allowed to be in only certain orbits. Taking our solar system as an analogy, it's as if it were possible to have planets where they are, but it is *impossible* to have planets between them. If we launched a probe to Mars, it would either be near Earth or near Mars, but never in between. Bohr's hypothesis was not rooted in any deep underlying theory, rather it was more of a "if this were true, it would explain a lot" kind of idea. And explain it did, as now each element could emit only specific colors of light. As electrons jumped from an outer orbit to an inner one, they would emit a single photon of light, with the color uniquely determined by where the electron began and where it ended up. Bohr's theory was merely an educated guess and not rooted in a deeper theory. It was quantum mechanics that finally provided the explanatory theoretical framework.

Beta Radiation and the Neutrino

Of course, with quantum mechanics explaining the colors (more technically the energy) of light emitted by atoms, physicists naturally turned their attention to the myriad of types of radiation that had been recently discovered. X-rays and gamma rays were now understood to be very energetic photons... one could think of them as colors not visible to the naked eye. The source of x-rays was the now relatively well-understood electron cloud, while gamma radiation originated in the nucleus of the atom. Alpha radiation was understood to be the emission of a helium nucleus by a much heavier element, while beta radiation was simply the emission of an electron by the nucleus of an atom. And, of course, cathode rays were now known to be electrons ejected from the cloud surrounding the nucleus. The nature of the various types of radiation seemed clear.

In order to further our discussion, we need to know about an important physical principle; the law of the conservation of energy. The story of this law is probably worth a book or at least a chapter by itself, but we'll only discuss the highlights here. Although a commonly used word, energy is a somewhat abstract concept. There are many forms of energy which, on first inspection, could not appear to be more different. The first kind of energy that we will discuss is kinetic or moving energy. A baseball thrown through the air carries kinetic energy, because it's moving. There are a number of kinds of kinetic energy: rotational, vibrational or translational energy. Thus anything that is vibrating, rotating or simply moving carries energy. Because the total amount of energy can't change, the energy may change forms, but not increase or decrease. A hammer hitting a bell is an example of converting translational energy to vibrational energy. The hammer stops moving and the bell begins to vibrate. We will return to this transmutation of energy soon.

The second type of energy is a little more difficult to visualize. This type is called potential energy. Such energy is explicitly *not* energy of motion; rather it is energy which could potentially cause an object to move. Thus a ball lifted above the floor will fall (i.e. move) if you let it go. So the ball has potential energy. Similarly, if you put a pebble in a slingshot and pull the slingshot back, it's not moving. But it *will* move when you let it go, so the rubber band in the slingshot has potential energy.

There is a third type of energy that is even trickier to appreciate, that of mass energy. Einstein's theory of special relativity, the famous $E = mc^2$, says that matter is a form of energy. Such a contention was truly revolutionary as it implies that one can convert moving energy into mass energy and back again. Relevant details of Einstein's theory are given in Appendix D.

The last idea that one needs is the law of conservation of energy. This law states that energy is neither created nor destroyed, but can only change forms. Thus one can see how the further back you pull the slingshot (the more potential energy you have), the faster the

Table 2.3 An illustration of how there can be many different combinations of kinetic, potential and mass energy that all sum to a single value.

Energy Type	Example			
	1	2	3	4
Kinetic	2	0	8	6
Potential	4	0	1	4
Mass	4	10	1	0
Sum	10	10	10	10

pebble will eventually move (the more kinetic energy it has). In this example, "before" and "after" refer to before and after the slingshot is released. Finally, one can work out the total energy by adding up the amount of each type of energy, for example the total amount of energy is simply the amount of kinetic, potential and mass energy added together. You can do the sum at anytime and you will find that the three numbers always add to the same amount. Let's illustrate this idea with a particular example. Suppose the total energy of some system or situation is some arbitrary amount, say 10. If you add the three kinds of energy, they must always add to 10. In Table 2.3, I show four completely arbitrary examples of this principle.

Now that we know something about the law of conservation of energy, we return to the idea of radiation and nuclear decay. Suppose you start with an atomic nucleus that isn't moving. In this case, you have no kinetic energy, no potential energy and only mass energy. The nucleus then decays into two fragments, which can in principle be moving. These two fragments have mass energy and kinetic (moving) energy, but no potential energy. Thus we might write that a little more clearly as

Mass (original nucleus) = Mass (fragment 1) + Mass (fragment 2)
+ Kinetic Energy (fragment 1)
+ Kinetic Energy (fragment 2)

Since we know the masses of the original nucleus and all of its fragments, our unknowns are only the moving energies of the two fragments. Since one fragment is usually enormously more massive than the other, one of the fragment's kinetic energy is very small and can be ignored (i.e. we call it zero). Thus, we have only one unknown. Since total energy is constant, this means that the kinetic energy (and thus speed) of the light fragment is completely determined. Making up some numbers for fun, say the mass of the original nucleus is 11, and the masses of the two daughter nuclei are 9 and 1. Since the kinetic energy of the massive daughter is about zero, each and every time the kinetic energy of the light fragment must be 1. There is no alternative.

So let's turn our attention to particle decay. As an example of an alpha particle decay, we'll discuss the situation when a uranium nucleus decays into a thorium nucleus and an alpha particle ($U^{238} \rightarrow Th^{234} + \alpha^4$). Because the alpha particle is so much less massive than the thorium nucleus, its moving energy is completely determined. When the experiment is done, one gets what one expects. One only measures a single and unique value for the kinetic energy of the alpha particle. Theory and experiment are in agreement.

The situation should be even better with beta radiation, as the mass of an electron is so much smaller than that of an alpha particle. Thus the theory should work even better. One could take the example of radium decaying into actinium via beta (i.e. electron) decay ($Ra^{228} \rightarrow Ac^{228} + e^-$). Again the kinetic energy of the electron should be completely determined. When measured, one should only get one particular value. However, when this experiment is done, we find that the electrons never have the predicted energy; they always have less … sometimes much less.

Further, it turns out that the kinetic energy of beta radiation can take any value, as long as it is less than that predicted. This is most mysterious. Figure 2.6 shows how the energies of alpha and beta radiation differ.

The original evidence that beta rays were emitted with the "wrong" energy was first observed by Chadwick in 1914. As you

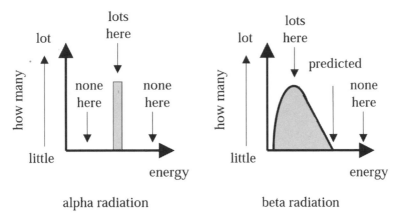

Figure 2.6 Alpha and beta radiation release particles with a different range of energies. In alpha particle emission, all particles are emitted at a specific energy. In beta particle emission, particles are emitted over a range of energies, all of which are lower than what one would predict. This observation led to the hypothesis of the neutrino.

recall, he had gone to Berlin in 1912, in order to work with Hans Geiger. A few months before the outbreak of World War I, he was able to measure the energy distribution for beta rays. Initially nobody believed him, but his results were eventually confirmed by Ellis in 1927 and Meitner in 1930.

The question of how to explain this mystery perplexed physicists for some time. It seemed as if the law of the conservation of energy was invalidated. Such a possibility would turn the entire world of physics on its head. So gifted a scientist as Niels Bohr suggested that perhaps the law of the conservation of energy was not respected in radioactive processes. While possible, this would be counter to all prevailing thought and experiment (beta decay aside).

Finally Wolfgang Pauli suggested an alternate explanation. Energy conservation could be preserved if instead of two particles after the decay, there were three. Then one would have kinetic and mass energy of the third fragment to add to the energy balancing equation. If energy were stored in moving a third particle, then there would be less to move the electron from beta decay (recall that the sum of the energies had to be always constant). Then the earlier prediction would be in error and this would explain why the electron didn't always have a specific energy after the decay. In order to be compatible with known measurements, the hypothetical third fragment would have to be very light and electrically neutral (or it would have been easily detected long before). Also, the third particle should not interact very strongly with matter, for the same reason.

Pauli announced his idea on December 4, 1930 to some colleagues at a conference via a letter that he had read for him in his absence. The letter read

> Dear Radioactive Ladies and Gentlemen,
>
> As the bearer of these lines, to whom I graciously ask you to listen, will explain to you in more detail, how because of the "wrong" statistics of the nitrogen and lithium-6 nuclei and the continuous beta spectrum, I have hit upon a desperate remedy to save the "exchange theorem" of statistics and the law of conservation of energy. Namely, the possibility that there could exist in the nuclei electrically neutral particles, that I wish to call neutrons, which have spin 1/2 and obey the exclusion principle and which further differ from light quanta in that they do not travel with the velocity of light. The mass of the neutrons should be of the same order of magnitude as the electron mass and in any event not larger than 0.01 proton masses. The continuous beta spectrum would then become understandable by the assumption that in beta decay a neutron is emitted in addition to the electron such that the sum of the energies of the neutron and the electron is constant...
>
> I agree that my remedy could seem incredible because one should have seen those neutrons very early if they really exist. But only the one who dares can win and the difficult situation, due to the continuous structure of the beta spectrum, is lighted by a remark of my honoured predecessor, Mr. Debye, who told me recently in Bruxelles: "Oh, it's well not to think of this at all, like new taxes." From now on, every solution to the issue must be discussed. Thus, dear radioactive people, look and judge. Unfortunately, I cannot appear in Tübingen personally, since I am indispensable here in

Zurich because of a ball on the night of 6/7 December. With my best regards to you, and also to Mr. Back.

Your humble servant,

W. Pauli

Note that Pauli used the word "neutron," but this was before Chadwick discovered the real neutron in 1932. With Chadwick's discovery that the real neutron was actually quite massive, it was clear that the neutron was not the particle described by Pauli. It was some time later when the noted Italian physicist, Enrico Fermi, gave Pauli's hypothetical particle a new name, neutrino, an Italian diminutive for "little neutral one." The neutrino was an intriguing idea. But was it real?

After the discovery of the neutron, some physicists turned their attention to the question of whether or not the neutrino had a physical reality. Just because it explains things doesn't make it true. In the intervening three years, several properties of the neutrino had become apparent. Speaking at the Solvay conference in Brussels, in October 1933, Pauli said

> ... their mass cannot be very much more than the electron mass. In order to distinguish them from the heavy neutrons, Mister Fermi has proposed to name them 'neutrinos.' It is possible that the proper mass of neutrinos be zero ... It seems to me to be plausible that neutrinos have a spin of 1/2 ... We know nothing about the interaction of neutrinos with the other particles of matter and with photons: the hypothesis that they have a magnetic moment seems to me not founded at all.

By the end of 1933, Enrico Fermi devised the first "real" theory of beta decay, including the hypothesized neutrino. In fact, he had devised the first theory of the weak force. We've not yet considered the forces encountered in particle physics, but we will remedy that deficiency soon. The weak force basically has the property that it allows a particle governed only by it to penetrate great distances of material, even greater than the width of the Earth, without interacting. It's no wonder that the neutrino had not been detected.

By this time, the neutrino, if it actually existed, had many of its properties determined. The most notable ignorance was of the neutrino's mass. Was it zero, or just very small? If the neutrino could not be directly observed, due to the tiny chance that it would interact with the detector, one could determine its mass in a beta decay experiment by measuring the motion energy of the larger daughter fragment after the decay. Recall that earlier we assumed it was zero, when in fact it was just really small. If we could measure this tiny number, we could determine the mass of the neutrino in much the same way Chadwick measured the mass of the neutron. After much effort, all attempts were unsuccessful.

As time progressed, nuclear fission was discovered, culminating with the first controlled nuclear reaction under Stagg Field at the University of Chicago and its deliberately uncontrolled cousin at Alamogordo, New Mexico. In such large nuclear reactions, uncounted atoms undergo beta decay and thus emit neutrinos. In 1951, a physicist by the name of Frederick Reines had the idea to place a detector near a nuclear detonation site. With so many neutrinos coming from the detonation, at least a few might be detected in a suitably designed experiment. Such an experiment is daunting and after some thought, he and Clyde Cowan, another physicist, decided that perhaps it would be a bit more civilized to do the proposed experiment near a tamer nuclear reactor. They proposed the experiment in February 1953. They would place their detector near the nuclear reactor at Hanford, Washington and try to directly detect neutrinos from the intense source. By late spring, the detector was built and by summer they had their results… inconclusive. It was very hard to distinguish between when the reactor was on versus when it was turned off. But they had learned valuable lessons and realized that the Hanford site was not suitable, due to an excess of fake signals in their detector (i.e. positive indications of neutrinos when it was known that there were nearly none present). So it was back to the drawing board.

In 1956, they tried again, this time near the reactor at Savannah River, South Carolina. This time, they had reduced the fake signals and they could clearly see when the reactor was running as compared to when it was off. They had directly observed the neutrino, more than 25 years after it was proposed. For this difficult and successful experimental result, Reines was awarded the 1995 Nobel Prize. (Cowan had died in 1974 and the Nobel Prize cannot be awarded posthumously.) We will return to neutrinos again towards the end of this chapter and again in Chapter 7.

However, let's consider what we've found. The beta decay of an element, say the before-mentioned radium (with its 88 protons and 140 neutrons) to actinium (with its 89 protons and 139 neutrons) involves changing the number of protons and neutrons. Thus, essentially this can be seen as a neutron (n^0) decaying into a proton (p^+), an electron (e^-) and a neutrino, for which we use the Greek letter "nu" (i.e. ν) ($n^0 \rightarrow p^+ + e^- + \nu$). An elemental particle (the neutron) is transmuting into other elementary particles (the proton, the electron and this new particle). Hmm ... we've heard this transmutation story before. Like an artful fan dance, have we been given a flirtatious peek at the next layer in the cosmic onion?

The answer is a most titillating "yes." With the observation of the transmutation of what had appeared to be elementary particles, we begin to see evidence that these particles are not elementary at all. Rather than the history that we've discussed to this point, over which we've traveled a comfortable path leading towards an understanding of the nature of the atom, we now enter into truly uncharted territory. From this point on, we will be investigating physics for which we've had very little foreshadowing.

More Forces

Before we continue on with the explosive discovery of particles unimagined by the very best physicists at the turn of the 20th century, we must pause for a moment and discuss some ideas that will significantly improve our understanding of just how confusing this discovery process was. We need to talk about the clear evidence that there were different types of forces acting on the particles, which had an important impact on their behavior. In addition, we need to introduce an important concept from quantum mechanics ... that of quantum mechanical spin. Armed with this clarifying knowledge, we will be ready to plunge ahead into extraordinary and dizzying discoveries that marked the middle decades of the 20th century.

Physicists knew of "the force" long before George Lucas usurped the term for his own use. An understanding of the concept of force and the types of forces present in our universe is crucial for one hoping to fully appreciate just how interesting is the world in which we live. The concept of force is tied to our language in ways that are less precise than those we use in physics, but we will start with these common meanings. One facet of the definition of force is the following. We say an object feels a force if it is attracted to or repelled by another object. Examples include our attraction to the Earth or the Earth's attraction to the Sun. An example of repulsion is provided by two magnets which, when oriented correctly, will repel each other. We also speak of force as something that can effect change, as in military force changing a regime or a political force changing a law. Force, as physicists define it, can also have a similar meaning, after all *something* causes a neutron to decay into a proton in beta decay or causes a helium nucleus to be ejected from a larger nucleus in alpha decay. Force thus becomes the agent of change, either through attraction or repulsion (and changing an object's motion) or through changing an object's identity.

Force is so crucial an idea that I devote all of Chapter 4 to it. But our modern understanding of the nature of force differs somewhat from the ideas held in the early part of the 20th century. To further our discussion of the early history of particle physics, we need to understand forces in a way similar to that of the early physicist pioneers.

People had been aware of forces since time immemorial. Even thousands of years ago, people knew of the forces of gravity and lodestones (magnetism), of static electricity and the orderly progression of the heavens. As discussed in Chapter 1, the second half of the last millennia led to the realization by the two great minds of Newton and Maxwell that all phenomena observed by the late 19th century physicists could be explained as manifestations of the forces of electromagnetism and gravity. Gravity explained our weight and the motion of planets. Electromagnetism was a much newer theory, but it explained static electricity and all of the other electrical phenomena, magnetism and indeed light itself. An understanding of atoms was not complete, but it was clear that electricity played a role, as one could break up molecules into their constituent atoms using electricity. Two forces explained it all.

With the advent of Rutherford's atom, this simplicity changed. To appreciate this, we must recall two facts. The first thing is that Rutherford had shown that the nucleus of an atom included as many as 100 positively-charged protons, packed into a small spherical volume with a radius of approximately 10^{-14} meters. The second fact one must recall is that two positively-charged objects will feel a repulsive force. In the nucleus of an atom, with its dozens of positively charged protons, each repelling the other, it's clear that the tendency must be for the nucleus to blow itself apart. However, we know it doesn't. With the exception of the radioactive elements, we know that the nucleus of an atom is stable, lasting essentially forever. So three possible solutions must be considered: (1) the idea of protons is wrong; (2) the theory of electromagnetism doesn't work at such small distances; or (3) another force must be present to counteract the electromagnetic repulsion.

Explanations 1 and 2 were excluded by experiments, leaving physicists with the inescapable conclusion that a new force had been discovered ... one that held the nucleus of the atom together. This force (illustrated in Figure 2.7) was called the nuclear force or occasionally the strong force, to highlight its strength as compared to the electromagnetic force. The first evidence for the strong force interaction (besides the elementary discussion given in the last paragraph) was found in 1921. Chadwick was scattering alpha particles (i.e. helium nuclei) from a target and found that more particles scattered into certain angles than could be explained by the electric force between the alpha particle and the nucleus. A proper theory for the behavior of the

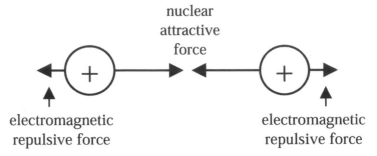

Figure 2.7 Cartoon describing the necessity of a strong nuclear force. In the absence of a counter-balancing force, the electromagnetic repulsion between two protons would cause them to accelerate away from one another. The fact that they don't points to a stronger attractive force. The arrows denote each force, with the length of the arrows indicating the forces' respective strength.

nuclear force was not available until 1935, when Japanese physicist Hideki Yukawa had some interesting ideas on the problem.

A few years earlier, Werner Heisenberg had an idea on how the strong force might work. He knew that one could think of beta decay as a neutron emitting an electron, thus turning itself into a proton. Similarly, a proton, hit by an electron, could turn into a neutron. He therefore hypothesized that the nuclear force could be explained by electrons jumping back and forth between protons and neutrons. As long as the electron didn't escape the nucleus, the total number of protons and neutrons wouldn't change, but the identity of a particular particle could. This basic idea is illustrated in Figure 2.8.

Heisenberg knew that his theory was wrong for reasons of quantum mechanical spin (more on that very soon). For purposes of our discussion here, think of spin as something that each particle has and each electron, proton and neutron carries a spin of value 1/2. To find the total spin, you simply add them together. So before and after the exchange, there is just $1/2 + 1/2 = 1$ units of spin, while during the exchange there is $1/2 + 1/2 + 1/2 = 3/2$ units of spin. Since the total amount of spin can't change (like the law of conservation of energy, there is a similar law of conservation of spin), Heisenberg knew that his idea was wrong, but it was interesting.

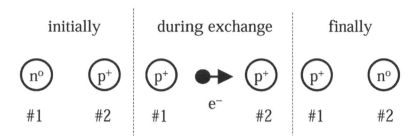

initially · during exchange · finally

Figure 2.8 Heisenberg's theory of nuclear force. A neutron emits an electron, changing into a proton. The electron travels to an adjacent proton, changing it into a neutron. The number of protons and neutrons do not change, but the electron exchange was thought to be the cause of the strong force. This idea was discredited because it did not conserve angular momentum, in the form of quantum mechanical spin.

Yukawa liked Heisenberg's idea, but he was just as aware as Heisenberg was with the theory's problems. He then spent a long time, trying to work out a similar theory, but without the problems. When he read Fermi's paper on beta decay, he had an epiphany. Fermi's theory required a particle that no one had seen (the mysterious and ghostly neutrino). Yukawa realized that he was taking a wrong tack. Rather than trying to shoehorn his theory into known particles, he would create a theory and see what particles were needed to make the theory work. As long as the predicted particles were not too outlandish, maybe they would be real.

Combining some of the various ideas he had read, Yukawa realized that his hypothetical particles could have positive electrical charge (the emission of which would convert protons into neutrons), negative charge (to convert neutrons to protons) or be electrically neutral (to create a force between two protons or two neutrons). He knew it needed to have zero spin, to fix up the problems with Heisenberg's attempt. Further, he knew the force was strong but it was weird because the force didn't extend very far. He knew this because, although the force was strong, no effects of it were apparent when one looked at the electrons surrounding the atom. In this, the force is a little like Velcro. It's very strong when two things are in contact, but essentially zero when they're not. Using this information, he found that his theory required a particle with a mass of about 1/10 that of a proton and about 200 times larger than an electron. The problem of course, as Yukawa himself stated in his paper, was that no such particle had been observed. Yukawa called his proposed particle the U-particle, a term which never caught on. Subsequent scientists proposed the name "Yukon," in honor of Yukawa, and the "meson" or "mesotron" (meso being Greek for "middle," i.e. having a mass between that of the proton and the electron). While Yukawa's particle had never been observed and so his theory was thus suspect, there was no disputing the existence of a mysterious new strong force.

With our force count now at three, we need to reconsider the phenomenon of radioactivity. We said earlier that forces could govern change and that radioactive decay is indisputably the change of a nucleus of an atom. However, the strength of the force is also related to how quickly it can effect change. It was known that the characteristic time involved in radioactive decay was very long and spanned a large range, from fairly small fractions of a second to many millions of years. Since electromagnetism and the strong force react on timescales tiny compared to a second, they were not likely candidates for the force that caused radioactivity. Thus, it was clear that there was a fourth force, much weaker than electromagnetism and the strong force and much stronger than gravity. In 1934, Enrico Fermi published his paper on beta decay which, as we now know, evolved into the theory of the weak interaction. The weak force appeared to be thousands of times weaker than electromagnetism, although incomparably stronger than gravity, and very mysterious. Just how mysterious will become apparent a little later.

So we see that by the mid 1930s, it was known that there were four forces with which to contend, each with different strengths and behaviors. As we continue our discussion of the discovery of new particles, it will become clear that not all particles interact equally with each of the forces. This cacophony of particles and forces begged for some sort of deeper understanding. Our modern view of what is going on is given in Chapters 3 and 4. But first some more history.

Something to Make Your Head Spin

The next principle with which we must become familiar is a subtle one … that of quantum mechanical spin. Spin is a concept that is fairly easy to state, but one that can quickly become quite difficult. We will restrict ourselves to the minimum appreciation necessary to continue with our understanding of particle physics. Rather than telling the entire historical tale of spin, we'll concentrate on what it is and why it's important.

All known fundamental particles act as if they were tiny spinning tops. A short calculation shows that they can't be spinning in the usual way, for instance when one asks just how fast the electron must move in order to account for its experimentally determined spin, one finds that the surface of the electron would have a velocity exceeding that of light. So the vision of an electron as a tiny ball of electrical charge, spinning furiously, is not quite right, although it is a good enough working picture that we can use it, as long as we remember some facts.

The laws of quantum mechanics are weird and wonderful and completely counterintuitive. We are all familiar with a spinning top. While we don't have a number whereby we can quantify spin in the same way we quantify weight, it is somehow intuitive that a fast-spinning top should have a big spin, while a slower spinning top should have a smaller spin. Further, the top can spin at all speeds, from its maximum down to zero. In this, quantum mechanical spin is different. Each particle is allowed to take on only specific discrete values of spin. It's as if when you stepped on a scale, you could weigh only exact values of pounds, say 1 or 2 or 3, but it was *impossible* to weigh 2.5 pounds. Only integer values of pounds are allowed. The units of spin are arcane and given in units of a thing called $\hbar$ (pronounced h-bar). $\hbar$ is simply a unit, like pound. If someone asks you your weight and you say 160, pounds are understood. Similarly, if one asks what spin is carried by a particle, we never say $\hbar$. We just say the number. The only allowed values of spin are ($\ldots -5/2, -2, -3/2, -1,$

−1/2, 0, 1/2, 1, 3/2, 2, 5/2, ...). (Note "..." means "and the pattern continues.") That's it. Spin 1/4 simply isn't allowed. Thus the only types of spin allowed are the integers (... −2, −1, 0 1, 2, ...), and half-integer (... −5/2, −3/2, −1/2, 1/2, 3/2, 5/2, ...) values.

Of the elementary particles we know so far, the electron, neutron, proton and neutrino all carry half-integer spin (specifically 1/2). The photon is different with a spin of 1, as is Yukawa's proposed particles, with their spin of 0. During the period of 1924–1926, it became clear that integer and half integer particles were fundamentally different and acted in very different ways. Half-integer spin particles are called fermions after Enrico Fermi, while the integral spin particles are called bosons, after Indian physicist S.N. Bose. Bosons are fundamentally gregarious and it is possible for more than one to be in the same place at the same time. Fermions, on the other hand, are loners of the atomic world and it is *impossible* to get two identical fermions in the same place. This fundamental difference has significant consequences for their behavior. In Chapter 8, we discuss some new theoretical developments that might bridge the divide between fermions and bosons, but as of this writing, they remain distinctly different kinds of particles.

As we return to our study of the particle discoveries in the first 60 years of the 20th century, we need to remember that there are several important properties that one must determine for each particle discovered. Of course, the mass and electrical charge of the particle is important, but of equal import is whether or not it is a fermion or boson. It's also necessary to determine which of the four forces affect each particle and a related question is how long does the particle live and into which particles does it decay? If a particle can decay in several different ways, which types of decay are more likely and which are rare? It was answers to these and other questions that allowed physicists to unravel the confusing situation that was about to confront them.

Cosmic Rays: Particles from the Heavens

Our continued voyage into the world of modern particle physics requires that we return to the very first days of the 20th century.

Earlier in this chapter, we read of how Marie Curie used an electroscope to precisely measure the amount of radioactivity present in various elements. Her reason for using this method was that an electroscope is an extremely precise method for measuring radioactivity. With such a precise instrument available, other experimenters were quick to adopt it for their own use. The electroscope soon became ubiquitous among early physicists.

One thing troubled the electroscope users. Technically, what an electroscope measures is the conductivity of the air surrounding it. The conductivity of the air is increased by the presence of radioactivity, but also by other things like moisture in the air (recall Curie's damp potato cellar). If you're trying to measure the radioactivity of a substance, anything that alters the conductivity of the air (except for the radioactivity in which you're interested) is undesirable. Therefore, physicists went to great pains to do their experiments under ideal circumstances. This involved using perfectly dry air and otherwise isolating their apparati from anything that might affect the air's conductivity. In order to verify that they had isolated their equipment adequately, they would charge their electroscopes and watched to see that their readings remained unchanged for a long period of time.

However, no matter how carefully they shielded their experiment, they found that it always acted as if there was radiation or moisture present. Since they had very carefully arranged to remove all moisture, they were led to the inescapable conclusion that there was a tiny, yet constant, presence of radioactivity here on Earth. Such a supposition was not so silly, as it was known that uranium ore was radioactive and it came from the Earth. So, perhaps trace amounts of radioactive elements were everywhere. Experiments were done to shield the equipment from the Earth's latent radioactivity. While shielding from alpha and beta radiation and x-rays was straightforward, gamma radiation, with its much more penetrating nature, was more difficult to accomplish. It became clear that if one could not easily shield the electroscopes from the Earth's gamma rays, the next best thing to do would be to move the equipment away from the source of the radioactivity. Of course, the only way to do this was to go straight up.

In 1910, a Jesuit priest named Theodor Wulf took an electroscope to the top of the highest man-made structure at the time, the Eiffel Tower. He was surprised to find that he measured *more* ambient radiation at the top of the tower than at the bottom. He checked that the tower itself was not radioactive and thus he was confused. The result was not at all as expected. Perhaps there was a type of radiation from the Earth that could penetrate the 300 meters of air separating Wulf's electroscope from the ground? Of course, what was needed was another experiment with even greater separation. Since the Eiffel Tower was the tallest man-made structure, another approach was called for.

In 1782, the Montgolfier brothers did something never before accomplished. They made the first balloon flight. Here was a way to lift an electroscope to a great altitude. Following Wulf's observation, several scientists attempted to repeat his experiment in a balloon, but the vagaries of the pressure and temperature variation with height proved to be challenging. The early measurements were not precise or reproducible enough for anyone to make firm conclusions.

In 1911, Austrian physicist Francis Hess entered the fray. He took a balloon to 1,100 meters and observed no decrease in radiation. In April of 1912, Hess made several different trips, rising to a height of 5,350 meters. He found the most amazing thing. Above 2,000 meters, he found that the amount of radiation increased rather than decreased. It was as if the source of radiation came not from the Earth but rather from the sky. An obvious source of energy in the sky was the Sun, but subsequent flights at night and during a full solar eclipse on April 12, 1912 showed no decrease in radiation. As Hess wrote later,

> The discoveries revealed by the observations here given are best explained by assuming that radiation of great penetrating power enters our atmosphere from the outside and engenders ionization even in counters lying deep in the atmosphere … Since I found no diminution of this radiation for balloon flights during an eclipse or at night time, we can hardly consider the Sun as its source.

Hess' observation was not immediately accepted by the majority of physicists but further research, interrupted by World War I, supported his results. While the radiation was originally named for Hess, in 1925 American physicist Robert Millikan termed the new phenomenon "Cosmic Rays," a manifestly more poetic title. The name stuck. For his careful study of cosmic rays, Hess shared the 1936 Nobel Prize in Physics with Carl Anderson, another pioneer of cosmic ray studies (and one of whom we will soon hear again).

Further study of the nature of cosmic rays required an improved detector. In 1911, Scottish physicist Charles Thomson Rees Wilson invented the cloud chamber. This new technology revolutionized the study of cosmic rays. The cloud chamber was basically a clear container consisting of moist air. It is a little-appreciated fact that the formation of clouds requires a trigger, say a speck of dust on which water molecules can condense. The interesting thing is that when a radioactive particle, say an alpha or beta particle, crosses the water vapor, it can knock electrons off the air molecules and provide a cloud formation site. Thus an electrically charged particle crossing a cloud chamber would leave a little trail, looking like nothing more than a tiny jet contrail. The contrail could be viewed or photographed.

While modern cloud chambers are constructed a little differently (one can easily get plans on the Internet to construct one at home using readily available materials), the principle is the same. If one takes a radioactive source and places it near a chamber, one sees little contrails form and fade away, to be replaced by the contrails from new particles crossing the chamber. For the casual science enthusiast, building a cloud chamber is a great project.

Even in the absence of radioactive material, contrails will form, indicating the presence of cosmic radiation. Armed with this marvelous new device, physicists further investigated these peculiar cosmic rays. In 1929, D.V. Skobelzyn took a cloud chamber and surrounded it with a magnetic field. Since charged particles will move in a circular path in the presence of a magnetic field (with the radius of the circle proportional to the energy of the particle), this allowed him to measure the energy of the cosmic rays. He took 600 photos and found that in 32 of them, there were cosmic rays that originated outside the chamber and passed through it essentially undeflected, indicating that some cosmic rays carried a great deal of energy, much more than that typically seen in radioactive decay. In addition, he saw cosmic rays enter the chamber and hit a nucleus. After the collision, several particles were created. Because early cloud chambers consisted mostly of moist air, it was realized that similar behavior would exist in the atmosphere at large. Since each particle leaving the collision could in turn react with more molecules of air, it was clear that a single particle, through its initial interaction and emission of secondary particles and their subsequent interactions, could result in many particles hitting the Earth's surface. This phenomenon was called a cosmic ray shower.

In the same year, W. Bothe and W. Kolhorster used another technology to study cosmic rays…the Geiger-Müller tube. As you may recall, Geiger was a Rutherford protégé who helped establish the nuclear atom. After many untold hours, staring in the darkness and looking for a small blink of light, Geiger invented a device that would generate an audible click when a charged particle crossed it. His eyestrain was over. Bothe and Kolhorster took two Geiger tubes to study cosmic rays. They noticed that when one tube clicked, it was frequently true that the other did as well. Whatever was firing one tube seemed to be observed in the other one too. Bothe and Kolhorster found that the frequency of the "coincidences" (i.e. times when both Geiger tubes subsequently fired) depended on the relative orientation of the two tubes. When they were near one another, with one above the other, they observed the maximum number of coincidences. As the two tubes were separated, either vertically or horizontally, the coincidence rate dropped off. Such behavior shed light on the nature of cosmic radiation. It appeared to be charged particles that could ionize the gas contained in each Geiger tube. It was a mystery how charged particles could penetrate so far in material, exceeding the behavior seen by alpha and beta particles. Thus, another curious question was raised. For Bothe's discovery of the coincidence method and subsequent use of it to make measurements, he shared the 1954 Nobel Prize in Physics. (The guy with whom he shared the prize that year received it for completely unrelated work.) ·

While Bothe thought he had shown that cosmic rays were "corpuscular," which means that they act like a little "bullet" carrying electrical charge, really he had only shown that the Geiger counters fired simultaneously. What was needed was a combination of the cloud chamber and Geiger counter techniques. In 1932, Patrick Blackett and Giuseppe Occhialini came up with a clever method. Rather than just randomly taking photographs of a cloud chamber, which results in mostly a bunch of empty photographs, they rigged the Geiger counters to send out an electrical signal to simultaneously snap a photograph. They showed that the electrical signal in the Geiger tubes was accompanied by one or more tracks in the cloud chamber. For his work in cosmic rays, Blackett received the 1948 Nobel Prize.

Despite the realization that cosmic rays consisted of charged particles, the mystery was far from resolved. The charged particles known at the time were protons, electrons and various atomic nuclei. None of these particles could possibly pass through the atmosphere from outer space all the way to the ground. While the neutral particles had a superior penetrating ability as compared to the charged ones, they too could not explain cosmic rays. To properly understand the results of the cosmic ray experiments spanning the crucial period of 1932–1947, we must briefly return to the two seminal theories of the first part of the 20th century: special relativity and quantum mechanics. While neither topic is central to this book, a brief foray into both topics is warranted. Essentially, quantum mechanics is the description of physics at very small distance scales; say approximately the size of an atom ($\sim 10^{-10}$ meters and smaller). In contrast, Einstein's theory of special relativity deals with objects moving very fast. Since particle physics deals with very small particles moving at extreme speeds, it is clear that a theory describing them must include both of these ideas.

The Antimatter Electron

Prior to 1928, such a theory was not forthcoming. However Paul Dirac was finally able to synthesize these ideas during the years of 1928–1930 and provided the seeds of a successful theory, which one can call relativistic quantum mechanics, but is now, in an extended form, more commonly called quantum electrodynamics, or QED. His theory did a great job describing the interactions between two charged subatomic particles at any speed, although it was derived mostly to clarify the behavior of electrons. The only problem was that it predicted another unknown particle that seemed to be an opposite analog to the electron; that is, a particle that is identical to the electron in every way except that the charge would be positive, in contrast to the negatively charged electron. Just for fun, we can get an idea of how the theory predicted this. While the mathematics needed to solve this question is pretty complex, in the end the equation looked something like: $x^2 = 1$. This equation is true for two values of x, they are $x = +1$ and $x = -1$. The first solution was easy to understand, as it described the electron. However, the second equation seemed to indicate a positive particle, although with the same mass as an electron. In Dirac's first paper, he indicated that this particle would probably be a proton. The fact that the proton and electron had such different masses just meant that the theory needed a little additional work. The problem with the theory was underscored when Russian physicist I.E. Tamm published a paper in 1930 which showed that if the proton were the positive particle predicted by Dirac's theory, atoms would not be stable. Nonetheless, it seemed pretty clear that Dirac's theory, suitably modified, would reconcile the interactions between protons and electrons.

In 1932, Carl David Anderson of the California Institute of Technology built a Wilson cloud chamber at the Guggenheim Aeronautical Laboratory at Caltech. He observed what appeared to be positively charged electrons among the other particles he recorded. Robert Millikan, or "The Chief" as Anderson called him, was very skeptical about Anderson's result and offered several alternate explanations. Finally, Anderson put a lead plate in his chamber. A particle passing through the lead plate would lose energy in the passage. Such a photograph was able to unambiguously rule out Millikan's alternate explanations. A positively charged electron had been observed.

The discovery of Anderson's positive electron, or positron as it is now more commonly known, was soon followed by the realization that an electron and positron can annihilate one another and convert their entire energy into two photons. This opposite of matter is now called antimatter and it is one of the scientific terms that have been commandeered by science fiction writers. However, unlike its cousin terms of warp drive, wormholes and hyperspace, antimatter is a firmly established, utterly inarguable, phenomenon. While Anderson only discovered the antimatter electron, the antimatter analog of the proton, the antiproton, would require the use of powerful particle accelerators to create. In the ensuing years, it has become clear that for every particle discovered, there is an antimatter analog. For a few neutral elementary particles, the photon for example, the particle and anti-particle are the same.

With the discovery of the positron, the number of known elementary particles had again increased by one, bringing the count to six: electron, proton, neutron, photon, neutrino and positron. With the discovery of the positron, the thought that there might be other types of antimatter to be discovered in cosmic rays was at the forefront of many physicists' minds. Because of Bothe's innovation using Geiger tubes to trigger cloud chambers, scientists would typically take thousands of data-rich photographs. Each photograph would have tracks bent by the magnetic field to either the right or left, signifying positively or negatively charged particles. Each track was analyzed, with the degree of curvature indicating the energy carried by the particle and, as important, one could determine the mass of the particle. Eventually the question of what particles were created in concert with which others became important (i.e. were electrons produced singly or in pairs, or did the presence of a positron mean that an electron must also be present, etc.) These more complex questions were still a ways off. However, the mid to late 1930s were heady times in cosmic ray physics.

Who Ordered That?

In August 1936, Carl Anderson, the discoverer of the positron, along with his graduate student Seth Neddermeyer lugged the cloud chamber to the top of Pikes Peak. By going to the top of a mountain, they would climb above much of the atmosphere that shielded so many of the cosmic rays. They brought with him a large magnet, so they could deflect the particles and thus ascertain more information about them. In driving the heavy load to the top of the mountain, they managed to blow the engine in their 1932 Chevrolet truck. They could have been stranded but, as luck would have it, they bumped into a vice president of General Motors who was General Manager of the Chevrolet truck division. He had been driving Pikes Peak with the intention of having an advertising campaign focusing on how fast a Chevy truck could climb the mountain. He spoke with the local Chevrolet dealer and had the engine replaced at 14,000 feet above sea level.

With their apparatus in place, Anderson and Neddermeyer noticed some tracks that seemed to best be explained by a previously unknown particle, with a mass somewhere between that of an electron and a proton, lighter rather than heavier, but in any event something new. However, one of Anderson's senior colleagues, Robert Oppenheimer, of American atomic bomb fame, remained unconvinced. He maintained that these highly energetic particles could be electrons and that any deviation from Dirac's theory of quantum electrodynamics indicated a limitation of the theory, rather than a new particle. Somewhat intimidated by Oppenheimer's exceptional command of mathematics, Anderson and Neddermeyer published their photos with little comment and less fanfare.

However low their confidence, Anderson and Neddermeyer's paper traveled to Japan, where it was read by none other than Hideki Yukawa, the architect of the U-particle which, as we recall, was an attempt to explain the force that held together the atomic nucleus. The U-particle was supposed to have a mass midway between that of the proton and electron. Needless to say, Yukawa's ears perked up. The 1930s were a time of rampant nationalism in Japan and one of Yukawa's colleagues, Yoshio Nishina, decided to try to find and measure the properties of some of these mid-massed particles, before the westerners appreciated their discovery. While the Japanese team knew what they were doing and for what they were looking, bad luck plagued them and they were able to record only one photograph that contained a U-particle candidate. With some more time, they would have solidified their effort but, unfortunately for them, time had run out. In the spring of 1937, Anderson had visited the Massachusetts Institute of Technology, where he learned that two physicists there, Jabez Street and E.C. Stevenson, had data similar to that of Anderson, but that they were considering announcing the discovery of a new particle. Not wanting to be scooped, Anderson wrote a quick article to the journal *Physical Review*, in which he claimed discovery of the particle, the existence of which Oppenheimer's earlier disbelief had caused him to soft peddle just a year earlier. Anderson's paper was published in May, with Street's paper presented at a meeting of the American Physical Society in late April, with final submission in October 1937. A new particle was added to the particle pantheon. In fact, two particles were added, as it was soon clear that this new particle came in both a positive, as well as a negative, variety. As is usual in the case of a discovery, it was soon evident that people had been photographing these new particles for years, without appreciating their significance. Many physicists went to sleep with the final words in their minds "If only …"

With the observation of a new particle, a name was needed. Tradition required that Yukawa, as the person predicting the particle, or Anderson, as the discoverer, name it. Yukawa's term "U-particle" never was popular and, after many candidates, the term "meson" (Greek for "middle one") was adopted and retained to this day. It seemed that the particle mediating the strong force had been observed (although Yukawa's neutral meson was still missing). Eventually the term meson became a generic one like "automobile," rather than a specific one like "1964 Volkswagen Beetle (my first car)." While a meson initially meant a particle with a mass midway between that of a proton and an electron, this was not the whole story. In Chapter 3, we'll explain what actually constitutes a meson and we'll see that the early physicists had it mostly correct, but not perfectly. However, in 1937, when only one type of such a particle had been discovered, the term meson meant Anderson's discovery.

With the discovery of the meson, the next order of business was to study its properties, in part so that physicists could verify that Anderson's and Yukawa's meson were one and the same. In 1939, Bruno Rossi published a paper in which there was a hint of evidence that mesons could decay and a stab was taken at the lifetime of the meson. It was the following year that E.J. William built a large Wilson cloud chamber. He recorded events in which the meson was clearly seen to decay. The meson disappeared and an electron with the same charge appeared in its place. In addition, because the direction of the electron was different from the meson, it was clear that the decay included neutral particles in the final state. The decay was thought to be "beta decay like," with the meson decaying into an electron or positron along with an accompanying neutrino. In 1941, F. Rasetti carefully measured the lifetime of the meson and found that it was some few millionths of a second, far longer than the lifetime predicted by Yukawa.

If the newly discovered meson was indeed Yukawa's U-particle, physicists should be able to demonstrate this by seeing it interact with the atomic nucleus. A very clever approach was followed during World War II by teams in Japan and Italy, working under unimaginably poor conditions. They reasoned that a negatively charged meson would be attracted to the positively charged atomic nucleus. The meson would be quickly absorbed by the nucleus and thus participate in the nuclear force. Positive mesons, on the other hand, would be repelled by the nucleus. These mesons would not enter the nuclei and thus they would decay instead. Of course, the difference between the decay rates of positive and negative mesons became the interesting measurement to make.

In 1947, after the war, an Italian team including M. Conversi, E. Pancini and O. Piccioni showed that positive and negative mesons decayed in roughly equal numbers. Thus it appeared as if the meson discovered by Anderson and Neddermeyer was not the "U-Particle" predicted by Yukawa.

While the measurements that clearly indicated that the meson was not Yukawa's predicted particle were not definitive until 1947, it was clear even before the beginning of World War II that this might be true. While physicists treat a result as official only after it has appeared in a refereed journal, the truth is that physicists love to talk shop. So many physicists knew that there was a problem with the hypothesis that the cosmic ray mesons were the source of the nuclear force. The particles lived too long and interacted in the cloud chambers too rarely. However, while most physicists knew that there were problems with the hypothesis, they usually had heard rumors from only one source. The war had played havoc with the normal international scientific lines of communication. The active groups in America, Italy and Japan could not confirm each other's result. Thus it was frequently true that many ideas were independently conceived throughout the world, after all genius does not respect the vagaries of geography nor the whims of temporary geopolitical realities. For instance, in June 1942, Japanese physicist Shoichi Sakata made the somewhat reckless and ultimately correct proposal that perhaps there were two mesons, one like the one proposed by Yukawa and the other one observed by cosmic ray physicists the world over. While at a different time such a proposal might have made it to the West, in a month that included the Battle of Midway, it is unsurprising that Sakata's idea went unnoticed by American and European physicists until much later. Tokyo Rose never even mentioned it. The journal containing Sakata's idea didn't make it to the United States until December 1947, six months after Robert Marshak of Cornell University independently had the same insight.

Marshak described his idea at a conference on Shelter Island in June of 1947. The gist of Marshak's idea was that a particle from outer space would hit the Earth's atmosphere and make Yukawa's meson, which would in turn decay into the meson seen first by Anderson. To distinguish between the two particles, he called Yukawa's particle a pi-meson (π) and the decay product meson a mu-meson (μ). In subsequent years, these terms have been contracted to pion and muon. While his idea was well received, physicists couldn't help but note that no such decays had been observed. When Marshak returned home, he opened his copy of *Nature*, an exceptional British science journal, and found a photograph that had many of the characteristics that he had described.

Don Perkins' team had done an experiment that was rather interesting. He had taken plates covered with a photographic emulsion in a plane that flew at 30,000 feet for several hours. Previous experimenters had found that cosmic rays would leave tracks in photographic emulsion. When the plates were later developed, the behavior of the track could be measured. Frequently the final moments of the particle would be observed, be it a decay or an interaction with an atom in the emulsion plates. One of the beauties of this technique was that while the magnetic field helped determine the momentum of the particle, the thickness of the track revealed information on the charge and the mass of the particle.

In one of Perkins' photographs, he observed a cosmic ray meson slowing in the emulsion until it came to rest. The particle then decayed into another slightly less massive meson. Perkins' team's photograph consisted of a single event. Confirmation was clearly required. Since the war was over, the lines of international collaboration were once again open. A team of British and Italian physicists, including Giuseppe Occhialini, C. Powell and C. Lattes performed a similar experiment in the Bolivian Andes, whereby they found 40 examples of one charged meson decaying into a second (and slightly less massive) type of meson. It was in their 1947 paper in which they suggested that events, which they described as "explosive disintegration of nuclei", were consistent with the production of these new charged mesons and another neutral particle "... the results are consistent with the view that a neutral particle of approximately the same rest mass as the μ-meson is emitted."

You might recall that Yukawa's theory called for three types of mesons, in principle with the same mass, one positively charged, one negatively charged and one electrically neutral. The positive and negative pi-mesons (π^+ and π^-) seemed to be present in cosmic rays and here was a paper suggesting that a neutral meson of similar mass might be observed. It took another three years for the existence of the neutral pi-meson (π^0) to be found, this time not in a cosmic ray experiment, but rather in a particle accelerator at Lawrence Berkeley Laboratory. The era of machines was soon to dawn. The π^0 was confirmed in 1950 in a cosmic ray experiment that had lifted photographic emulsion plates to the dizzying height of 70,000 feet in an unmanned balloon. Yukawa's triad of U-particles had been found.

If the three pi-mesons (π^+, π^-, π^0) fit into the order of things by providing the mechanism that held the nucleus of the atom together, what was the purpose for the mu-mesons (μ^+ and μ^-)? The muon did not seem to be affected by the strong or nuclear force, but was affected by the electromagnetic and weak forces. Basically, the muon seemed to be a fat electron (carrying about 200 times the electron's mass), but without a purpose. Since all other particles could be fit neatly into a niche, the muon was especially disconcerting. Physicists' confusion was exemplified by I.I. Rabi's oft quoted remark "Who ordered that?" when the non-utility of the muon became apparent to him. Indeed the muon was a mystery. To add to the conundrum, while the mass of the muon was such that it could be considered a meson, it turns out that the muon wasn't a meson at all. Don't worry. If this all seems confusing, it's only because it is. However, in Chapters 3 and 4, the whole situation will be vastly simplified.

Strange "V" Particles

But the late 1940s provided more than one mystery, for in October 1946, G. Rochester and C. Butler had finally been given permission to turn on Blackett's large electromagnet at the University of Manchester (they had been forbidden to do so during the war for reasons of power consumption). This magnet surrounded their counter-triggered cloud chamber. They could again do research, and so they did; placing a lead sheet above the chamber to absorb uninteresting low energy cosmic rays so they would look at only the high energy particles. Of their 5,000 photographs, two pictures contained events "of a very striking character." These were what were called "V" events and they appeared to be neutral particles decaying into two particles, one positively charged and the other negatively charged. Another class appeared to be a charged particle decaying into another pair of particles, one charged and the other electrically neutral. Rochester and Butler wrote in their 1947 paper in *Nature*,

> ...We conclude, therefore, that the two forked tracks do not represent a collision process, but do represent spontaneous transformations. They represent a type of process with which we are already familiar in the decay of a meson with an electron and an assumed neutrino, and the presumed decay of the heavy meson recently discovered by Lattes, Occhialini and Powell.

These events were entirely unexpected.

Rochester and Butler's results were so controversial that they required confirmation. The two physicists continued to take data for another year and found no additional similar events. People began to suspect that their initial results were somehow flawed. In 1948, Rochester met in Pasadena, California with Carl Anderson who was excited at the prospect of being involved in the discovery of yet another mystery from space. Anderson took his best chamber to the top of a mountain in order to get an increased flux of cosmic rays. Rather quickly, Anderson obtained another 30 photographs of events with the same character as that described by Rochester and Butler. It appeared that whatever was causing the "V" events was a new meson, with a mass somewhere between that of the pion and the proton. Just like the discovery of the muon, physicists went scurrying back to their cache of photographs taken over the previous decade, only to find that "V" events had been showing up, unappreciated the entire time. Yet another generation of physicists got a case of the "If only's…"

With the observation in cosmic rays of the positron, the muon, the pion and now the "V" particle, it was clear that the mountaintop was the place to be. Physicists the world over ascended to great heights. The Rockies, the Andes, the Alps and the Himalayas all provided marvelous, if Spartan, laboratory conditions. Like the gods dwelling high atop Mt. Olympus (how's that for delusions of grandeur?), physicists sat on their mountaintops and contemplated the meaning of life and the very nature of reality. Oh yeah, and froze their butts off too. In fact, many lives were lost by physicists who had not enough respect for the danger of great heights, where a hidden crevasse could swallow the unwary physicist, out for his morning stroll.

Nevertheless, while the conditions were hard, the rewards were great. Within five years of the discovery of the "V" particle, dozens of other rare types of events were observed. Because they were rare, each scientist would find only one or two of his particular discovery and often none of the type reported by his competitors. There were reports and discoveries and retractions galore. These were giddy times for cosmic ray physicists although, as we will see, their days were numbered.

While it was now thought that there were several subclasses of "V" particles, as well as other newly discovered particles with names like the theta (θ), the tau (τ) and even K particles, scientists kept coming back to the "V" events. The particles created in "V" events were now thought to be made rather easily in cosmic ray collisions and observed with their spontaneous two-particle decay. Because the particles were created so easily, it was known that the force that created the particles was the strong force. The weird thing was that the particles didn't decay rapidly. This suggested that the decay of the particle was caused by a force other than the strong force. This was unusual. Ordinarily if a particle could be created by a particular force, the same force could cause it to decay. Since this seemed to not occur in this case, it was apparent that something else was going on … something was inhibiting the decay. This was strange and consequently these particles were called strange particles. The answer to the question lay in the fact that these "V" particles were created in pairs. It became apparent that strangeness was something like electrical charge. Ordinary matter had none of this property now called strangeness. Further, strangeness appeared to be mostly conserved. Thus if a particle carrying strangeness was formed, then at the same time an antiparticle carrying anti-strangeness had to be created. If one could write the idea of strangeness numerically, we could say that a particle carrying strangeness could be represented by ($S = +1$) and one carrying anti-strangeness by ($S = -1$). Taken together, they have no strangeness $(+1) + (-1) = 0$. In many ways, this concept is analogous to electric charge. A neutral particle (say a photon) can be converted, under the right conditions, into two particles carrying electrical charge (say an electron and positron (e^- and e^+)). This point is detailed in Figure 2.9.

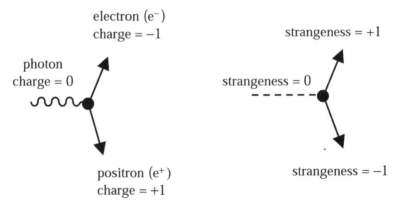

Figure 2.9 Electrically neutral photons can split into positively and negatively charged particles. Similarly, particles containing no strangeness can split into two particles, one carrying strangeness and the other carrying anti-strangeness.

So far this is a pretty easy idea, but the question of what makes the particles in the "V" event live so long remains. This is explained by the idea that each particle contains some amount of strangeness ($S = \pm 1$). Except. In 1953, Abraham Pais hinted at and a young hotshot theorist named Murray Gell-Mann explicitly stated the solution to the problem. While the strong force and electromagnetic force could not explicitly change a particle's strangeness, the weak force can. Thus, strange particles can be produced copiously in pairs, via the strong or electromagnetic force, but for single particle decay, they needed the weak force and thus the particle would have a long life. This strangeness was a new quantum number, the first proposed in about 25 years. (Quantum numbers are those properties that describe a particle, with mass, charge and spin being more familiar examples.) It is similar to the quantities like charge or spin that can only come in discrete quantities. Strangeness is our first really foreign concept, so let's recap. Strange particles are made easily, indicating that the strong force governs their creation. They live a long time (i.e. don't decay easily), showing that the strong force doesn't govern their decay, rather the weak force is the culprit. The fact that particles could be created by a particular force, but were not allowed to decay by the same force is what was considered to be strange. Normally in the particle world, if you live by the sword, you die by the sword. Eventually, it became clear that the strong force could easily create strange particles in pairs, but only the weak force could let individual strange particles decay.

With the addition of the new particles, it became clear that what was needed was some order. Like any scientific field where understanding remained elusive, the first order of business is the measurement of the properties of the various particles and some sort of classification scheme so that one can understand how the particles are similar and how they are different. As summarized in Table 2.4, we see that particles that can be affected by the strong or nuclear force are called hadrons (protons, neutrons, pions and the "V" particles). The hadrons are subdivided into two classes. The first, baryons, were

Table 2.4 Summary of the knowledge of particles and their defining properties as of about 1960. If a review of the particle names (symbols) given in the above table is necessary, the reader is invited to read over Appendix C. (*Note: the muon didn't fit well, as it had a meson's mass, but a lepton's indifference to the strong force.)

Major Class	Minor Class	Forces Felt			Example	Strangeness Possible?	Mass
		S	W	EM			
Hadron	Baryon	Y	Y	Y	$p^\pm$, n^0, Δ, Λ	Yes	Large
	Meson	Y	Y	Y	$\pi^\pm$, π^0, $K^\pm$, K^0		Medium
Lepton	Charged	N	Y	Y	$e^\pm$, $\mu^\pm$	No	Light*
	Neutral	N	Y	N	ν_e, ν_μ		Massless

initially defined as those particles with a mass greater or equal to that of a proton. The second class is called mesons and was initially defined to be those hadrons lighter than a proton (pions and "V" particles, as well as the newly discovered θ, τ, and K-particles were all mesons). The names θ and τ were not long-lived and the τ symbol was eventually used to name an entirely different particle (see Chapter 3). Also in Chapter 3, we finally understand the true differences between baryons and mesons, so the initial mass-based definition should be understood only for its historical context. In addition, there were the particles that were not affected by the strong force. These particles were called leptons and examples are the electron, positron, neutrino and both types of muon. Thus the muon, which had initially been called a meson on mass terms, was now understood to be a lepton on grounds of the types of interactions that it feels. Such was the confusion of the mid 20th century.

In addition to the above classification, the properties of each particle were interesting. The charge, lifetime and spin, as well as the more abstract strangeness and even more esoteric parity, were important to measure. In addition to the lifetime, the various possible ways each particle could decay provided crucial information. The field was in a bit of a turmoil and it really was great fun.

The chaos that was particle physics needed order so, in the summer of 1953, a conference was held at Bagnères-de-Bigorre, a small town in the foothills of the Pyrenees. Physicists the world over came to exchange information and ideas. Strategies were mapped out to classify particles and to set new paths of investigation. The study of cosmic rays had revealed the existence of particles not imagined here on Earth. There were new frontiers to explore, new truths to discover.

Also reported at the conference was data taken on particles created, not from space, but rather from monster particle accelerators that could generate more new and esoteric particles in minutes than a cosmic ray experiment could do in months or years. The scientists at the conference could not but feel the heat. Particle accelerators could be built near their offices, rather than at the top of mountains. The era of cosmic ray physics had passed, although they hoped to continue for a while. In his closing remarks, Louis Leprince-Ringuet said

> We must run without slackening our pace; we are being pursued, pursued by the machines!... We are, I think, a little in the position of a group of mountain climbers climbing a mountain. The mountain is very high, maybe almost indefinitely high, and we are scaling it in ever more difficult conditions. But we cannot stop to rest, for, coming from below, beneath us, surges an ocean, a flood, a deluge that keeps rising higher, forcing us ever upward. The situation is obviously uncomfortable, but isn't it marvelously lively and interesting?

But cosmic rays were no longer the place to be. Aggressive physicists turned to new monster accelerators for their studies.

The first particle accelerator of note was the early Crookes tube. The next big accomplishment was made in 1931 by two protégés of Ernest Rutherford, John Cockroft and Ernest Walton, who made a huge scaled-up version of the Crookes tube (although the technical details were considerably different). Beams from the Cockroft-Walton accelerator were energetic enough to split the nucleus of an atom and it is here where the term "atom smasher" comes from. In the same year, Ernest Lawrence and his student E. Stanley Livingston made a nearly modern particle accelerator called a cyclotron. This accelerator was only 27 centimeters in diameter and could accelerate a particle with an equivalent voltage of 1,000,000 volts. But bigger things were not too far in the future.

In 1947, physicists at Lawrence Berkeley Laboratory (LBL), perched high above the University of California, Berkeley, built a large cyclotron accelerator. This accelerator was 184″ (4.67 meters) in diameter and could accelerate alpha particles to the extraordinary energy of 380,000,000 electron volts. An electron volt is a measure of energy useful in particle physics. A proton or electron accelerated by one volt will have an energy of 1 electron volt (or 1 eV). An electric field of a million volts will accelerate a proton or electron to 1 million eV or 1 mega eV or 1 MeV, three ways to write the same thing. A television accelerates electrons to a few tens of thousands of electron volts, so Berkeley's 380 million electron volts was quite an accomplishment. After some tuning up, they accelerated alpha particles to 380 MeV and directed them onto a carbon target. Among other things, what came off the target were a huge number of pions. Rather than waiting for cosmic rays to randomly make pions under non-ideal conditions, physicists could make them at will. In science there are two phases of research. Initially, one begins with observational science, in which scientists look around at phenomena but can do little to change the conditions of what they are observing. The fields of cosmic ray physics, astronomy and biology as it is taught in high school are examples of this type of science. Later, when the field is more advanced (and typically following a technical discovery that provides new tools) science enters an experimental phase. In this phase, scientists have considerable control over their experiments. Accelerators provided that control for particle physicists. Physicists could generate pure beams of a specific type of particle (electrons, protons, alpha particles, pions, muons, etc.) at a well determined energy and aimed at a carefully prepared target, surrounded by optimized instrumentation. With this degree of control, it is no wonder that particle physicists left the stark high mountains and returned to their university campuses, at which cyclotrons were popping up all over. Cosmic ray physics experienced a long decline in favor of the more controllable accelerators and is only now experiencing a renaissance because cosmic rays occasionally generate a collision with an energy that exceeds that available to even today's great accelerators.

By 1949, Berkeley could make pions at will. A cyclotron at the University of Chicago was claimed to have created "V" particles by bombarding a metal target with high energy protons. This claim was retracted, but it was no matter, as experimenters at an accelerator at Brookhaven National Laboratory (BNL) on Long Island, with the marvelous name "Cosmotron" announced firm evidence of "V" production in 1953. In addition, in the period 1952–1953, the Cosmotron yielded yet another surprise. Two physicists, Luke Yuan and Sam Lindenbaum, shot a high energy beam of pions at a hydrogen target. They varied the beam energy and found that when the beam energy was in the rather tight range of 180–200 MeV, the number of pions passing through the target dropped, indicating that they had somehow interacted in the passage. This type of interaction is called a resonance, because of the fact that the reaction occurs at a "magic" energy. Examples of resonance in ordinary life occur when pushing a child on a swing. In order to get the swing to go high, one must push the child at a particular rate. If the pushes come much faster or much slower, no big movement occurs. Another example of resonance occurs when you drive and the front end of your car is slightly out of alignment. At low speeds, everything is OK. However, as you increase your speed, you feel a vibration in the steering wheel. This vibration is a maximum at a particular speed and decreases once the speed exceeds the "magic" speed.

In particle physics, a resonance implies that a particle is being created. Yuan and Lindenbaum called their new particle the Δ (delta) and it quickly became evident that there were several types of Δ particles, each with similar mass, but varying electric charge. The Δ^-, Δ^0, Δ^+, Δ^{++} were eventually all observed. As is frequently the case, the observation of a new phenomenon sent others scurrying to look for other examples of the same behavior. Soon many resonances were observed, with all sorts of properties. Each resonance had its own mass and electric charge. Some had strangeness, some didn't. There was a broad range of particle lifetimes and different ways each could decay. Both baryons and mesons were discovered. There were new particles everywhere! It was an unlucky physicist who couldn't point to a particle discovery of his own.

The period from about 1950 to 1963 was a time where the experimentalists reigned supreme. Because the existence of these particles was not expected from other, earlier experience, it took a while to absorb the information and begin to see the patterns. It was the 1960s in which understanding became possible. This modern understanding is discussed in Chapters 3 and 4. But in 1957 there were about 30 unique particles, while in 1964 there were more than 80. Particles that were clearly variants of other particles increased the count. The particles named so far were referred to as the particle zoo. A partial list of the particles discovered by that time is: π, μ, Λ, Σ, Ξ, ν, η, η', K, K*, Ω, ρ, ϕ, and this is just some of the cool ones with Greek letters for symbols (look at Appendix A for the proper pronunciation of the various symbols). Many of the particles came with different charges, some positive, some negative, some neutral. As mentioned before, the Δ had four distinct charge possibilities. The whole situation was a deeply glorious mess.

While the 1950s were characterized by a frenzied search for new particles, a few discoveries stood out. The first was the discovery in 1955 of the antiproton by Owen Chamberlain and Emilio Segrè, using the enormously powerful Bevatron accelerator at LBL. With the discovery of the antiproton, which was expected, the understanding of antimatter became murkier. Originally, the theory treated matter and antimatter on an equal footing. Further, in particle physics experiments matter and antimatter were created in equal quantities. However, the world in which we live consists exclusively of matter. So where is all the antimatter? The question is an ongoing mystery and we will revisit it in detail in Chapter 7.

Neutrinos Get Even More Complicated

Another interesting experiment, which had consequences far beyond what is apparent from its initial description, came in the first few years of the 1960s. While an understanding of the strong and electromagnetic forces had been obtained by experiments at a number of different energies, the theory of the weak force was derived on the basis of the low energy phenomena, in beta decay and assorted nuclear reactions. Enrico Fermi's theory of beta decay, despite its groundbreaking nature, was known to have problems. The most dramatic problem was that while it worked well at low energy, it predicted an impossibly strong behavior as one raised the collision energy. The theory predicted that interactions governed by the weak force became more and more likely, until they became more common than the much stronger strong force. Taken seriously, this meant that the force that was weak at low energies, became stronger than the strong force at high energies. Such a behavior is in principle possible, but is at least suspicious. However, even more deadly to the theory was its behavior if one raises the energy even more. Eventually the theory predicted that each particle had a greater than 100% chance of interacting. Such a prediction is manifestly nonsense and shows that the theory needed work. However, since the theory predicted the low energy behavior so well, what was required was data about the behavior of the weak force at high energy. With such data, theorists could test their ideas and thus receive much needed guidance.

The problem with trying to measure the behavior of the weak force is that it is…well…weak. The effects of the strong and electromagnetic force are so much larger that the weak force doesn't contribute much. It's like two people trying to talk to one another at a rock concert. Their voices contribute little to the overall noise level. Clearly, what was needed was a particle that could only feel the weak force. Luckily, such a particle existed, the enigmatic neutrino.

Postulated by Pauli in 1931 and observed by Reines and Cowan in 1956, the neutrino is the only known particle that is unaffected by both the strong and electromagnetic forces. But since what we wanted was highly energetic collisions between neutrinos and the target, one had to figure out how to obtain high energy neutrinos. By the late 1950s, physicists knew that pions decayed into muons and neutrinos ($\pi \rightarrow \mu + \nu$) and muons decayed into electrons and neutrinos ($\mu \rightarrow e + \nu$). (Note: we now know that these ideas were incomplete and we will see how as we proceed.) Since one could make beams of pions and muons using accelerators of the day, perhaps if one allowed a beam of pions to travel a long distance, some would decay and produce a beam of both muons and neutrinos. Now comes the tricky part. Since neutrinos only interact with matter via the weak force, they can pass freely through material. To give a sense of scale, a highly-energetic neutrino could pass through millions of miles of solid lead with essentially no chance of interacting. All other particles can only penetrate tiny fractions of that distance and so to make a neutrino beam, one aims the beam containing muons and neutrinos at a bunch of iron and earth. The only particles that come out the other end are neutrinos. With a beam of neutrinos, one could direct them at an immense target weighing many tons. While neutrinos rarely interact with matter, they do react occasionally. A tiny fraction of the neutrinos would interact in the detector and their behavior could be measured. Finally one would have measurements that would illuminate the behavior of the weak force at high collision energies.

What was needed was the right accelerator and the right guys to do the work. In 1960, an extraordinary accelerator was commissioned at Brookhaven National Laboratory. This accelerator was the Alternating Gradient Synchrotron (or AGS) and it could accelerate protons to the unheard of energy of 30,000,000,000 electron volts or 30 Giga electron volts (30 GeV). The necessary people were Leon Lederman, Mel Schwartz and Jack Steinberger, all professors at Columbia University who jointly shared the 1988 Nobel Prize for their results.

The premise of the experiment was simple. Neutrinos were expected to do two different things. A neutrino would hit an atomic nucleus and emit either an electron or a muon and, incidentally, knock the bejeezus out of the nucleus. Never mind what happens to the nucleus, what was really interesting was the ratio of electrons produced compared to muons. Predictions varied, but something approximating half of each seemed reasonable. Lederman and company turned on their detector, told their accelerator colleagues to turn on the beam and waited. They expected about one neutrino interaction in their detector per week. To accomplish even such a tiny rate, the accelerator would shoot 500 million billion (5×10^{17}) particles in their direction. Neutrinos *really* don't interact very often.

Their first neutrino interaction produced a muon, as did their second. The third was a muon too, followed by the fourth. As the muon events tumbled in, the experimenters made a brilliant observation (hey, there were three future Nobel Laureates involved). There were no electrons created in their detector. Neutrinos were known to interact with electrons and here were ones that refused to cooperate. After they thought about it for a while, they recalled that all of the neutrinos that hit their detector were created in tandem with a muon (recall $\pi \rightarrow \mu + \nu$?). They interpreted their results as the neutrino retaining some knowledge of its history, some sense of "muon-ness." It seemed as if there were two classes of neutrinos, one muon-like and one electron-like. The two types of neutrinos were called muon neutrinos (ν_μ) and electron neutrinos (ν_e). The particle zoo had again increased by one.

With the discovery of the muon neutrino, a curiosity was apparent. In the leptons, there appeared to be two different sets of particles that were very similar and yet somehow different. Physicists took to writing them in pairs

$$
\begin{array}{cc}
\text{electron} & \\
\text{electron neutrino} & \begin{pmatrix} e \\ \nu_e \end{pmatrix} \\
\hline
\text{muon} & \\
\text{muon neutrino} & \begin{pmatrix} \mu \\ \nu_\mu \end{pmatrix}
\end{array}
$$

Just why this pattern should be repeated was not understood, but it was clearly a clue of some kind. A related clue was that such a pattern was not observed in the mesons and baryons. What it meant still needed to be worked out. In addition, Lederman and company showed that the probability that an interaction would occur because of the weak force did indeed increase with collision energy, in agreement with the theory for the range of energies for which they had data. Therefore, that particular mystery remained to be solved.

So in the last days of the 1950s and the first few years of the 1960s, the situation in particle physics was rich. Rich, in this context, can be defined as a totally chaotic mess. There were nearly a hundred particles known, leptons and hadrons, which were further subdivided into mesons and baryons. The particles' mass ranged from zero to about 60% more than the proton. The particles had different spins; integral and half-integer, thus being bosons and fermions. They had vastly different lifetimes and were affected by different mixes of forces. Each decayed in unique ways. Some particles had strangeness and some didn't. Somehow, order had to be made out of the chaos. The time for an answer had come.

chapter 3

Quarks and Leptons

Daring ideas are like chessmen moved forward; they may be
defeated, but they start a winning game.

— Goethe

Given the hundreds of particles discovered in the 1950s and the
preceding decades, what was clearly missing was a unifying principle ... some idea that would bring order out of the chaos that was particle physics at the time. There was ample precedent for this desire. For
example in the field of chemistry, first the Periodic Table and then
quantum mechanics explained the many previously mysterious patterns
observed in atoms. The 1960s was the decade where physicists began
to achieve the clarity of vision for which they had worked so long. The
following decades provided even sharper focus to these ideas and now,
in these first years of the 21st century, physicists can successfully predict most of the data that they observe in experiments. In this chapter
and the following one, we will learn in detail about how physicists now
view the world. We will see how the hundreds of particles discovered
in the early accelerator and cosmic-ray experiments can be explained as

various combinations of twelve much smaller particles; six particles called quarks and six called leptons. In the following chapter, we will see how only four forces are needed to describe the behavior of these particles and, by some accounting, only two. The progress in our understanding of the world over the past few decades has been noth ing short of astounding. We call the ensemble of theories and ideas dis cussed in these next three chapters the Standard Model of Particle Physics (or just the Standard Model for short). Standard implies that it works well and Model reminds us that it is still incomplete. While it is well known that the Standard Model doesn't answer all questions, it does a fantastic job of explaining all measurements made thus far. It's an extraordinary accomplishment and, if questions remain, that just leaves opportunities for further study and investigation. We'll discuss in Chapter 8 some of the questions on which the Standard Model remains silent. But, even incomplete, the Standard Model provides deep insights into the nature of the universe and a strong base from which to launch aggressive sallies against the remaining mysteries.

Quarks and Mesons

The situation in particle physics in 1960 was confusing. Many hun dreds of particles had been observed: the heavy hadrons, the lighter mesons and the even lighter leptons. What was missing was a unify ing principle. In 1964, two physicists, Murray Gell-Mann of California Institute of Technology and George Zweig of CERN inde pendently proposed a model that provided the guidance that has thus far been missing. They proposed that the pattern of hadrons and mesons could be explained if there existed even smaller particles contained within them. While Zweig called these particles "aces," Gell-man's name is the one that has been adopted by the field. He called these particles "quarks," after a line in James Joyce's *Finnegans Wake*: "Three quarks for Muster Mark…" This choice of name was unusual and may have set the custom of having rather fanciful parti cle physics language (as we shall see).

How one pronounces "quark" is a topic of some lively debate (usually involving beer). When one looks at the passage from which the name was taken, one might expect it to be pronounced so that it rhymes with mark, dark and park. However, most of the people I know pronounce it "kwork," rhyming with fork.

Initially, the quark model stated that there were three kinds of quarks. They were called: *up, down* and *strange*. While a bit odd, these names actually had some meaning. The proton and neutron have a similar mass and it was possible in earlier nucleon physics models to treat them as two manifestations of the same particle. This particle was called the nucleon and had a property called isospin. Isospin is a complicated concept and we won't pursue it further, except to say that there are exactly two kinds of isospin for the nucleon, which one can call up and down (one could have called these two types type 1 and 2 or cat and dog or Steve and Mary, but up and down were chosen). Protons have up isospin and neutrons have down. A nice analogy would be men and women, manifestly different (vive la différence!), but who can be treated as two aspects of a unifying object called a person. A man has the male property, while the woman has the female property.

Getting back to quarks, the proton contains more up quarks and the neutron contains more down, which is why they have their respective isospin. The name of the strange quark was chosen because it was thought that this quark carried the "strange" property that caused some particles to exist for a longer time than one would ordinarily expect. So the names, while somewhat obscure, have a historical basis.

Quarks were predicted to have some unusual properties. The proton and electron have equal and opposite electrical charge and further, they were understood to have a fundamental (that is, the smallest possible) electrical charge. The charge on a proton is $+1$ unit, while the electron carries -1 unit of electrical charge. However, quarks, as originally imagined, were thought to have an even smaller charge, a somewhat heretical postulate. Up quarks were to have a positive electrical charge, but only two-thirds that of the proton

(+2/3 charge). Similarly, the down and strange quarks were thought to have a negative electrical charge, but one-third that of an electron (−1/3 charge). Antimatter quarks have opposite electrical charge as compared to their matter counterparts (anti-up has a −2/3 charge, while anti-down (and anti-strange) have +1/3 electrical charge).

Another property of quarks is their quantum mechanical spin. As discussed in Chapter 2, particles can be broken down into two different spin classes: bosons, with integer spin ($..., -2, -1, 0, +1, +2, ...$) and fermions with half-integer spin ($..., -5/2, -3/2, -1/2, 1/2, 3/2, 5/2, ...$) (where "..." means "The pattern continues"). Quarks are fermions with spin $\pm 1/2$.

While quarks have some other properties that we will discuss later, we now turn to how quarks combine to make up many of the particles described in Chapter 2. To begin with, let's discuss mesons, the medium mass particles.

Gell-Mann and Zweig decided that mesons consisted of two objects: a quark and an antimatter quark (called an antiquark). For instance, the π^+ meson (pronounced "pi plus") consists of an up quark and an anti-down quark, which we write as $u\bar{d}$. (Note that up, down and strange quarks are written as u, d and s. An antiquark has a bar written over the letter so anti (up, down and strange) are written $\bar{u}$, $\bar{d}$ and $\bar{s}$.) We can see that the electric charge works out correctly: $u(+2/3) + \bar{d}(+1/3) = \pi^+(+1)$. Similarly, we can look at the quantum mechanical spin of the quarks and meson. The quark and antiquark both have (+1/2) spin, but they can be in the same direction or in the opposite one. In this case, the spin of the quark and antiquark are in the opposite direction $u(spin = +1/2) + \bar{d}$ ($spin = -1/2$) = $\pi^+(spin = 0)$, which is the spin of a pion (Yukawa's particle), as we saw in Chapter 2. It doesn't matter if the quark or antiquark carries +1/2 spin (in fact they can switch), but it is important that they are in opposite directions. Figure 3.1 illustrates this and the other spin states discussed in the text below.

Given the fact that any particular meson can have one of three different quarks and three different antiquarks, this implies that one

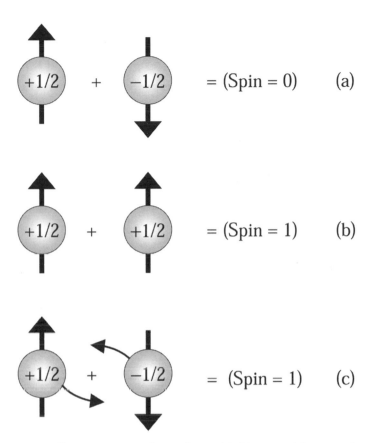

Figure 3.1 Different spin configurations. (a) Two particles carrying equal spin, pointing in opposite directions, have zero net spin. (b) Two particles carrying equal spin, this time in the same direction have a net spin. (c) In this case, somewhat similar to (a), the particle's spin points in opposite directions, yielding zero net spin. However, in this case, the particles orbit around a central point and the particle's motion contribute to a net spin.

can make up $3 \times 3 = 9$ different mesons (as you can pick from three types of quarks and three types of antiquarks). All possible combinations are: u$\bar{\text{u}}$, u$\bar{\text{d}}$, u$\bar{\text{s}}$, d$\bar{\text{u}}$, d$\bar{\text{d}}$, d$\bar{\text{s}}$, s$\bar{\text{u}}$, s$\bar{\text{d}}$ and s$\bar{\text{s}}$. The reality is a little bit trickier, as one never sees a meson that is only u$\bar{\text{u}}$ or only d$\bar{\text{d}}$ or only s$\bar{\text{s}}$. Recall that *identical* matter and antimatter particles (e.g. u$\bar{\text{u}}$ or d$\bar{\text{d}}$, but not u$\bar{\text{d}}$) can annihilate when they touch. Thus an up quark and an anti-up quark would touch and disappear, changing into energy. This energy will eventually turn into a $q\bar{q}$ (quark-antiquark) pair and, while it could turn back into a $u\bar{u}$ pair, $d\bar{d}$ is also possible, as is $s\bar{s}$. One might write this as $u\bar{u} \rightarrow$ energy $\rightarrow u\bar{u} \rightarrow$ energy $\rightarrow d\bar{d} \rightarrow$ energy and so on. We'll be better able to tackle this idea after Chapter 4 and you might want to make a mental note of this point and wait until we get there. The impatient reader might flip ahead to the discussion surrounding Figure 4.24.

There is a mathematically technical way to write this, but for our purposes, it's OK to write "mixture($u\bar{u}$ & $d\bar{d}$)" which is how we'll indicate "This particle contains a quark and antiquark pair, but sometimes it's $u\bar{u}$ and sometimes it's $d\bar{d}$." Note that the two quark combinations listed as mixture ($u\bar{u}$, $d\bar{d}$ & $s\bar{s}$) are different in a technical way. We will ignore that difference here. Trust me. You don't need to know. If you must, look at the suggested reading, concentrating on the pro suggestions.

So nine mesons are listed in the Table 3.1, in the column titled ↑↓. For all of these mesons, the spin is 0. While we can now see how the quark model can simplify our understanding of the world (nine mesons can be explained by three quarks), the real story is even better. While the above discussion talked about mesons in which the spins of the quark and antiquark were in opposite directions, it is also possible that the quark and antiquark's spins could be pointing in the same direction. This would result in a meson with a different spin, as illustrated in Figure 3.1b and in Figure 3.2. Take for instance the same quark combination we used as an example before, $u\bar{d}$. If the spins now are in the same direction, we make u(spin $= +1/2$) $+$ $\bar{d}$(spin $= +1/2$) $= \rho^+$(spin $= +1$). So one can use the same quark-antiquark combinations, but require the spins to be aligned and we can make more particles, this time listed in Table 3.1, in the column headed by ↑↑.

The strength of the quark model becomes even greater when we realize that for the two instances listed above, the quark and antiquarks pairs weren't moving (not true in the strictest sense, but close

Table 3.1 Quark combinations that describe many mesons. Each symbol in each of the "Meson" columns is the name of a specific type of meson. For instance, the first row shows that an up and antidown quark can make a pi-plus (π^+) meson, a rho-plus (ρ^+) meson or a b-plus (b$^+$) meson, with the type of meson being created depending only on the spin orientation or the motion of the quark and antiquark.

Quark Combination	Charge	Meson ↑↓ Spin = 0	Meson ↑↑ Spin = 1	Meson ↑↓ + Movement Spin = 1
u$\bar{\text{d}}$	+1	π^+	ρ^+	b$^+$
d$\bar{\text{u}}$	−1	π^-	ρ^-	b$^-$
mixture (u$\bar{\text{u}}$ & d$\bar{\text{d}}$)	0	π^0	ρ^0	b^0
mixture (u$\bar{\text{u}}$, d$\bar{\text{d}}$ & s$\bar{\text{s}}$)	0	η	ω	h
mixture (u$\bar{\text{u}}$, dd & s$\bar{\text{s}}$)	0	$\eta^{'}$	ϕ	h$^{'}$
s$\bar{\text{u}}$	−1	K$^-$	K*$^-$	K$_1^{*-}$
d$\bar{\text{s}}$	0	K^0	K*^{0}	K$_1^{*0}$
s$\bar{\text{d}}$	0	$\overline{\text{K}}^0$	$\overline{\text{K}}$*^{0}	$\overline{\text{K}}_1^{*0}$
u$\bar{\text{s}}$	+1	K$^+$	K*$^+$	K$_1^{*+}$

enough for illustration). More correctly, we say that they are in the ground state, which is physics-ese for lowest energy configuration. However, the quark and antiquark pair can move around one another, somewhat like the Earth and the Moon. This is one of those times where the weirdness of quantum mechanics pops up. When the q$\bar{\text{q}}$ pair moves, they are required to move so that the "spin" that their motion gives to the meson is an integer (i.e. +1, +2, +3, …) So now one can construct 9 new mesons, each of which contain the same quark content as listed in Table 3.1, but with the quark-antiquark pairs moving and contributing +1 (for instance) to the spin of the meson. These mesons are listed in Table 3.1, in the column headed (↑↓ + Movement). In Figure 3.2, we show a few specific quark configurations for representative mesons. We further can now see what

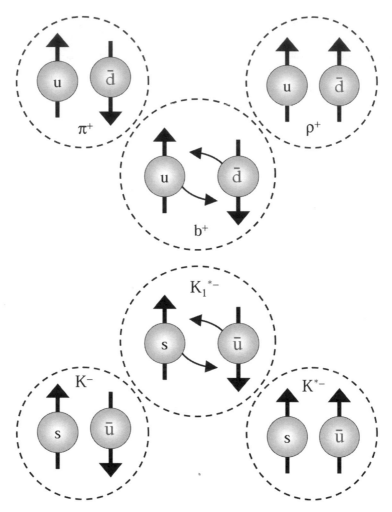

Figure 3.2 Examples of three different mesons that can be built with an up and anti-down quark or a strange and an anti-up quark. Additional motion configurations would yield additional mesons with no additional quarks.

was going on with the strange particles of Chapter 2. They simply contain a strange quark. From our earlier discussion, we recall that it seems that the strong force can easily make *pairs* of strange and anti-strange quarks, while it is only the weak force that can decay *single* strange quarks. We'll return to this in Chapter 4.

Table 3.2 Basic properties of the three originally postulated quarks.

Quark	Symbol	Charge	Decay Properties
up	u	+2/3	Stable
down	d	−1/3	Stable under many circumstances
strange	s	−1/3	Can decay via the weak force

Thus we see that for each configuration of quark and antiquarks, with differing contributions to the meson's spin from the pair's movement, we get nine new mesons. While we have listed 27 mesons in Table 3.1, there are in fact many more. But all of these particles can be described as different combinations of three quarks (and their corresponding antiquarks)! So the quark model greatly simplifies the understanding of mesons. Table 3.2 summarizes our knowledge thus far.

Quarks and Baryons

While mesons were some of the particles discovered in the early cosmic ray and accelerator experiments, there were also the much heavier baryons (the most familiar of which are the proton and neutron). The quark model would be even more powerful if it could explain the pattern seen in the baryons and, of course, it does. While mesons consist of a quark and antiquark pair, baryons are made of three quarks. Protons consist of two up quarks and one down quark (written uud) and neutrons consist of two downs and an up (udd). We can double check that these combinations of quarks properly predict the correct electrical charge: for the proton $u(+2/3) + u(+2/3) + d(-1/3) = \text{proton}(+1)$ and for the neutron $u(+2/3) + d(-1/3) + d(-1/3) = \text{neutron}(0)$. Figure 3.3 illustrates the familiar baryons, along with the most common type of meson.

While the proton and neutron are familiar baryons, other baryons were discovered before the quark model was understood. Before we show how quark models greatly simplify our understanding of

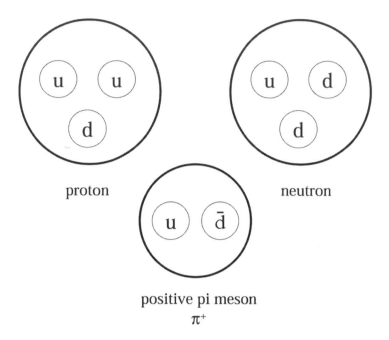

Figure 3.3 Baryons like protons and neutrons contain three quarks. Mesons contain one quark and one antiquark.

baryons, we must recall some important facts about quarks. Quarks have many properties, but we concern ourselves mostly with mass, charge and quantum mechanical spin. The mass of an up quark is very small (although we will return to this later), it has an electrical charge of $+2/3$ and it has a quantum mechanical spin of $1/2$. Recall from earlier discussions that, unlike the other attributes, a quark's spin can be either $+1/2$ or $-1/2$. We can write an up quark with spin $+1/2$ as $(u\uparrow)$ and an up quark with spin of $-1/2$ as $(u\downarrow)$.

Given this information, let us consider a hypothetical baryon containing three up quarks (uuu). Such a baryon must have an electrical charge of $(2/3 + 2/3 + 2/3 = 2)$, twice that of a proton. However, because each quark can have a spin of $(u\uparrow)$ or $(u\downarrow)$, we see that when spin is considered, we have two general spin cases. The first is when all of the quark's spins are in the same direction $(u\uparrow\ u\uparrow\ u\uparrow)$. Since the spins are all in the same direction, such a baryon would have a spin of

$(+1/2 + 1/2 + 1/2 = +3/2)$. The second case is when two quarks have spins that are aligned, while the third quark's spin points in the opposite direction (u↑ u↑ u↓). In this case, since the spins of two quarks cancel each other out, the baryon's spin is $(+1/2 + 1/2 + (-1/2) = +1/2)$. Thus there are two baryons consisting of three up quarks, but differing by their spin.

All of the possible combinations of quarks and spins that can be present in hadrons are much more complicated than in mesons. It turns out that there are eight unique ways in which one can combine the three different kinds of quarks (u, d and s) into a baryon in such a way that the final baryon has a final spin of 1/2. In addition, there are 10 different ways in which the same three quarks can make up a baryon with final spin of 3/2. Note that this is for the case when the quarks are not moving much within the baryon. Of course, the quarks are allowed to orbit within the baryon according to the strict rules imposed by quantum mechanics. These motions are restricted such that they can add only integer spin to the baryon $(0, 1, 2, ...)$, just like the meson case. Thus for each possible movement configuration of quarks, the quark idea can explain 18 different baryons.

We should recall that mesons were allowed to contain both quarks and antiquarks. Baryons contain only quarks. So how do antiquarks fit into the baryon picture? Well, antiquarks are not allowed in baryons. However, one can make anti-baryons using three antiquarks, for example an anti-proton (written $\bar{p}$) consists of ($\bar{u}\bar{u}\bar{d}$), while the anti-neutron consists of ($\bar{u}\bar{d}\bar{d}$). And this pattern is true for all baryons.

Thus we are now able to appreciate some of the genius of the quark model. Given the three kinds of quarks (and their corresponding antiquarks), we can explain 18 mesons, 18 baryons and 18 antibaryons ... 54 particles in total. And if the quarks revolve around one another within the particle, we can explain many more particles for each additional allowed movement configuration. So instead of hundreds of unexplained particles, we have reduced the complexity to three quarks and their associated antiquarks.

Often when one uses a model to greatly reduce the complexity observed in the world, one must pay a price, because the theory introduces its own complexity (although the complexity introduced is much less than the complexity explained). To see why we must introduce something new to our thinking, we must consider a particular baryon, the Δ^{++} (delta double plus). The Δ^{++} has a mass somewhat greater than that of a proton, has twice the electrical charge of the proton and has a spin of 3/2. From what we now know about quarks, we see that this object must consist of three up quarks in the $+1/2$ spin state ($u\uparrow\ u\uparrow\ u\uparrow$). But we now must recall something that we learned in Chapter 2. Any object with half integer spin is called a fermion and it is impossible for two identical fermions to exist at the same place. And yet here we have 3 up quarks, all with the same mass, electrical charge and the same spin state. This is very bad. With what we now know, a particle containing ($u\uparrow\ u\downarrow$) is OK, as is a particle containing ($u\uparrow\ \bar{u}\uparrow$). But ($u\uparrow\ u\uparrow$) is a no-no and ($u\uparrow\ u\uparrow\ u\uparrow$) is definitely a no-no. So either the quark model is wrong, or we need to do something to rescue it.

A Colorful World

Given that quarks with identical properties are not possible, but the three quarks in the Δ^{++} are "obviously" identical, a new property of quarks was proposed. Oscar Greenberg of the University of Maryland made the daring proposal that perhaps the quarks contained some previously unknown property that distinguished them. It was presumed that the quarks in the well-studied protons would have this same new property. But the proton itself did not (or else it would have been observed earlier). It's easy to see how two objects could cancel out to make nothing, just like adding $+1$ and -1 yields 0. So the quarks in mesons didn't seem so tricky, but baryons contain *three* quarks. Thus the new property needed to be such that when all three quarks were added together, the result was zero (or equivalently, the baryon did not contain the property).

There is a more familiar field of physics that had a similar property. When one takes three light bulbs, each of a specific color, in particular red, green and blue, and shines them on a white wall, the three lights combine to produce white light (try it!) Thus in analogy, this new property of quarks is called color. So in the Δ^{++}, the quarks can now be called (u↑ red) (u↑ blue) (u↑ green) and they produce a (Δ^{++} white). This is true of all baryons. The three quarks each carry a particular color, just like they do an electrical charge, but the baryon must be "white" or color neutral (which is just physics-ese for it has no net color, because the color of the quarks cancels out). Figure 3.4 shows the quark content for a real Δ^{++} particle.

It should be emphasized that quark color has nothing to do with visible color. A "red" quark is not red in the way we normally mean red. The dazzling blue of my wife's eyes is not because they are covered with blue quarks. We say that a quark has red, green or blue color to remind us that we need three of them to produce the color neutral objects we observe in the world. Please do not go to your local particle physics laboratory and ask for a bucket of green quarks because you think that the color of your living room needs a change. They'll think you're foolish and immediately put you to work.

Since we now know that the quarks carry color, we have solved the problems of identical quarks in the Δ^{++}. While it does contain

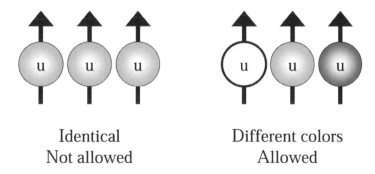

Identical
Not allowed

Different colors
Allowed

Figure 3.4 The fact that two identical fermions cannot exist in the same place led to the hypothesis of color. As long as the fermions have something that make them different they can be together.

three quarks of the (u↑) variety, identical in all respects except for color (i.e. (u↑ red) (u↑ blue) (u↑ green)), the fact that each quark carries a different color means that the quarks can be distinguished (at least in principle). So now there does not exist two or more identical quarks (i.e. fermions) in the Δ^{++} and the theory works. Phew!!!

If quarks carry color, then antiquarks carry anticolor. Like electrical charge where a positive charge can cancel out a negative charge of equal size, quark color charges can cancel each other out. Since baryons and mesons have no net color, we can work out the cancellation rules. Take any particular meson, which as we know consists of a quark and antiquark pair. If the quark carries a particular color charge (say red or R), the antiquark must carry antired (or $\bar{R}$). The R and $\bar{R}$ cancels out so that the meson has no net color (or is color neutral or is white, all the same thing said in different ways). The quark could carry blue (B) or green (G) color, in which case the antiquark would have to carry antiblue ($\bar{B}$) or antigreen ($\bar{G}$) color.

The situation with baryons is more complicated (as always). Each baryon carries three quarks, each carrying a different color (RGB). Then, if the resulting baryon is color neutral (i.e. white), we can say that (R) + (GB) = (W). So compared with the discussion of mesons, we see that (GB) must be equivalent to ($\bar{R}$). Similarly (RB) is the same as $\bar{G}$ and (RG) is the same as $\bar{B}$. This is a little hard to get your head around, but it's just a consequence of having to add three equal things together to get zero.

First Evidence for Quarks

Up to this point, everything that has been presented is theoretical. We know that mesons and baryons exist, but quarks are just hypothetical. On the face of it, there is an obvious experimental path. We should try to extract a quark from a baryon (say from a proton, as we have many of these). All of this is analogous with the experiments done to show that atoms contain electrons. Energy is added to the atoms, the electrons leave the atoms and we collect and study the electrons.

Similarly, we can try to break up a proton by adding energy to it. The easiest way to add energy to a proton is to use a particle accelerator to smash the proton into a target. Presumably, when the proton is hit hard enough, it will break up and we will be able to see the three quarks coming out.

As we recall from Chapter 2, what typically happens when a proton is smashed into a target is that a bunch of pions (pi mesons) are made. When we do such an experiment and try to identify quarks (easily identifiable due to their fractional electric charge), the result of the experiment is that *no* free quarks (i.e. quarks that are not carefully ensconced in a hadron) are observed. As an experimental scientist, one needs to be honest in stating what this result means. True, an observation of no free quarks could mean that free quarks don't exist. It could also even mean that the quarks themselves were an interesting, but ultimately false, idea. Really the experiments say that less than one free quark is observed for every 10^{10}–10^{11} pions. The interpretation of this observation was a topic of debate.

The non-observation of free quarks is extremely serious. It could have signaled the death knell of the quark model. Still, the elegance and predictive power of the quark model was compelling and physicists needed to offer a dirty and inelegant hypothesis: quarks could exist only inside mesons and baryons. This is the so-called confinement hypothesis. When made, this hypothesis was rather ugly and offered in order to save the quark model. But no one liked it. We now know that this hypothesis was actually correct (and we will discuss the reasons in Chapter 4), but it was touch-and-go there for a while.

There was one fact that allowed physicists to stomach the confinement hypothesis. Prior to the proposal of the quark model, many mesons and baryons had been discovered. Specifically, baryons had been discovered that the quark model explained as having 0, 1 or 2 strange quarks contained within them. However, no baryon containing three strange quarks had been observed. If the quark model was correct, the baryon (sss) had to exist. In addition, there was a pattern in the baryons as they contained more strange quarks. Baryons carrying one strange quark have a mass of about 150 million electron Volts (MeV) more than baryons carrying no strange quarks. Further, baryons carrying two strange quarks have a mass of about 150 MeV more than ones carrying one strange quark. (As a reminder, an electron Volt is a unit of energy, but since energy and mass are equivalent, we can express mass in energy units.) So, if the mass difference between the baryons was due to the mass of the strange quark, the quark model predicts the existence of a baryon containing three strange quarks and having a mass of 150 MeV more than baryons carrying 2 strange quarks.

Late in 1964, the Ω^- was discovered in a bubble chamber experiment at Brookhaven National Laboratory on Long Island in New York. This particle decayed in a way consistent with having three strange quarks and had a mass 140 MeV more than that carried by baryons with two strange quarks. Since the particle was predicted (and with very specific properties) before it was discovered, this was regarded as a singular triumph of the quark model. Concerns with the confinement hypothesis were put aside for the moment while physicists tried to work out the confinement mechanism.

It is one of the ironies of modern physics that while the quark model had great explanatory and predictive power, even the architects of the quark model initially did not think of quarks as actual constituents of the mesons and baryons. The quark model was just thought of as simply a mathematical organizing principle. However Gell-Mann and Zweig were more prescient than they knew. Experiments performed in the late 1960s could not free quarks from protons, but they did reveal that the proton had a small but finite size and that there appeared to be something inside the proton, as the much more massive proton would scatter the incoming electron more violently. The objects contained within a proton were poorly understood in the beginning, as their properties had not been measured. However, once their existence was proven, the objects were named "partons" as they were part of the proton. Initially it was not possible to identify partons with quarks (although we are now able to prove this). At this point in our story, we do not have enough information to properly discuss these ideas (we will resume this discussion towards the end of Chapter 4), but we can roughly understand this experiment by analogy.

The early experiments accelerated an electron to high energies and aimed them at a chunk of material (often hydrogen cooled until liquefied). Since hydrogen atoms consist of an electron, a proton and no neutrons, this experiment was essentially one of firing an electron at a proton. These particles both carry electric charge and thus they interact via an electric field. Since the electric force was quite well studied, the different possible behaviors of the electron in the scattering process were well known and they beautifully described the experimental data. However, as the electrons were accelerated with ever increasing energy, they could approach ever closer to the proton. When electrons were made to approach within about 10^{-15} meters of the center of the proton, the scattering pattern abruptly changed. Something was different. Some new physical process was beginning to come into play. An analogy might be a comet, which passes through the solar system again and again. According to the laws of gravity, one can treat the comet, the Sun and all of the planets as having no size (i.e. as point-like particles). The calculations work out perfectly and the motion of the comet is accurately described. However, if the comet passes so close to a planet that it hits it (as comet Shoemaker-Levy hit Jupiter in 1994), why then the physics changes. This is a reasonable way to measure the size of planets, although it's a bit hard on both the planet and the comet.

Just because something has a size, doesn't mean that it contains smaller particles. Compare a beanbag and a billiard ball. Ignoring for the moment what we know about atoms, the billiard ball is a uniform and solid structure with no internal features. The beanbag however is a loose aggregate of smaller objects held intact by the "force" of the outer cloth skin. When each of these objects is hit by something, they react differently in the collision. The billiard ball can have its speed and direction changed, but since there is nothing inside the ball, the insides are unaffected. The beanbag is quite different. During the collision, the beans can move around. So, in addition to having the speed and the direction of the bean bag change, the beans can move with respect to each other. Moving the beans takes energy. Since it is one of the fundamental tenets (and observations) of physics that energy is conserved (physics-ese for "doesn't change"), if energy can go into swirling the beans, there is less energy in the motion of the beanbag. So one measures the energy of the projectile before and after the collision and finds that they aren't the same. This means that something inside the target jiggled. And that means that it has structure (i.e. contains something within it).

When electrons (which to the best of our knowledge have no size or structure) are made to hit stationary protons, one can measure the electrons' energy before and after. When the electrons pass at distances more than about 10^{-15} meters from the center of the proton, the incoming and outgoing energies are the same. But when the smallest distance between the proton and electron becomes 10^{-15} meters or less, the outgoing energy can be less than the incoming. This means that the innards of the proton are jiggling. When the experiments were being done in the late 1960s, it was thought that the structure of the proton could be viewed as several particles (in analogy with the electron's making up an atom). The constituent particles of the proton were called partons. We will return to this topic at the end of Chapter 4.

While the quark model did a brilliant job of explaining the myriad of baryons and mesons that had been discovered, it also raised new questions. In addition to the question of quark confinement and the need to prove that the quarks were physical, and not just mathematical, entities; there were at least two additional questions that kept physicists awake at night. The first problem is the easiest to explain. Basically the question was "Why are there two $-1/3$ charge quarks and only one carrying $+2/3$ charge?" Physicists love symmetry (largely because the universe seems to). When you find an odd man out, it often indicates that your understanding is incomplete and certainly warrants further inspection. Thus it was obvious (obvious is so easy to say in retrospect) to speculate that perhaps there might be another, as yet undiscovered, quark with $+2/3$ electrical charge. I really want to emphasize that such an argument is somewhat religious at this point. This is just a gut feeling about how the world must be. But science on the frontier is often driven by gut feelings. Sometimes they're right. Sometimes they're not…experimental evidence is the final arbiter.

The second question concerned specifically the strange quarks. Recall from Chapter 2 that strange particles were unstable, eventually decaying into other, more familiar, particles. Once one believed in the strange quark, it was natural to believe that it was the strange quark that was unstable. Because strange particles live a long time (recall that this is why they were called strange in the first place) the force that caused them to decay had to be very weak (we'll talk more about forces in Chapter 4). Since the various ways in which the assorted strange particles could decay had been observed in experiments and further we knew by that time the quark content of both the parent strange particle and the daughter decay products, it was possible to understand the ways in which a strange quark could decay. Because the strange quark carried the strange quantum number, when a strange particle decayed into a particle containing no strange quarks, the "strangeness" changed (how very strange…er…I mean peculiar…) The real mystery was why there appeared to be no decays which were both caused by the weak force and changed the strange quark into a down quark. This should have been possible, but it simply wasn't observed. This was weird and not initially understood.

In 1970, Sheldon Glashow (Shelly to friends and rivals alike), John Iliopoulos and Luciano Maiani (eventually lab director at CERN, the premier European paricle physics laboratory) proposed a way to solve the problem. They rather cleverly showed that if another quark existed with electrical charge $+2/3$, the theory was modified so that the strangeness-changing interactions of the type described above were now forbidden by the theory. Experimental observation and theoretical predictions were once again in agreement. This new quark was called *charm* and was conceptually paired with the strange quark. Now there were two pairs of quarks (up & down) and (charm & strange). There was only one problem … no charm quarks had been observed.

Discovery of More Quarks and Leptons

This troubling set of circumstances changed in 1974, when two experiments jointly announced the discovery of a new long-lived particle. This particle was about three times heavier than a proton and was quite a surprise. Further investigation showed that the "sharpness" of the mass was extremely narrow. One of the architects of quantum mechanics, Werner Heisenberg, devised an uncertainty principle (detailed somewhat in Appendix D) that said that if a particle exists for a long time, its mass was extremely well determined (i.e. it has little uncertainty in energy), but if it decays quickly, it does not have a unique mass, because the energy uncertainty is large. For instance, a particle that decays quickly (say in 10^{-23} seconds) typically has a range of masses of 100 MeV. This new particle had a mass spread on the order of 0.063 MeV, so it lived about 10^{-20} seconds or about 2000 times longer. When a particle lives longer than it should, this implies that something is keeping it from decaying; for instance, a new type of quark is being produced. In essence, this is strange particle production redux.

This new particle had two names for a while. One experiment was performed at Brookhaven National Laboratory (BNL) and headed by Sam Ting. They were smashing protons on a beryllium target and looking for particles decaying into two muons. When they saw their evidence, they named their new particle the "J," I'm told because a Chinese character similar to J is used to represent Sam Ting's family name. The competitor experiment at Stanford Linear Accelerator Laboratory (SLAC) was headed by Burton (Burt) Richter. They were looking for particles by smashing electrons and antielectrons (also called positrons) together at different energies. Because they could choose the energy of their beams very precisely, they could look for particles at very specific energies. As they scanned the energies, the electrons and positrons collided at a fairly predictable rate until they hit the "magic energy" of 3100 MeV. At this energy, the number of interactions increased dramatically. Depending on which types of particles they were producing, the rate went up by a factor of 10–100. Presto. A new particle. The SLAC guys called this new particle the ψ (psi).

It is traditional in science for the discoverer of something, be it a particle or new species, to name it. Yet here were two very competitive groups of physicists essentially simultaneously announcing the discovery of a new and spectacular particle. After some, um, … spirited … debate on the question of who got there first, it was finally resolved that the two groups would jointly be declared discoverers and the particle was called the J/ψ (Jay-sigh). And, in 1976 when Richter and Ting jointly shared the Nobel Prize for the discovery of the J/ψ, amity returned. Mostly.

The J/ψ was eventually shown to be a new meson containing both a charm quark and anticharm quark ($c\bar{c}$). Soon after the discovery of the J/ψ, many other mesons and baryons were discovered. These were the ones that could be made now that four different quarks were known, for example the D$^+$ meson, consisting of a charm quark and a down antiquark ($c\bar{d}$) and the D$^-$ meson, containing a down quark and a charm antiquark ($d\bar{c}$). One question remained … why was the charm quark so heavy? The charm quark was about three times heavier than the strange quark and even 1.5 times as heavy as a proton or neutron. This question was just a shade of things to come.

While the quarks make up the mesons and baryons, there remain the much lighter leptons, which we will discuss presently. They are relevant here, because they seem to be related to the pattern of quarks. Prior to 1974, four leptons were known to exist: two charged leptons, the electron and muon, and two neutral leptons, the electron neutrino and the muon neutrino. A mysterious pattern seemed to be present. For each pair of quarks (for instance up and down), there appeared to be a corresponding pair of leptons (for instance the electron and electron neutrino). In 1974, this symmetry between quarks and leptons was very apparent, although not understood.

In 1975, an experiment at SLAC, headed by Martin Perl, announced that their data showed that another charged lepton existed. This lepton was called the τ (tau) lepton. Of course, this neatly destroyed the comfortable symmetry observed between the quarks and leptons. Or did it?

One way to restore that comfortable symmetry would be if another pair of quarks existed. While there really wasn't any evidence for a new pair of quarks, the very possibility excited experimental physicists. Like bloodhounds after a wounded fox, they set off in pursuit. In 1977, an experiment conducted at Fermilab, headed by Leon Lederman, saw what looked like a signal for a new particle with mass of 6 GeV, only to see the signal disappear like a mirage as more data came in. This was not an error or carelessness on the experiment's part. Often one sees clusters of data that initially look like patterns, only to have the pattern disappear as more information is obtained. Luckily for Lederman's group, while the cluster at 6 GeV became less interesting, a new cluster at 9.5 GeV started to look appealing. Now a little more gun shy, they waited and watched. Unlike their early experience, as the data came in the signal looked even more solid. In June of 1977, Lederman's experiment announced the discovery called the Υ (upsilon). (Lederman took some good natured ribbing about the non-particle at 6 GeV, which some wags termed the "oops-Leon.") Ribbing aside, the Υ was a marvelous discovery. Like the earlier J/ψ, the Υ particle had a very well determined mass, indicative of a long-lived particle. And, like the J/ψ, the reason this particle was long-lived was because a new quark was being created. This fifth quark was called *bottom* (although for a while, the term *beauty* was competitive). This new quark was massive, about 4.5 GeV (about three times heavier than the charm quark) and had an electrical charge of $-1/3$. Given our previous experience, it seemed that there probably was another quark to be discovered, this one with electrical charge $+2/3$. Even before this quark was discovered, it was named. It was called the *top* quark, counterpart to the bottom quark. (Note that for a while, alternate names for the pair of quarks were *truth* and *beauty*, but these names have fallen out of favor.)

Discovery of the Top Quark

The search for the top quark was long and arduous. In 1984, an experiment announced that they had perhaps observed the top quark, with a mass of about 8 times that of the bottom quark (and about 40 times that of the proton). Further experimentation revealed that this result was in error. The search continued. There was indirect evidence gathered by clever experiments that supported the existence of the top quark, but indirect evidence is often problematic. Direct evidence is preferred. In 1992, two huge, leviathan experiments got underway at Fermilab, their primary purpose to find the top quark or as one physicist of my acquaintance said "Bag it, tag it and take it home…" The two experiments, one called DØ (pronounced D-Zero) and the other called CDF (for Collider Detector at Fermilab), were friendly, but deadly serious competitors. Each experiment consisted of large detectors of approximate dimensions $(30' \times 30' \times 50')$ and weighed about 5000 tons. Each is housed in its own building and took years to build. Both experiments involved about 400 physicists of which about 100 on each experiment were directly working on trying to discover the top quark. They worked feverishly, days, nights, and weekends; each worried that the other experiment might get there first. In high stakes science, there is first and not-first. There is no second. (With apologies to Yoda.) Finally in March of 1995, both experiments simultaneously announced that they had firm evidence for the top quark. The chase was over.

We will learn more in Chapter 6 about the accelerators and detector techniques necessary to make this discovery, but even without that information, the story of the last days in the search for the top quark is pretty interesting. The reader should realize that the two experiments had a total of 800 individuals on them so there are about 800 slightly varying stories. The one given here is my take on it, with all of my own biases. I joined DØ in the spring of 1994 and the excitement was palpable. CDF was an older experiment and had taken data before. While the amount of data that they had taken was small, it was the most data ever taken at such a high energy. The experience that they had gained was invaluable and some thought that it would provide an edge that would be hard to overcome. DØ, on the other hand, was a newcomer and had never seen colliding beams before. Our detector was in some ways significantly superior (being built later and thus having the advantages of being able to use newer technologies), because our energy measuring equipment was superior and because more of each collision was recorded. (If you think of a particle collision as an explosion and your detector as a sphere that wraps around the explosion, the detector that covers more of the angle will have the advantage. DØ had the edge by this measure.) On the other hand, CDF had a detector that DØ didn't. They had a silicon vertex detector (discussed in Chapter 6), which had the ability to measure the trajectories of particles very near the collision point to a precision much smaller than a fraction of a millimeter. They also had a magnetic field in the region where the collision occurred, which allowed them a second way to measure the energy of many of the particles exiting the collision. Both of these components provided capabilities that DØ lacked. If the truth be known, both detectors were superb collections of technology, as carefully designed for their job (particle detection) as the combined brainpower of 400 really smart people could make them. Any attempt at design involves compromise and choice, and the differences between the detectors reflected each group's best guess on making the crucial elements work the best, while realizing that this necessarily meant that other, less crucial, elements might not work as well as they might, had other choices been made. As they say, time would tell who had designed correctly.

While both detectors were quite evenly matched, each with their strengths and weaknesses, the sociologies of the two experiments were really quite different. CDF was older, more established and had the advantage of years of experience. It was my impression that they were confident that this experience would keep them ahead of DØ while DØ worked to catch up. DØ, on the other hand, was the new kid on the block. We were brash, driven, talented, but unproven. This is not to say that we did not have experienced people on the experiment; we certainly did. But the experiment itself was new and to shake it down would take time.

In 1992, both experiments hit the ground running. You have to be around scientists to fully appreciate how intensely driven and hard working they can be, especially when a crucial discovery is on the line. Slackers worked sixty-hour weeks. Hard workers lived on the passion for the hunt (although gallons of truly-awful coffee helped too). Both experiments had significantly different approaches, each tailored to their respective strengths. Over the years, the data came in, although at a rate slower than the experiments had hoped. Even so, collisions were recorded by both experiments that looked promising. Of the zillions of collisions that were inspected and the millions that were recorded, each experiment had a handful of events that "smelled" like top quarks (we will revisit what this means in Chapter 4).

DØ released a paper in January 1994 (just prior to my arrival) in which they discussed several collisions recorded by their detector that were consistent with top quark production, with one particularly interesting event. While interesting, one event usually proves nothing, no matter how tantalizing. Because so few events were observed, this implied that top quarks were even heavier than originally thought. DØ said that their data suggested that if the top quark existed (which was not established at this time), then its mass exceeded 131 GeV, or 140 times that of a proton.

In April of 1994, CDF released a paper entitled "Evidence for Top Quark Production in $p\bar{p}$ Collisions at $\sqrt{s} = 1.8\,\text{TeV}$." DØ was a bit incensed by this action. The paper technically didn't claim that they had discovered the top quark (otherwise "Evidence for" would have been "Observation of") and DØ viewed the paper as a preemptive claim of precedence. If later the top quark was discovered, then

CDF could claim to have seen it earlier, while if top did not material-ize, CDF could correctly state that they never claimed discovery, so they had nothing to retract. I'm sure there are those on CDF who see it differently. As usual, the truth probably lies somewhere in the mid-dle. In all fairness, in their paper CDF got the mass of the top quark about right, although their measurement for how likely it is that a top quark would be produced was about twice as much as the correct answer. This over-estimate presumably was what made them feel com-fortable with their paper.

After the flurry of publication in the spring of 1994, it was back to the grindstone, as data continued to flow in. When an additional amount of data was taken, identical in size to the amount used to support the spring of 1994 papers, an identical analysis was performed on both CDF and DØ's new data. The significance (i.e. solidness) of CDF's data went down slightly, while DØ's increased by a similar amount. The work continued and the analyses became more sophisticated.

The first real announcement of the discovery of the top quark occurred in March 1995. The events that lead up to the joint announcement are rather interesting as a study of the sociology of competitive science when so much is at stake.

Fermilab has weekly particle physics presentations, officially called the "Joint Experimental–Theoretical Physics Seminar," but known to one and all as a "Wine and Cheese." Started in the early days of Fermilab by Marty Einhorn and J.D. Jackson (the author of *the* grad-uate level book on electricity and magnetism and bane of young physics graduate students everywhere), these seminars were designed to mimic the seminars regularly held at research universities and to bring the theorists at Fermilab together with the experimentalists who were alternately freezing or sweating their butts off, trying to turn the Illinois prairie into a world-class physics laboratory.

Wine and Cheeses (which were really Juice and Cheeses for a few years, although I'm glad to say they're now Wine and Cheeses again), occur each week on Friday at 4:00 P.M., as a nice end to a usually hectic week. A one-hour talk is given, typically by a junior scientist, on some measurement or discovery that they have made. In the fall of 1994, DØ decided that we needed to start letting others at the laboratory know of all of the seriously cool work that was being done (after all, the top quark frenzy consumed only about 1/4 of the physicists on DØ, the other 3/4 were busy on other things). So the DØ brass scheduled a Wine and Cheese every 6–8 weeks, with the idea that a young researcher would give a talk on their work. Since the reservations were done half a year in advance, the name of the actual speaker was not given, but rather a placeholder name.

On both DØ and CDF, the analysis efforts are organized into groups. While a single graduate student or post-doc often does each analysis, there are often sufficient commonalities between analyses that physicists with similar interests band together to share knowledge. On DØ, during the period of 1992–1996, there were five groups: Top, Bottom, Electroweak, QCD (Quantum ChromoDynamics) and New Phenomena. (CDF's organization was similar.) Top and Bottom were groups concentrating on their respective quark. Electroweak studied how quarks and leptons interact with each other. QCD was interested in how the less exotic quarks behaved and New Phenomena was concerned with unexpected physics. Each group had some few-dozen members and during the top quark search frenzy, both the DØ and CDF Top groups each had about 100 members. Each group has two co-leaders called conveners and in the case of the Top group of DØ, these were Boaz Klima of Fermilab and Nick Hadley of the University of Maryland, while their counterparts on CDF were Brig Williams and Brian Winer.

In February of 1995, one of the Wine and Cheeses had been scheduled and the placeholder name was Boaz Klima. Since Boaz was a Top group co-convener (and thus moderately senior), he was not a natural choice for a Wine and Cheese speaker. That is, unless a big announcement was about to be given. Someone on CDF looked at the Wine and Cheese schedule, saw Boaz's name, put two and two together and said "Poop!" (although I'm told that the actual reaction was somewhat stronger). If DØ was going to announce the discovery of the top quark, CDF didn't want to be caught flatfooted. They stepped up their already insane pace to try to firm up their analysis. Meanwhile DØ, who did not intend to announce anything so big just yet, merely continued to work at the usual insane pace.

Some three years prior, early in 1992, the two experiments had made a gentlemen's agreement that if they were going to make a big discovery announcement (say of the top quark or even bigger), they would give the other experiment a warning of one week, so as to prepare a response. The response could be "We agree," "You're full of it" or "We don't know." So, with the upcoming DØ Wine and Cheese, CDF went into overdrive to finish up their results, which, as we recall, were already promising. On February 17, in an attempt to preempt DØ's Wine and Cheese, CDF notified DØ and John Peoples, then Fermilab's director, that they were going to announce the discovery of the top quark on February 24th. DØ was caught unaware. DØ had a result that looked promising and that we believed that we would announce eventually but perhaps a little later, when all of the consistency checks had been done. But now this complete set of tests had to be done *now*. We had one week. It might be just a coincidence, but coffee stock prices jumped quite a bit that week.

Well the tests were done and, on February 24th, the two experiments simultaneously submitted their papers at 11 A.M. to *Physical Review Letters*, America's most prestigious physics journal. Both papers were accepted within a week, after the necessary and proper peer review. Unlike less responsible researchers, both experiments submitted their results to a refereed journal before calling *The New York Times*. John Peoples was out of town at the time and out of respect for the director, the formal presentation to the Fermilab scientific staff was deferred until his return. John returned on March 2 and both DØ and CDF presented not a mundane Wine and Cheese, but rather a special joint seminar to the scientists then resident at Fermilab. The Fermilab auditorium is fairly big and seats 847 people. There were a whole lot more people than that in the auditorium to see the talks. While I'm not sure, I suspect that a Fire Code or two were bent slightly that day, irrespective of Fermilab's safety group's best efforts. Some things you simply can't miss. We even had a live video feed to Fermilab's second largest conference room and it was packed too.

D∅ went first. Paul Grannis of the State University of New York, Stony Brook, then one of D∅'s co-spokesmen (which means supreme leader, rather than something like press secretary), gave the talk, meticulously going over D∅'s case. CDF's spokesmen followed, with Georgio Belletini of the University of Pisa and Bill Carrithers of Lawrence Berkeley Laboratory giving their case. When both presentations were done and the hard questions over, a short hush fell over the auditorium. Then applause and cheers thundered through the room. It was an impressive day.

Following the scientific presentation was a two-day media frenzy. Reporters from all over the world came to see what the hoopla was all about. Luckily, Fermilab is a multinational laboratory, so usually language wasn't an issue. But for two days, the conveners of CDF and D∅'s Top groups, the respective experimental spokesmen, as well as Fermilab's management, didn't get much rest. The rest of us basked in the glow.

So what was announced? Each experiment announced the mass that they measured (with an estimate of the experimental uncertainty) and their measured top quark production cross-section (which is a number that is proportional to how often top quarks are made). Now that nine years have passed since that frenzied week, it is interesting to ask: How accurate were we?

CDF said that they thought the mass of the top quark was 176 ± 13 GeV, or about 188 times as heavy as a proton. D∅'s mass measurement was much less precise; we quoted a mass of 199 ± 30 GeV. The little "$\pm$" means something important. It's an estimate of how uncertain we are. For instance, D∅'s top mass of 199 ± 30 GeV means "We think that the most likely answer is 199 GeV, but it could be 30 GeV larger or smaller without any problem." (It's kind of as if I asked you how much money you have in your pocket right now. You likely have a good idea, but aren't sure to the cent. So you'd tell me a range that you believe to be likely. Then we'd count the money and see what the real number was.) More technically, the "$\pm$" means that we were 70% certain that the real answer lies between $199 - 30\,\mathrm{GeV}$ and $199 + 30\,\mathrm{GeV}$. (Yes, that means that there is about one chance in three that the real number isn't in that range.) But bottom line is that the number after the "$\pm$" is a statement of our uncertainty in our measurement, i.e. of how far from our best estimate of the top mass that the real value can reasonably be. DØ's uncertainty of $30\,\mathrm{GeV}$ means that it is unlikely that the real top quark mass would be $50\,\mathrm{GeV}$, because that is too far from our best guess.

When we return to DØ's and CDF's estimate for the mass of the top quark, we see that CDF was more confident of their answer than DØ was, although the two estimates didn't disagree. Now, in the fullness of time, both experiments have improved their measurements and have comparable errors of about $7\,\mathrm{GeV}$. When we combine both experiment's measurements, currently our best estimate for the mass of the top quark is $174.3 \pm 5.1\,\mathrm{GeV}$. Data taking currently underway is expected to appreciably reduce this uncertainty.

So it seems that CDF made both a more accurate, as well as a more precise, first measurement of the mass of the top quark than DØ did. We see the situation is somewhat different when we look at the cross-section measurement. We might recall in CDF's earlier "Evidence" paper, that they said that the data supported a large cross-section (top quark production probability). However, in March of 1995, both experiments announced a similar (and much smaller) cross-section, with DØ having the smaller uncertainty. You win some and you lose some.

Any story involving over 800 people will have little nuances, depending on who is telling it. I believe this account accurately reflects how events unfolded. Others may differ slightly in their emphasis or on their take on certain events. But this version is consistent with the one reported in the journal *Science* immediately after the measurement. When others write their book, they can tell their view. The thing that you should take from this story is that science is a very human endeavor, although held to much stricter rules than most. There is *an* answer and we try to find it. You can be right or you can be wrong. There isn't much room for "We disagree, but we're both right," unless it turns out that you're talking about different things. In this story, there were many heroes and very few villains. Two intensely motivated and extraordinarily competent groups of scientists chased a discovery hoping to get there first and the contest was a draw. Both felt compelled to announce their results just a little bit prematurely (although not much…the fact that either experiment would announce was never in doubt by that time), rather than coming in second. It's a little like the psychology that sometimes drives countries into conflicts that neither wants. However, this is science and only reputations, rather than lives, were on the line. This was one of those times when you shake hands and admit that the contest was fair and that the competition was good. The two experiments made the same discovery at the same time. It was a tie. The next time however....

Exactly how the top quark was discovered requires some knowledge from Chapter 4. Since I would like to discuss in some detail the technical aspects of how the top quark was observed, I will defer this until a little later in the book. However, for now we can take on faith that the top quark has been discovered and now look at the quark's properties. The top quark looks much like a charm quark which, in turn, looks a lot like an up quark: electrical charge +2/3, spin 1/2 and associated with a partner quark of charge −1/3. The most remarkable thing about the top quark is while the mass of the up quark is currently unknown (although known to be very small), and the mass of the charm quark is about 1.5 times that of the mass of a proton, the top quark has a mass of 175 GeV, fully 187 times the mass of a proton and even 40 times the mass of its partner, the bottom quark. To give you perspective, we often say that this single quark has a mass similar to that of an entire gold atom. (In fact, it's more similar to that of an ytterbium atom, but ytterbium doesn't have gold's cachet.) And this, as my teenage daughter says, is *soooo* weird. What makes a top quark have a mass about 100–200 times more than the average of the other 5 quarks? We don't know. But we do have some ideas and we will discuss the possibilities in Chapter 5.

The top quark does have an additional property that makes it unique. It is so massive that it decays very rapidly. In fact, it decays before it has time to combine with an antiquark and make a meson. So there will be no study of mesons and baryons containing top quarks. But a positive side effect of this fact is that the mass of the top quark has been the most precisely measured of all the quarks. This is because the other quark types (for example, the charm quark and antiquark in a J/ψ) have time to form mesons. Thus, in addition to the energy going into the masses of the quarks and antiquarks, there is energy going into the force holding them together. This confuses the issue and makes it difficult to unambiguously determine the quark's mass. The rapid decay of the top quark sidesteps the whole problem, which is why we can measure it to an accuracy of 5 GeV or 3%. Experiments are currently underway to reduce this already-impressive uncertainty.

So now you know much of what there is to know about quarks. There exist six types of quarks, arranged into pairs. We say that there are six flavors of quarks, where, as usual in particle physics lingo, flavor doesn't have the usual meaning. In this context, flavor means "type." Three of the quarks have electrical charge of +2/3, the others have charge of −1/3. All of the quarks have associated antiquarks (which have been observed), each with the opposite electrical charge and identical mass of their related quark. All of the quarks are fermions with spin 1/2. All quarks carry color charge (and antiquarks carry anti-color). Quarks can combine in quark-antiquark pairs to form mesons and quark triplets to form baryons. All the mesons and baryons have no net color, which sets restrictions on the possible quark combinations allowed. And, perhaps most importantly, literally hundreds, if not thousands of particles can be explained as various combinations of the six flavors of quarks.

Return of the Leptons

While quarks are fascinating objects, there exists a type of particle that can't be understood as combinations of quarks. These are the leptons. Leptons are both more and less complicated than the quarks. Their more complicated nature mostly concerns the forces that dominate their behavior and thus we defer a discussion of this for Chapter 4. For this chapter, we concentrate on the physical properties of the leptons. Physically, leptons are much simpler than the baryons and mesons in that they do not appear to have any internal structure. Currently we know of six leptons, three carrying electrical charge and three electrically neutral. Like the quarks, we can group the leptons into pairs, each consisting of one charged and one neutral lepton. Each of the charged leptons carries the same amount of electrical charge, specifically negative charge equal in magnitude to that of the proton. We write this charge as -1. The charged leptons are: the chemically-important electron (e), the muon (μ^-) and the tau (τ^-).

While the charged leptons do carry the same amount of electrical charge, they do not carry the same mass. The electron is the lightest charged lepton, with a mass of $0.511\,\text{MeV}$, just about 2000 times lighter than a proton. The muon has a mass of $106\,\text{MeV}$, or about 200 times that of an electron, while the tau's mass is even higher at $1784\,\text{MeV}$.

An elementary particle long before anyone knew that elementary particles existed, the electron was the first real subatomic particle discovered. Discovered in 1897 by J.J. Thomson at the famous Cavendish Laboratory at Cambridge University, the electron also made up the first controlled particle beam. While we now know much about the electron, perhaps the most critical observation of the electron's nature was when Thomson asserted that the electron was a particle with mass much smaller than that of the hydrogen atom. Since prior to this discovery, atoms were the smallest particle of nature thought to exist and further hydrogen was the smallest atom, all of the understanding of the atom, so painfully gained in the 1700s and 1800s, was called into question. Thompson's discovery that the electron was a component of the atom started physicists down the path which led first to the tricky world of quantum mechanics and finally to the field of particle physics that we study today.

Cosmic ray experiments were initially intended to try to understand their most basic properties. As described in Chapter 2, it was discovered that air was more ionized as the altitude increased, leading to the hypothesis that perhaps the cause of this phenomenon came from outer space. With the invention of the cloud chamber (a cloud-filled device which would display a track if crossed by a charged particle) one could see distinct tracks of particles rather than a diffuse radioactive glow. These tracks were then photographed for further analysis. One study that was natural to perform was to surround the detector with a magnetic field to ascertain the particles' energy and also to insert metal plates in the cloud chamber to ascertain the degree to which the particles interacted with matter. (As a rule, if they could cross several plates, they were fairly energetic.) Both electrons and positrons were observed (and identified by their rather poor penetrating power, even for fairly energetic examples). But there existed a type of particle that had significant penetrating power, even for relatively low energies, and further this type of particle did not seem to interact very much with the material that made up the cloud chamber. These measurements were accomplished in 1937 by two groups: Anderson and Neddermeyer & Streets and Stevenson. It was originally thought that this particle, which had a mass of about 100 MeV, was the one predicted some years earlier by Hideki Yukawa as a particle essential for explaining nuclear physics. Consequently, this particle was called at various times a *yukon* after Yukawa or a *mesotron*, for its medium mass (meso = medium). However the fact that this penetrating particle interacted so weakly proved that it was not the particle that Yukawa had predicted. Its ability to penetrate matter was so unusual that when I.I. Rabi heard of the particle's existence, he is reported to have said "Who ordered that?" After further study, this particle was identified as the μ (mu) lepton, or simply muon.

The discovery of the tau lepton in 1975 was a truly superb bit of scientific deductive work. A group of scientists at the Stanford Linear Accelerator Center (SLAC), headed by Martin Perl, was colliding electrons and positrons that annihilated and, by all expectations, this energy should have reappeared into a particle and its antiparticle. However, they saw 24 events in which the collision resulted in two particles, one an electron, the other a muon. Somehow they deduced that they were making a new pair of leptons (now known as the tau and antitau), which were each decaying into a different lepton. Diagrammatically, what they were saying was:

$$e^+ + e^- \rightarrow \tau^+ + \tau^-$$

$$\mu^- + \bar{\nu}_\mu + \nu_\tau$$

$$e^+ + \nu_e + \bar{\nu}_\tau$$

This is a nice way to say: "A τ^+ and a τ^- were created. The τ^+ decays into an e^+, an electron neutrino and a tau antineutrino, similarly the tau decays into a muon and an antimuon neutrino and a tau neutrino." As we discuss below, neutrinos essentially do not interact (and therefore cannot be detected). So what they were saying was that they were making two never before observed particles that decayed into a total of six particles, of which four were invisible. Further, since the charm quark and tau lepton masses are very similar *and* given that the charm quark had only been recently discovered in a similar energy region, the confusion had to be immense. Yet they claimed that they had found a new charged lepton and also inferred a new neutrino. And they were right. And I'm impressed. The Nobel Prize for this discovery was well deserved.

Much of the story of the discovery of the neutrinos was given in Chapter 2, but I briefly recap it here. The electron neutrino (although at the time, they didn't know that there was more than one kind) was inferred in 1930 by Wolfgang Pauli and observed in 1959 by Frederick Reines and Clyde Cowan. The muon neutrino (and just as important was the fact that there were at least two different kinds of neutrinos) was discovered in 1961 and Leon Lederman, Jack Steinberger and Mel Schwartz shared the Nobel Prize for that discovery in 1988 (the impromptu party that we had at Fermilab for Leon when the prize was announced was a lot of fun). As we have noted above, the existence of the tau neutrino was inferred in 1975, but not experimentally observed until 2000 at Fermilab by an experiment lead by Byron Lundberg and Vittorio Paolone.

The electrically neutral leptons are intriguing. Collectively, they are called neutrinos, although since each neutrino is paired with a charged lepton, they are called: the electron neutrino (ν_e), the muon neutrino (ν_μ) and the tau neutrino (ν_τ). Neutrinos are fascinating in that they interact very weakly with other types of matter. Neutrinos are prodigiously created in nuclear reactions. Neutrinos from the biggest source around (the Sun) interact so weakly with matter that it would take about four light-years (about 20 trillion miles) of solid lead to reduce the number of neutrinos by a factor of two. The probability that a neutrino will interact with matter goes up with the energy carried by the neutrino (and thus the amount of material that can be penetrated goes down), but even the vastly higher energy beams of neutrinos available at modern accelerator facilities can penetrate approximately 200 million miles of lead before losing half of their number.

It's a good thing that neutrinos interact so weakly. Six hundred fifty million million (6.5×10^{14}) neutrinos from the Sun pass through every person on Earth every single second. To give a sense of scale, 6.5×10^{14} BB's weighs about 20 billion (2×10^7) tons, yet with all of these neutrinos hitting you every second, on average only about thirty interact in your body each year and no more than one with "real" energy. This number sounds respectable until you fold in the amount of energy deposited by each neutrino. You then find that taking all of these neutrinos, it would take 60 billion (6×10^{10}) years to deposit as much energy in you as generated by a typical sneeze. Neutrinos *really* don't interact with matter very much. And just to make sure that you know that you can't get away from neutrinos, each adult person contains something like 20 milligrams of Potassium 40, a radioactive isotope which decays with a neutrino as a final product. Because of this Potassium, each person emits approximately 340 million (3.4×10^8) neutrinos per day.

Neutrinos are extremely light particles and it is not too much of an approximation to say that they are nearly massless. In Chapter 7 we will discuss the possibility that (and consequences of what would happen if) neutrinos have a small mass. But we have been able to set limits on the neutrinos' masses. When a physicist says, "set limits," he really means "I don't know what the real answer is, but it's smaller (or bigger) than X." In this case, we know the mass of the electron neutrino is smaller than 15 eV, the muon neutrino's mass is smaller than 0.17 MeV and the tau neutrino's mass is lower than 24 MeV.

The notation for denoting particles can be maddening for the non-expert. Appendix C gives a more detailed description of the naming rules, but we can give a brief description here. Like quarks, for every lepton, there exists a corresponding anti-lepton. The antilepton for the electron (e^-) is the positron (e^+) (e plus). The antileptons for the muon (μ^-) and tau (τ^-) are the antimuon (μ^+) (mu plus) and antitau (τ^+) (tau plus) respectively. Note that the little "+" and "−" in the superscripts indicate the electric charge of the lepton (or antilepton). We see that leptons have -1 charge, while the antileptons have $+1$ electrical charge. It's also true that, unlike quarks, we usually do not write an antilepton with a bar over it. The information that distinguishes the leptons and antileptons is the charge in the superscript. In principle, $\bar{\mu}$ can mean antimuon, but this convention is rarely used.

Neutrinos are electrically neutral and their corresponding antineutrinos are also neutral. The neutrinos use the "overline" convention to indicate antiparticles. The electron antineutrino is written ($\bar{\nu}_e$), while the muon antineutrino and the tau antineutrino are written ($\bar{\nu}_\mu$) and ($\bar{\nu}_\tau$), respectively.

Like the quarks, both the charged and neutral leptons are fermions. Recall that fermions have half-integer spin. The electron, muon and tau are conventional in that they can have either $+1/2$ or $-1/2$ spin. But the neutrinos are different. Neutrinos can only have one kind of spin. Specifically, all neutrinos have spin $-1/2$. All antineutrinos have spin $+1/2$. This seemingly innocuous property is truly earth shaking. It means that one can, in principle, distinguish between anti-matter and matter, rather than simply choosing which is by convention. We will discuss this property of neutrinos more when we talk about the forces that govern how neutrinos interact.

Leptons also differ from quarks in that they are colorless. This is evident in part due to the fact that leptons have no internal structure. Since color is not observed as a property of leptons and further we do not believe that leptons have structure, we can also conclude that even inside them, leptons do not contain color, in stark contrast to the mesons and baryons. Color charge plays a role in the strong force and consequently, the leptons do not interact strongly. All of this will be addressed in more detail in the next chapter.

Table 3.3 lists all of the particles that we currently think of as fundamental (i.e. have no smaller particles contained within them). The table is organized to show the repeating structure of the particle generations. Generations II and III appear to be carbon copies of the first generation, except for the ever-increasing mass. The reasons for this repeating structure are currently unknown and consequently a focus of active research.

While we now know of these 12 particles, it is noteworthy to remark that the quarks and charged leptons of generations II and III are unstable and decay in fractions of a second into generation I particles (the neutrinos are a special case and will be discussed in Chapter 7). Because these particles disappear essentially instantly, this means that all of creation can be constructed from the four particles of the first generation. It still amazes me that everything that we see: you, me, the Earth, Moon and stars, *everything* (well not certain Hollywood celebrities, who often seem to be from an entirely different dimension) are simply endless combinations of four miniscule and point-like particles. Since the generation II and III particles can exist

Table 3.3 Organization of quarks and leptons. The numbers in parentheses are the mass of that quark or lepton in GeV. (Note: "<" means "less than.")

Charge	Generation I	Generation II	Generation III	Particle
+2/3	up (u) (small)	charm (c) (1.5)	top (t) (175)	quarks (q)
−1/3	down (d) (small)	strange (s) (0.7)	bottom (b) (4.5)	
−1	electron (e) (0.0005)	muon (μ) (0.1)	tau (τ) (1.7)	leptons (ℓ)
0	electron neutrino (ν_e) (<0.000000015)	muon neutrino (ν_μ) (<0.00017)	tau neutrino (ν_τ) (<0.024)	

for only a very short time and are created in only the most energetic collisions, they can only exist under very special conditions. With a little thought, one can recognize that the last time that these conditions existed generally in the universe was a tiny fraction of a second after the Big Bang, the primordial fireball that started it all. In this sense, particle experiments are a way to look back to the very dawn of creation.

[End note: Beginning towards the end of 2002 and becoming more solid during the summer of 2003, several groups have observed a new particle, called the Θ^+. Current thinking is that this particle consists of four quarks and an antiquark, specifically two ups, two downs and a strange antiquark (uudds̄). The data is fairly compelling and will probably stand the test of time. Such a particle was not unexpected, having been predicted essentially with the birth of the quark model. When you think about it, it isn't so odd, since it has the quark content of a proton (uud) and a K meson (ds̄). Physicists, both experimental and theoretical, are trying to reconcile the theory and data. Stay tuned, as with the next edition of this book, I may need to rewrite this chapter. Science is always exciting!!!]

chapter 4

Forces: What Holds it All Together

Research is to see what everybody has seen and to think what nobody else has thought.

— Albert Szent-Györgi

If the universe were only occupied by the particles described in the preceding chapter, the universe would be a very lonely place indeed. Particles would zip hither and yon, never giving one another so much as a "How do you do?" Electrons would not be bound to atomic nuclei and, with no atoms; there would be no molecules, no cells, no us. And since readers wouldn't exist, I wouldn't bother writing this book. Luckily, in addition to the interesting particles about which we are now familiar, there also exist forces that bind the particles together into useful configurations. As alluded to in earlier chapters, we know of four distinct forces with very different properties. The first thing that we will discuss is the character of the various forces, but then we will discuss a new and interesting idea. The existence of forces implies that new particles exist. These particles carry the various forces. This is a non-intuitive concept and we will discuss it in detail when appropriate.

At our present level of knowledge, there appear to exist four forces. These forces are gravity, the electromagnetic force, the strong (or nuclear) force and the radiation-causing weak force. Gravity is perhaps the most familiar. It keeps us on Earth and guides the stars and planets through the cosmos. Gravity is always an attractive force, which means gravity will always make two particles want to move closer to one another. When one thinks about forces, an important question is always "What governs the strength of the force?" For gravity, just three things are relevant: (a) the mass of each of the two objects, (b) the distance separating the centers of the two objects and (c) a constant factor which is related to how strong the gravity force is, once the other two factors are taken into account.

Gravity

Mass is a somewhat tricky concept, with which most of us have a mildly incorrect familiarity. Everyone is familiar with the concept of weight (in my case an often depressing familiarity). While weight is not mass (weight is really the force due to gravity), weight is related to mass. A person who weighs more also has a greater mass. However, while weight goes away in outer space, mass does not. Further, while your weight would change if you were to stand on a different planet, again your mass would remain unchanged. So this is a very important idea: weight can change, but mass doesn't. If it makes you more comfortable, you can use the two interchangeably as long as you stay on Earth (just don't tell my physicist colleagues that I said it was OK). Really weight is a force; a greater weight means that you experience a greater force. The reason that one's weight can change, while one's mass is unchanged is because of how gravity works. The force due to gravity is proportional to the mass of one object, multiplied by the mass of the second object. Since Jupiter is the biggest planet, it has a much larger mass than Earth. So, if you were standing on Jupiter, you would feel a greater force than on Earth because, while your mass is unchanged, Jupiter's mass is much greater. Since your weight is related to your mass multiplied by the planet's mass, voilà, you're heavier on Jupiter. And, when you're deep in outer space, there is no nearby planet, so the planet's mass is zero. Now you multiply your (unchanged) mass by the zero mass of the planet and the result is zero (recall that anything multiplied by zero is zero). So, no force in outer space.

Actually, what I just told you is a tiny lie. This is because two objects that have mass always feel an attractive force, no matter how far they're separated. The force due to gravity extends to the edge of the universe. So even if you're extremely far from Earth, Earth will always exert a force on you. So why don't we feel a force due to gravity from Jupiter if it is so much more massive than Earth? This is because the mass of the two objects is not the only thing that affects the force of gravity that an object feels. The distance that separates the two objects also affects gravitational force. Physicists say that the force goes down as the square of the distance (physics-ese for the distance multiplied by itself). So, if you have two objects which feel a particular attractive force, when you double the distance between them ($\times 2$), the force goes down by a factor of four ($2 \times 2 = 4$). Similarly if you increase the distance by a factor of ten ($\times 10$), the force goes down by a factor of a hundred ($10 \times 10 = 100$). Thus one sees that the force due to gravity drops off rather quickly; but while the force gets weaker, it never becomes exactly zero. However, as the distance increases, the force drops until it can be neglected (i.e. gets "close enough" to zero).

The final component relevant to determining the force due to gravity is a single, universal constant. While the amount of the mass involved is important, as well as the distance between the two objects, one also needs to include the strength of gravity itself. It turns out that gravity is really a very tiny force. The only reason that it appears to be so strong is that the force is always in one direction and further each proton or neutron (recall we call them collectively nucleons) in your body feels gravity from each nucleon that makes up the Earth; and that's a whole bunch of nucleons (you $\sim 10^{28}$ nucleons and

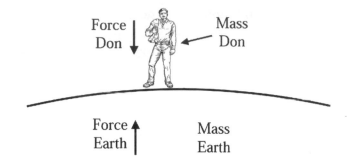

Gravity Force

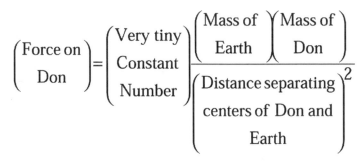

Figure 4.1 The effect of gravity depends on three things. The mass of the two interacting bodies, the distance that separates them and the fundamental strength of gravity. (Drawing courtesy of Dan Claes.)

Earth $\sim 10^{51}$ nucleons). When you think about it, with so many atoms involved, it's a good thing that gravity is so weak, otherwise we'd be squashed like an unlucky bug. The essential points in the preceding few paragraphs are illustrated in Figure 4.1.

The knowledgeable reader will recognize that we've been discussing Newton's theory of gravity. In 1916, Albert Einstein realized that there were situations where the theory breaks down.

If one has a huge amount of mass concentrated in a small space, then gravity is better thought of as a warping of space (which is a very cool idea!). So one might be inclined to say that Newton was wrong, but in fact he really wasn't. It's more correct to say that his theory was incomplete, that is, it applied only in limited circumstances. One can think of ample examples where a theory is correct and yet incomplete. If you punch a brick, what happens is the brick is unaffected and your hand hurts a lot. However, if a karate expert hits the brick, the brick breaks and his (or her!) hand doesn't hurt (well much anyway). So a hypothetical Newton's "law of bricks" might be something like "Hitting a brick doesn't affect the brick and hurts your hand, with the degree of pain proportional to how fast the hand was moving." This is a good theory, which works very well over a vast range of hand speeds. Einstein's theory would be more like "When you hit a brick, the brick flexes an imperceptible amount (i.e. little enough that zero flexing is a good approximation) and your hand hurts in an amount proportional to hand speed (although the flexing does reduce the pain by an equally imperceptible amount). As the speed of the hand increases, the amount that the brick flexes increases, although the flexing remains very small. At a particular hand speed, the flexing of the brick becomes large enough that the brick breaks and the hand no longer hurts." We see that as long as one's hand is moving slowly enough and one doesn't measure the flexing of the brick too precisely, that Newton's and Einstein's law of bricks are nearly identical. However, at high enough hand speed and for good enough brick flexing measurements, Newton's law is no longer sufficiently accurate. Newton's law of bricks should rightfully be called "Newton's law of brick hitting at low hand speeds." Similarly, with gravitation, it should be "Newton's Law of Universal Gravitation, as long as speeds aren't huge, masses aren't enormous and distances aren't galactic." Einstein's law of gravitation should be "Einstein's Law of General Relativity (i.e. gravity) unless the sizes involved are tiny." I'm sad to report that Einstein's theory, cool as it may be, also fails under particular circumstances. It's also true that nobody knows how to write a new theory that supersedes Einstein's theory as Einstein's theory superceded Newton's. We'll come back to this when we discuss the modern mechanism for forces and again in Chapter 8.

Electromagnetism

The force of electromagnetism, the reader will no doubt recall, is one that explains both the phenomena of electricity and magnetism. So let's start out with the electric force. The electric force is in many respects similar to the gravitational force. Instead of mass, the equivalent quantity for electrical force is electric charge. However, unlike the gravitational force, the electrical force can be either attractive or repulsive. This stems from the fact that there are two "flavors" of electrical charge, which have been named, positive $(+)$ and negative $(-)$. While the reasons for this naming convention are historical (and arbitrary, as any two names would do), it turns out that these names are handy when one is doing the math that one needs to do to calculate things. This is because if you put an equal amount of positive and negative charge in the same place, they cancel each other out, just like positive and negative numbers in math class, and the result is zero net charge.

So what governs how strong the electric force is between two electric charges and what direction the force points (i.e. attractive or repulsive)? Well the strength is governed by three things (which should sound familiar): (a) the amount of electric charge carried by each of the two objects, (b) the distance between their centers and (c) a constant which turns out to be vastly larger than the similar gravitational constant (about 10^{20} or 100 quintillion times greater, in fact). The direction depends on the flavor of not one, but both charges. If both charges are of the positive type, or if both are of the negative type, then the two charges will be repelled. If the two charges are of opposite flavor (that is, one is positive while the other is negative...it doesn't matter which), the two charges will be attracted. This is where the phrase "opposites attract" comes from (and not that old girlfriend or boyfriend about whom all of your friends asked "What *were* you thinking?" after the fact).

As illustrated in Figure 4.2, just like gravity, the electric force felt by *each* particle is dependent on the properties (in this case the electrical charge) of both. Increase either particle's electrical charge and

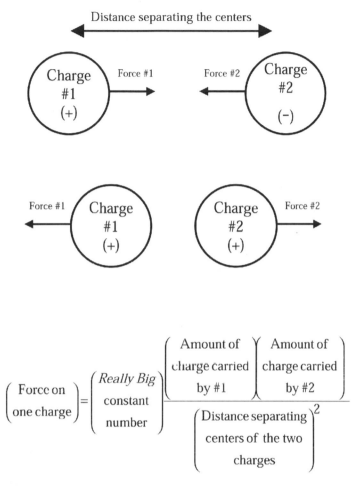

Figure 4.2 Like gravity, the electromagnetic force depends on three things. The electric charge of the two interacting bodies, the distance that separates them and the fundamental strength of electromagnetism.

the force on both increases. It's also true that how the electrical force varies with particle separation is identical to that of the gravitational force (e.g. double the distance, reduce the force by a factor of four; increase the distance by ten, reduce the force by a factor of 100). So except for the fact that the electrical force can repel as well as exhibiting gravity's attractive behavior, the two forces appear very similar.

The two forces also differ enormously in their strength. The electrical force is vastly stronger than the gravitational force. I cannot tell you in general how much they differ (remember that the forces also depend on mass and charge) but if one uses "obvious" units (one kilogram and one coulomb for the technically minded), the electrical force overpowers the gravitational force by that mind-boggling factor of 10^{20} (100 billion billion). Wow!

If you're still awake at this point, I hope your first reaction will be to join me in that "Wow!" You're second reaction should be "Wait a cotton-picking minute. That can't be right. If the electrical force is that much bigger, why doesn't *it* dominate the universe rather than gravity?" To this, I reply "Good question. I'm glad you're awake!"

The answer stems from the fact that most objects have a very small total electrical charge. Since each atom has the same amount of positive charge (in the nucleus) as negative charge (in the electrons surrounding the nucleus), the net charge is zero (remember that positive and negative charge cancel). So it doesn't matter how strong the electrical force *could* be, if one (or especially if both) of the charges were essentially zero, they would feel no electrical force.

So why talk about the electrical force? Because there are situations where it matters and where it matters a whole bunch. Recall that the electrical force gets much larger as the particles get closer together. Since the size of an atom is about 10^{-10} (one ten billionth) of a meter, it stands to reason that in this situation, the electromagnetic force must be very strong. This is because of the fact that the electrical charge of the atomic nucleus and the charge of the electrons "see" one another. Because the electrical force dominates the gravitational, it is electricity that holds the atoms together. If you put in the correct charges and masses of the electrons and atomic nucleus of a hydrogen atom, you see that in this case the electrical force is 10^{39} times larger than the gravitational force.

Since magnetism is just caused by electrical charges in motion, we do not go over this force in detail. Things are a little different, because velocity now matters too. However, as Maxwell showed, electricity and magnetism are two faces of the same phenomenon; as the magnetic force increases, the electric force compensates. How it does this is really interesting, but a little technical. So I do not discuss the details, but simply state that much of what was said about the electrical force also applies to the magnetic force.

We now know enough to be perplexed. If positive charges attract negative charges, why don't the negative electrons get sucked into the positively charged atomic nucleus, instead of swirling around the atom in a little "planetary system"? Similarly, why don't the planets crash into the Sun? Remember in Chapter 1, when it was revealed that Newton said that things moving would go in a straight line unless a force acts on them? Well, if the electric force was somehow magically "turned off," the electrons would instantly (ignoring for the moment a few of Einstein's ideas) start traveling in a straight line, with the directions determined by where they were going when the electric force was turned off. However, the electric force *does* exist and the electrons are always attracted to the atomic nucleus. As we see in Figure 4.3, the electrons do get pulled towards the nucleus, but since they're moving, they miss. A little later, they're still moving (but in a different direction), but still getting pulled towards the center. The net effect is that the electrons keep moving in a circle, always being pulled towards the nucleus, but always missing it.

So why is this interesting? The reason is it shouldn't be possible. In the 1860s, Maxwell (remember him?) showed that electricity and magnetism were the same. According to his theory, the electrons should lose energy (physics-ese for slow down) as they felt the electrical force and they should have spiraled down into the nucleus of the atom in a brief fraction of a second. So either Maxwell was wrong (heresy!) or something else was going on. Maxwell's equations have been heavily tested under lots of circumstances. They predicted radio and most of the electrical phenomena that makes our modern technology possible. So his theories obviously applied. Except. Just as Newton's laws made wrong predictions under some circumstances, Maxwell's theory only worked when the sizes involved were large.

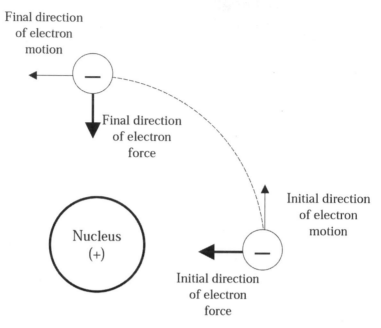

Figure 4.3 Even if two particles experience an attractive force, they will not necessarily come together and collide. If the particles have a velocity, they will orbit each other, much in the same way as the planets orbit the Sun.

(Note: large in this context means large when compared to the size of an atom. Maxwell's laws work rather well even for charges separated by distances that are so small that the eye can't see them.)

The fact that Maxwell's equations didn't work for atoms caused no end of consternation. How this quandary was resolved is a very interesting story, about which many books have been written. The birth and growth of quantum mechanics is a fascinating tale, involving some of the most brilliant and storied physicists of the 20th century. Bohr, Heisenberg, Schroedinger and Pauli, legends among physicists, are but a few of the people involved. This book is *not* about the story of quantum mechanics, but some of quantum mechanics' ideas are needed to further our tale. Bohr postulated that electrons could only orbit the nucleus only at fixed distances, although why this should be so he was quite uncertain. His postulate worked though and broadly explained why atoms gave off the light that they do (particular atoms, hydrogen for example, are observed to only give off certain discrete colors of light and no others). Clearly his idea had merit. It was the work of Schroedinger and Heisenberg that generated all of the fuzzy and counter-intuitive aspects of quantum mechanics.

There's a great story about how Schroedinger made his great contribution. He is said to have gone on a holiday (European for vacation) in the Austrian Alps, having brought some paper, a pen, two pearls and a mistress. He placed a pearl in each ear to screen out distractions, put the mistress in bed for inspiration and tried to work out the mystery of the atom. Somehow he had to keep his woman happy while simultaneously creating a new physics theory that explained many difficult mysteries. When I tell this story, I often add that, as a physicist, of course he was up to the challenge and succeeded at everything that he set out to do.

As I'm writing this, I'm in an airplane, returning from a physics conference that was held in the Italian Alps. I am returning without any new and brilliant theories. Of course, I was in Italy, not Austria, I brought no pearls and I was unaccompanied. As a scientist who really would like to make a great discovery, I really feel that I need to do the experiment to determine which factor was critical to Schroedinger's success. Luckily, my wife is a caring and understanding woman, so I'm sure she would agree to both Austria and the pearls.

The upshot of the theory of quantum mechanics is that electrons sort of orbit atomic nuclei. While it is in principle impossible to know where any particular electron is at any particular time, you can know where it is on average. Quantum mechanics also explained the particular colors of light emitted by each kind of atom. Perhaps most interestingly, physicists were finally able to explain Mendeleev's Periodic Table of the elements (introduced in Chapter 1). This fact gave substantial credence to the theory. Prior to the full understanding of quantum mechanics, scientists knew of about 100 atomic elements and knew of electrons and atomic nuclei. Now they knew of the rules governing electrons and the known elements were just a consequence of electrons, nuclei and quantum mechanics. The world was thereby greatly simplified.

While Schroedinger had extended physics understanding to the ultra-small, the theory had an obvious flaw. It had not included Einstein's special theory of relativity and thus was not guaranteed to work at speeds nearing the speed of light. Clearly an extension of quantum mechanics was needed. This melding of quantum mechanics and special relativity was accomplished in 1927 by Paul Dirac.

Dirac's notions were extremely impressive. However, as frequently is the case, subsequent experimentation showed the chinks in the armor. There is a property of the electron, called the magnetic moment, which Dirac calculated to be exactly 2. Knowing what the phrase "magnetic moment" means is not important for our discussion, except to remember that Dirac's theory precisely predicted a value of 2. However, in about 1948 experiments revealed that the correct number was close to 2.00236, with an uncertainty of about 6 in the last decimal place. Since the uncertainty is much smaller than the total deviation from 2, the experiment showed that Dirac's equation was wrong, as the measurement was clearly *not* 2. It was known that the electron interacted with photons and it was thought that a somewhat more sophisticated calculation would reveal the correct answer. But when the calculation was done, the result was not 2, not 2.00236, but rather infinity. And, as my son likes to say, that is *not* good. The reason was that at the very small scale of the size of the electron, the electron reveals itself to be a very busy object with photons swirling around it. As one gets closer to the electron, there are more photons swirling even faster. When the effects of the swirling were added up, the answer was infinity. Hmmmmm.

Luckily, in about 1948, three very bright guys, Julian Schwinger, Richard P. Feynman and Sin-Itiro Tomonaga, independently solved the problem with a particularly clever way of doing the math. This second quantum revolution has allowed unprecedented calculations of the properties of the electron, accurate to one part in a billion. The resultant theory is called Quantum ElectroDynamics or QED. The name comes from the fact that it concerns the quantum realm (Q) and electrical interactions at a small size (E). Further the particles described are not static, but rather dynamic (D), usually moving at nearly the speed of light.

With the advent of "second quantization," we have begun to see the beginning of how modern particle physics views forces. Forces are viewed as the exchange of a force-carrying particle. In the case of electromagnetism, the particle that is exchanged is the photon, which is the same particle (in many respects) as the photons that allow you to see. After we introduce the other forces, we will return to this idea of particle exchange.

The Strong Force

When you think about it, the nucleus of an atom should not exist. The nucleus consists of a bunch of protons and neutrons, all within a sphere of radius about 10^{-14} or 10^{-15} meters (that's about one quadrillionth of a meter and between ten thousand and a hundred thousand times smaller than an atom). While the neutron is electrically neutral, each proton is positively charged with a charge equal in number (but not in sign) to that of the electron. Since (a) like charges repel and (b) nearby charges feel a greater electrical force than distant ones, the various protons should feel a repulsive force. Doing a quickie calculation, I find that two adjacent protons feel a repulsive force of about fifty pounds. When one thinks of a large nucleus like uranium, a proton on the periphery of the nucleus feels a force of about 133 pounds. That much force is appreciable, even on the size-scale of a person, let alone for an object as unfathomably small as a proton. So an obvious question is "What keeps the nucleus of an atom intact?" There can be only one answer. If there is a force of approximately 50–100 pounds on each proton repelling them outwards, and they don't move, then there must be an even stronger force holding the protons together. This force was called the strong or nuclear force for lack of a better name, although little more was known about it. But definitely there must exist another force. This force has some peculiar properties. We know of only about a hundred different kinds of atoms (or elements). Since quantum mechanics puts no upper limit on the number of electrons one can put around a nucleus, something else must be limiting the number of elements. It turns out that if you put enough protons and neutrons together, eventually the nucleus becomes unstable and falls apart. So there is a maximum size of the nucleus. This occurs when there exists so many protons that the electrical charge actually overwhelms the strong force. Since the strong force really *is* stronger than the electromagnetic force, the only thing that can explain the facts is that the strong force must have a shorter range over which it operates than the electromagnetic force. If the strong force only extended to adjacent nucleons, but was zero at a distance of two nucleons away, while the electromagnetic force extended forever, one could explain the data. We can see how this could explain how atomic nuclei can be held together so stably for small nuclei and be unstable when the nuclei gets bigger. This point is made more visually in Figure 4.4. Suppose

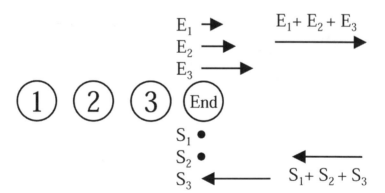

Figure 4.4 Cartoon showing how a strong force with limited range will eventually be overpowered by a weaker force that has a larger range. The cumulative effect of the many contributions to the smaller force eventually exceeds the stronger force that can only feel effects from immediate neighbors. Note that "E" denotes the electric force, while "S" denotes the strong force.

that you have four protons in a row. The end proton sees an electro-magnetic force from the other three protons (we denote the electric force from protons 1, 2 and 3 as E_1, E_2 and E_3, each of which drops off as the distance between that proton and the end proton increases). However, the end proton only sees the strong force from proton 3, while the distance between the end proton and protons 1 and 2 is so large that the end proton feels no strong force from them. Since forces add, you see at least in principle how eventually the strong force can be overcome. Since this does occur, we have demonstrated that the strong nuclear force is very strong, but only over a limited range, after which the force goes to zero. Essentially, the strong force can be thought of as "Velcro" between two nuclei. If the nuclei are touching (or nearly so), the force is strong. If they're not touching, they don't interact at all.

The properties that go into determining the strength of the strong nuclear force are somewhat harder to describe. Obviously, the distance matters although we have difficulty stating exactly how. Since two pro-tons cannot get any closer than surface contact, we don't know how the strong force acts for distances small compared to the proton (although in our particle physics experiments we can probe smaller dis-tances; more on that later). Similarly, we know that the force gets small quickly if we go to distances more than about three or four times the size of a proton. We also know that the direction of the strong force is attractive. We recall that the electric force depends on a constant, the distance separating the two objects and the electric charge. For the strong nuclear force, we have covered the constant and the distance behavior (big constant for smallish distances, zero for bigger ones), but not the charge. It turns out that the strong charge is a bit tricky. It is unrelated to the electric charge, as the electrically-neutral neutrons experience the same strong force as the proton. Further, no other nat-ural (i.e. stable) particle feels the strong force. One thing that protons and neutrons have in common (but not the electron or neutrino) is that they both contain quarks. So perhaps somehow the quarks are the source of the strong charge (which turns out to be true). We will talk a little more about this quark-strong force connection later when we discuss forces in terms of the particles that carry them.

The Weak Force

The weak force was discussed in Chapter 2, but it remains mysterious and will be clarified best when we discuss force-carrying particles (don't worry, that time is coming soon), but we can review some of the ideas introduced in Chapter 2. The weak force gets its name from the fact (you guessed it) that it is a very weak force. We know it's weak because while reactions caused by the strong force occur on a time scale of 10^{-23} seconds and the electromagnetic force reacts in about $10^{-20} - 10^{-16}$ seconds, there also exist reactions that take 10^{-8} seconds to 10^9 years. Recall that charged pions decay in 2.6×10^{-8} seconds, while muons decay in 2.2×10^{-6} seconds. Carbon fourteen (C^{14}, an isotope of carbon that is very useful for dating organic things that are a few tens of thousands of years old) decays with a characteristic time of 5730 years, while the decay time of uranium is 10^9 years. Since it takes a force to make a particle decay, if something takes that long to react, it must be weak. In fact, one might ask if maybe there are many more forces involved, after all 10^{-8} seconds is very much different than 10^9 years (10^{16} seconds), but in fact it's not true. One force spans this entire time scale.

The weak force has several notable features. The first one is that it is the only force that distinguishes between our world and a hypo-thetical world that you see when you look in a mirror. When you look into a mirror and raise your right hand, your mirror image raises its left hand. When you throw a ball towards the mirror, the ball goes away from you, but the image ball (i.e. the ball you see reflected in the mirror) comes towards you. If you define the direction in front of you as forward and behind you as backwards, the thrown ball is going forwards, while the image ball is going backwards.

We discuss in Chapter 7 the famous experiment of Wu, in which she showed that particles produced by the weak force knew about whether they were in our world or a mirror world. We became aware of this property of the weak force because we found that neutrinos could only spin clockwise compared to their direction of motion and anti-neutrinos spin counter-clockwise. This is in contrast to the more familiar top (the children's toy, not the quark), which can rotate in either direction. This special "knowledge" that the neutrino has of its direction of motion is unique in physics, as most interactions do not know which way a particle is moving (or spinning).

The weak force was poorly understood until Enrico Fermi proposed on January 16, 1934 a theory of weak interactions that was in many ways analogous to the earlier theory of electromagnetism. Weak charges replaced electric charges and a new (and much smaller) constant characterized the strength. The weak force differed from the electromagnetic force in that it had a much shorter range, much shorter even than the strong force. The characteristic range of the weak force is about 1000 times shorter than the strong force. Like the strong force, the strength of the weak force is inextricably intertwined with the distance between the two particles. A consequence of this fact is that the strength of the weak force increases as particles get closer together. Since the collision energy determines the minimum distance between the two particles, one can see that the weak force changes with collision energy. This fact is very interesting and will be discussed more in Chapter 8.

Another enormously interesting facet of the weak force is the fact that it alone can change the flavor of quarks and leptons. As we discussed in Chapter 2, matter and antimatter can completely annihilate each other into pure energy. But this is only allowed if the matter and antimatter are the same; i.e. an up quark can annihilate an anti-up quark, but an up quark won't annihilate an anti-down. However, that's not true for the weak force. For instance, a positive pion decays into a positive muon and a muon neutrino ($\pi^+ \rightarrow \mu^+ + \nu_\mu$). Recall that a π^+ contains an up quark and an anti-down quark. It can decay only if somehow the two different flavor quarks can combine, destroy each other, and be replaced by a lepton pair. Such behavior is not allowed by the strong or electromagnetic force, but the weak force allows it. But because the weak force is ... well ... weak, it takes a long time, so the π^+ can live longer than many particles.

The fact that the weak force takes so long actually shapes our view of its nature. We see it as the force that combines different flavor quarks, but this isn't its full nature. For instance, the neutral pion, π^0, which contains an up and anti-up quark pair, decays via the electromagnetic force into two photons (we say $\pi^0 \rightarrow 2\gamma$). It is also possible for a π^0 to decay via the weak force, but because the electromagnetic force works so much faster, the allowed weak decay never happens. It's kind of like dusting at my house. In principle, my teenage children will voluntarily dust. However, long before the dustiness of the house reaches their threshold to spontaneously dust, it has crossed mine, so I dust. And, I'm somewhat chagrined to say; my wife's threshold is lower than mine, so I rarely spontaneously dust either.

On the other hand, if something somehow prevented my wife from dusting (say a coma or a long visit to her parents), then the next higher dust threshold (mine) would dominate. And, if I too were forbidden to dust, then eventually the kids would (although experimental evidence suggests that this would occur only simultaneously with a distinct nip in Hell's morning air).

Getting back to forces, the reason that a π^0 decays via the electromagnetic force is because it is the lightest hadron and so it can't decay via the strong force (which must have hadrons as decay products) and you can't decay the lightest hadron into any other hadrons without using more energy than is present. So strong decays of the π^0 are forbidden.

The π^0 can decay via the electromagnetic force, as the up and anti-up quarks can annihilate to photons (and since photons are related to the electromagnetic force, this is proof that the electromagnetic force is involved). However, a π^+ or π^- cannot decay via the strong force for the same reasons that forbid the strong force decay of the π^0. Also now the electromagnetic force is forbidden as the electromagnetic force can only annihilate quark and antiquark pairs of the same flavor. So only the weak force, with its unique flavor-combining property can do the job.

Now that we know of the four forces, it becomes of interest to know which particles feel which forces. The quarks are the particles that have the richest force behavior and can be affected by all four forces. The charged leptons are not affected by the strong force, but feel the rest. The neutrinos are the least gregarious of the particles, affected only by the weak force and gravity. Because gravity is somehow a bit different, a word or two is in order. It is the mass (or equivalently energy) that causes gravitational interactions. Because gravity is intrinsically so weak and because the masses of the particles are so small, we are unable to do experiments to see how it really works at such small size scales. Since the strength of the gravity force is so small, we disregard it in the remaining discussions. The interrelationship between the particle types and the forces they feel are given in Figure 4.5.

Another important parameter when comparing the forces is their respective strengths. As we have seen in our earlier discussions, the strength of the forces depends on the distance between the two particles under consideration, which means that the results of the comparisons of the strengths of the various forces are distance-dependent. If we take the size of the proton (10^{-15} meters) as a good size, we

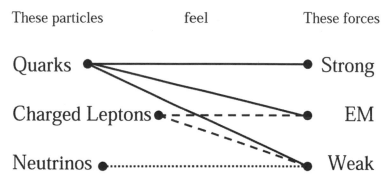

Figure 4.5 Diagram showing which particles are affected by which forces.

Table 4.1 Range and strength of the four known forces.

Force	Range	Relative Strength at 10^{-15} meters
Strong	$\sim 10^{-15}$ meters	1
Electromagnetic	Infinite	$\sim 1/100 \ (10^{-2})$
Weak	$\sim 10^{-18}$ meters	$\sim 1/100{,}000 \ (10^{-5})$
Gravity	Infinite	$\sim 10^{-41}$

can do the comparisons. In order to make the information easier to understand, we define the strength of the strong force to be unity (i.e. 1) and give the strength of all of the other forces in comparison, for instance a force of 0.5 is half as strong as the strong force, while 0.01 would be 100 times weaker. The strengths of the respective forces are given in Table 4.1. The numbers given in Table 4.1 are for matter under normal conditions, not in the massive particle accelerators in which we do experiments. As always, it's hard to represent the entire truth in a single table.

If you've been reading carefully, you can think of one thing that doesn't hang together with what is given in the table. This is the fact that the top quark decays *before* it has time to become part of a meson (meson creation is caused by the strong force). Since the top quark decays, in part, by turning itself into a bottom quark (i.e. changing flavor), this interaction can occur only by the weak force. So here is a clear example of the weak force interacting before the strong force. Sigh. This is a consequence of the mass of the top quark. It's huge. Since, as we have described, the strength of the forces depend on distance, energy and mass, it is possible for the relative strength of the forces to change. The fact that a top quark decays so rapidly doesn't really mean that the weak force has gotten that much stronger than the strong force. What it reflects is the fact that the top quark is extremely massive and, by the miracle of $E = mc^2$, that means that it contains a lot of energy. The probability that some interactions or behaviors will occur for a particular particle is related both to the strength of the force and the available energy.

$$(\text{Probability of behavior}) \sim (\text{Strength of force})$$
$$\times (\text{Energy available})$$

$$(\text{Probability of cool present}) \sim (\text{Cheapskate factor})$$
$$\times (\text{Available money})$$

One can draw an analogy between particle decay and the kind of present that a guy buys his girlfriend. Two factors come into play. One is the amount of money that he has available and the other is how fundamentally generous he is. Even if he's extremely generous, if he's a poor physics graduate student (a state of affairs with which I was once intimately familiar), he's going to have to work pretty hard to scrape up the money to take his girl to the zoo. On the other hand, if the guy is some generous and highly successful rock star, he might fly his girl to Paris for lunch on the Champs Elysées. He might even throw in a diamond necklace to boot. Available resources matter. On the other hand, even if the same rock star is inherently a cheapskate, he's still likely to at least spring for a dinner at a nice restaurant in Chicago. This analogy shows how both the natural tendency of a force and the available energy will both affect the probability that a particle will decay. With the top quark decay, the thing that's really increased is the available energy.

We now turn to another big question. Even though we know about the strengths of the various forces, we don't know how the forces actually work. How do particles see one another? We discussed in Chapter 1 the concept of fields, like a gravity field or an electromagnetic one and this idea is really good up to a point. However, the idea of a field works best when you are talking about "big" things, where big means big compared to an atom, say people-sized. But we've seen that the rules change when we look at things at a much smaller scale. This is true of fields as well. A very nice analogy of fields is a river. Everyone has seen a river ... a wide expanse of water, moving (sometimes fast and sometimes slow), but with a uniform liquid moving in one direction. This is a pretty good analogy to a field, say with all of the gravity force pointing downwards. Now let's take another look at the river. While we all have experience with water, we also know that when you look at it closely enough, you can see individual water molecules. You don't directly observe this aspect of water, but it's there. Just as the uniform water you see becomes individual molecules at a small enough scale, so it is with forces. At a small enough size scale, a field becomes a swarm of force-carrying particles, buzzing hither and yon. And, of course, this opens a new and interesting story.

Forces and Feynman Diagrams

As we may recall from Chapter 1, in the 1860s Maxwell showed that electricity and magnetism were two facets of the underlying force: electromagnetism. More importantly for this discussion, he also showed that electromagnetism and light were intimately related. And this starts us down the path of force carrying particles.

One question that absorbed the thoughts of many people in the centuries preceding 1900 was: "What is the ultimate nature of light?" As early as Newton, scientists discussed whether light was intrinsically a wave or a particle. Waves and particles are very different objects. A particle has a definite size and position, while a wave has neither. The story of that particular debate and subsequent resolution is inextricably linked with the saga of quantum mechanics and worth a book in and of itself. However, we start our story when it was resolved that the answer was "both." A particle of light could be localized like a particle, but had enough "waviness" that it can act like a wave when experiments sensitive to those qualities are done. We call the particle of light a photon. Figure 4.6 shows three models of light: particle, wave and photon.

We see how a photon has a reasonably well-defined position. We also see that the size of a photon is a few times its wavelength. For photons that are interesting (with energy larger than a GeV), we see

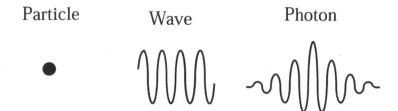

Particle   Wave   Photon

Figure 4.6 A particle exists at a specific point. A wave extends to great distances. In contrast, a photon exhibits characteristics of both; it can be more-or-less localized, but still has wave-like properties.

that the wavelength is less than 10^{-15} meters (or smaller than the size of a proton), which suggests that its position is quite localized. Thus we often treat the photon as a particle, but we need to never forget that it can act like a wave when necessary.

The mathematics of two electrons exchanging a photon and the electrons' subsequent behavior is really quite grim. In fact, if I ask my experimentally-minded colleagues to calculate the behavior of two electrons when they scatter by exchanging a single photon, their (and my!) response is usually something like "Well, yes ... ummm ... I used to know how to do this. Can I review a little and get back to you?" (Those pesky theorists, however, reply "Yeah, sure, piece of cake.") Luckily a truly gifted physicist, Richard P. Feynman, had a deeply intuitive insight. He worked out a series of pictures, which anyone can understand, but which precisely map to the mathematical equations. That way, you could quickly draw a clear diagram, translate the picture into a mathematical equation that a (sufficiently diligent) scientist can solve. These diagrams are called Feynman diagrams and they are pretty easy to write. Let's write down an example of an electron being made to hit another electron. To draw the Feynman diagram, we need to know how to draw an electron, a photon and an interaction point. The way to draw these objects is shown in Figure 4.7. So two electrons scattering by an exchange of a photon can be drawn in the way shown in Figure 4.8.

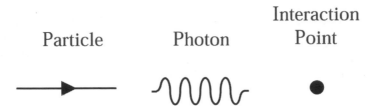

Figure 4.7 The symbols with which one can build up a Feynman diagram.

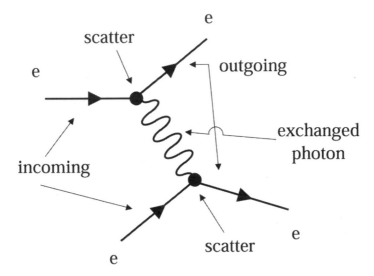

Figure 4.8 Feynman diagram showing two electrons scattering by exchanging a photon.

"Exchange" means a slightly different thing than is usually understood. In particle physics, exchange means one particle emits a photon and the other particle absorbs it. Exactly which particle does the emission and which does the absorption is not knowable, even in principle, so our theory must mathematically combine both possibilities. Just for fun, although we won't do anything more with it, let's take a quick look at how the pictures and the math relate. To do this, we need to realize that in Figure 4.8 there are two incoming electrons prior to the scatter and two outgoing electrons after. There is a single photon exchange and two vertices (places where an incoming electron emits (or alternatively gets hit by)) a photon and becomes an outgoing particle. We can then write down the rules for each piece. I'm not going to define the math symbols, as we're not going to do anything with them. These rules are shown in Figure 4.9.

So, with this information, we can translate a simple Feynman picture into mathematics. (For the mathematically anxious, don't

Object	Diagram	Symbol
Incoming Particle		I
Outgoing Particle		O
Photon		$\dfrac{-ig_{\mu\nu}}{p^2}$
Vertex		$ie\gamma^\mu$
Antimatter Particle		I or O

Figure 4.9 Table showing the correspondence of each piece of a Feynman diagram and their corresponding mathematical terms. By drawing the figure, one can easily write the correct equation. Matter and antimatter particles use different symbols. (Note to the purist, rather than the "open end" and "close end" arrowheads drawn here, in textbooks, the same "close end" arrow is used but for antimatter, the direction is reversed. For clarity, I have introduced a non-standard notation, as now all arrowheads point in the direction of motion.)

worry ... we're not going to solve or for that matter even try to really understand the equations.) An example is shown in Figure 4.10.

After this, the math gets hard. The heroic reader can look at books on the subject (see the bibliography, under super hero reading), but you will find the books daunting, my friend. The one that I learned from, written by Francis Halzen and Alan Martin, I still find imposing.

The point of this little jump into hard math is not because I want anyone to understand the equation. The real reason is that I want you to know that every Feynman diagram I write is really an equation in disguise. This is an *INCREDIBLY COOL* idea. There are many more rules and much fancier and more complicated diagrams, but each one is a clear way to write an equation. Once you draw the picture, writing the equations comes almost for free. Of course, even practicing scientists find pictures clearer than equations and we thus have Feynman to thank for making an intrinsically difficult problem more tractable.

So let's get back to physics. How does an electron exchanging a photon with another electron relate to the electric force ideas discussed previously? We can best explain this by analogy. Let's start out

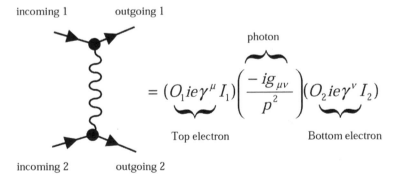

Figure 4.10 The correspondence between a particular piece of a Feynman diagram and the related mathematical expression. While we will not solve this equation, this figure underscores the fact that all Feynman diagrams are simple ways to represent equations. This insight makes calculations much easier.

with something simple. Suppose you are standing on a rowboat in a lake and you dive into the water. What happens to the boat? It moves in the direction opposite to the one in which your body went. So projecting an object can cause another object to move. Since the phenomenon that causes an object to move is a force, one can say that throwing an object from a boat results in a force on the boat. The next idea is to consider what happens if you throw a heavy sack of sand into a boat. The boat will move in the same direction that the sack was going before it hit the boat. By reasoning similar to that given above, if you throw something into a boat, the boat feels a force and then moves in the direction the sack was originally going. Now let's extend these ideas to two nearby and identical boats, each carrying a person of similar size, a situation shown in Figure 4.11. If one person throws a heavy sack at the person in the other boat, the throwing person's boat will move away from the other boat. When the other person catches the sack, that boat will move in the same direction that the sack was moving (and away from the initial boat). This is exactly analogous to the situation of two electrons which, because they both carry negative charge, feel a repulsive force through the exchange of a photon.

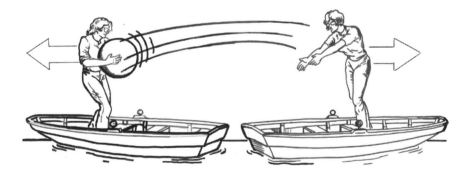

Figure 4.11 Interactions between particles can be thought of as an exchange of another particle. By tossing a ball back and forth, the boats experience a repulsive force. (Drawing courtesy of Dan Claes.)

We have to strain our analogy a little bit in order to think of the case of a positive and negative electrical charge (and thereby one in which an attractive force is felt). However, pretend we have our two brave mariners and one of them throws a boomerang in a direction opposite the other boat. The boomerang then swoops in a circle and is caught by the person in the other boat. Figure 4.12 depicts the situation when both boats recoil in such a way that the two boats move towards one another (are attracted).

So we see how the exchange of a photon can look like what we understand as a force. In order to become fully comfortable with this new idea, we need to consider a few other things. One real question might be "Does the exchange of a single photon fully account for the electrical force?" The answer to this question is, of course, "no," as the real answer is more complicated. In reality, two particles can exchange many more photons and they do. There is a continuous bombardment of photons being emitted by one and absorbed by the other (and vice versa). So why do we draw in our Feynman diagrams a single photon exchange? Well, one reason is mathematical simplicity; including more than one photon greatly complicates the calculations. But, by itself,

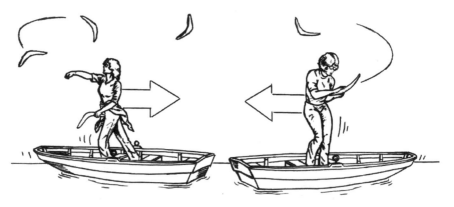

Figure 4.12 Two boats experiencing an attractive force through the exchange of a particle. In this case, the two people are throwing a boomerang, causing the two boats to feel an attractive force. While this analogy is a bit contrived, it illustrates the relevant behavior. (Drawing courtesy of Dan Claes.)

this is a lousy reason, because that approximation may poorly reflect reality. The physics-motivated reason is that in particle physics experiments, we shoot particles towards one another so fast that they are near one another for so short a time that they don't have sufficient time to exchange more than one highly energetic photon before they are past each other and then too far away to exchange another highly energetic photon. The two particles do exchange photons both before and after the "big exchange," but these are exchanges of lower energy photons and they don't change things much. In truth, we can now do experiments with sufficient accuracy that we can measure the effects of these additional photons, but for the purpose of this book, these are just little tweaks, which we now ignore.

The simple interaction between two electrons shown in Figure 4.8 is just one of the possibilities. In this particular situation, one must consider a subtle point. Initially, you have two electrons (call them one 1 and 2) entering the collision and two leaving (call them A and B). When you look at outgoing electron A, was this incoming electron 1 or 2? You can't know. So you need to add a new Feynman diagram to account for both cases. This point is illustrated in Figure 4.13.

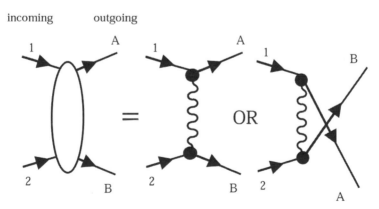

Figure 4.13 Subtle complication in scattering calculations. When two identical particles are collided, it is impossible to know which of the outgoing particles corresponded to which of the incoming ones.

While the situation discussed above is technical and only moderately interesting (although critical for accurate calculations), when one collides electrons and positrons (anti-electrons) together, something similar (but far more interesting) occurs. While an electron and positron can scatter in the simple way discussed earlier, the fact that an electron and positron are matter/antimatter pairs means that they can annihilate into pure energy (i.e. a photon) and then the energy can reappear as another matter/antimatter pair (although the new pair can be electron/positron, muon/anti-muon, quark/antiquark, etc.). It is this property (the conversion of matter to pure energy and then back to a new form of matter) that allows us to create matter and antimatter in the laboratory. For instance, a $\pi^{\pm}$ decays in 2.6×10^{-8} seconds, so if we want to make a $\pi^{\pm}$ beam, we need to create them from the energy carried by a beam of stable particles. The two Feynman diagrams that go into electron/positron exchange are given in Figure 4.14.

Note that the "switched scatter" Feynman diagram (Figure 4.13) cannot occur because it stemmed from the fact that a particular outgoing electron could have been either of the two original incoming

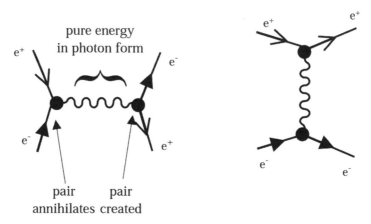

Figure 4.14 Two different ways in which an electron and positron can interact. While they can scatter in the traditional way by exchanging a photon, because they are identical forms of matter and antimatter, they can annihilate into energy before re-emerging as a new electron/positron pair.

electrons. Since one can always (in principle) identify electrons from positrons, this kind of confusion doesn't exist in this case. We will return to the pair annihilation diagram frequently (Figure 4.14).

So now that we know about Feynman diagrams, what about photons themselves? Photons are point-like particles (physics-ese for having a very small size); they have no mass and they jump from one electrically charged object to another. Photons are bosons (which we recall means that they have integer spin) and sound like new and exotic matter, but they're really not. Ordinary light is made up of photons and you can see because a photon jumps from an electron in (say) a flashlight to an electron in your eye. While the photons that occur in high-energy particle collisions are of considerably higher energy than those with which you see, except for this fact, they are essentially the same.

If the photon is the boson that carries the electromagnetic force, what of the other two forces? A particle called a gluon carries the strong force, while two particles called the W and Z bosons carry the weak force. Physicists speculate about a graviton that causes gravity, but there is no supporting experimental evidence nor any real prospect of any in the near future (although in Chapter 8 we discuss some efforts in this direction).

Feynman Diagrams and the Strong Force

The gluon has both many similarities and differences as compared to a photon. The gluon is a massless boson that couples to a charge and mediates a force, but here the similarities end. The gluon does not couple to the electrical charge, but rather to the strong force charge. Since the strong force is felt only by quarks, it is natural to ask what constitutes this charge. The one thing that quarks carry, but not any of the other particles thus far discussed, is color. As quarks carry color, it turns out that color is the strong charge. As explained in Chapter 3, color comes in three distinct types that add together to form a color-neutral (white) hadron. Thus, in analogy with

Quantum Electro-Dynamics (QED), we call the theory of the strong force Quantum Chromo-Dynamics (QCD). So we can start our discussions of gluons by drawing a simple Feynman diagram with two quarks scattering from one another by exchanging a gluon (Figure 4.15).

Note that we see that we have replaced the wavy line that denotes a photon with a "corkscrew" line that denotes a gluon. Rest assured that there is a mathematical analog to this corkscrew (and furthermore it's pretty tricky). In addition, when one collides an identical quark/antiquark (q$\bar{\text{q}}$) pair, as shown in Figure 4.16, one can also annihilate the two quarks into a gluon which can then reappear as another q$\bar{\text{q}}$ pair. Note that because the leptons are not affected by the strong force, neither charged leptons nor neutrinos can be the particles that come out of the collision. The gluon is forbidden to convert into

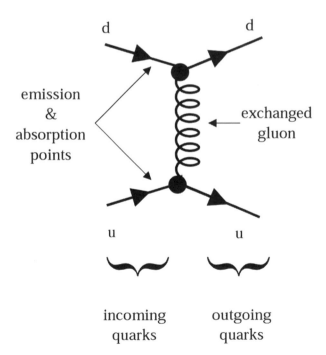

Figure 4.15 The scattering of quarks by the exchange of a gluon.

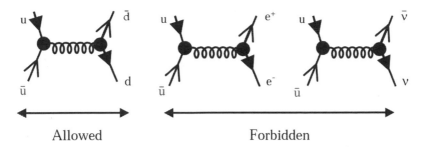

Figure 4.16 Gluons only interact with quarks (and other gluons). They do not interact with charged leptons or neutrinos.

those kinds of particles. We will return to this annihilation diagram eventually.

One point that is crucial in differentiating how photons and gluons act is the fact that gluons carry the color charge. Since photons are electrically neutral and photons are exchanged only by electrically charged particles, this means that one photon does not see (and cannot interact) with another photon. In Feynman diagram terms, this means that it is not possible to have a vertex with three photons. In contrast, since a gluon does carry the color charge, one gluon can "see" another gluon, as illustrated in Figure 4.17. One thing that this means is that one *can* have a vertex with three gluons connecting to it. A perhaps not-so-obvious consequence of the "triple gluon" vertex is that you could, in principle, get two gluons as measurable, post-collision particles.

Of course, this fact vastly changes how the strong force behaves as compared to the electromagnetic force. Because gluons are attracted to both quarks and other gluons, the strong force field is more concentrated along the line connecting the two quarks. To see the practical consequence of this fact, we first must turn to some phenomena with which we are more familiar: the force between two magnets and the force generated by a rubber band. With two magnets, the force gets stronger as the magnets are moved closer. When great distances separate the magnets, the force between them is very small. In contrast, a

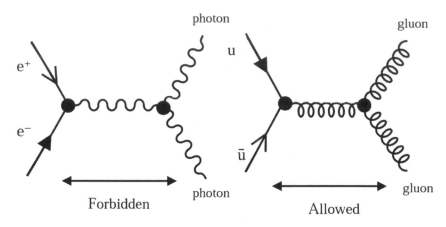

Figure 4.17 Photons only interact with particles carrying electric charge, thus a vertex with three photons is forbidden, since photons do not carry electric charge. In contrast, since gluons carry the strong charge, gluons can interact with other gluons, thus allowing a vertex with three gluons attached.

rubber band acts quite differently. When the ends of the rubber band are near one another, there is little force. However, as the ends become separated by greater distances (i.e. the rubber band gets stretched), the force increases. The electromagnetic force acts like pairs of magnets, while the strong force acts like a rubber band. As the distance between two quarks increases, the strong force between them also increases.

If we look at the Feynman diagrams given in Figures 4.15 and 4.17, we see that it is "obvious" that we can smash two quarks together and get two quarks out or possibly two gluons, which we can then measure in our detector. However, we've said in the past that one cannot see free quarks, so either we're wrong or there's more to the story. Of course, the reason is that there is more to the story. When a quark exits a collision something occurs which converts the quarks into other kinds of particles, typically mesons (and specifically π^+, π^- and π^0, although other mesons are possible). We'll discuss how this conversion occurs in a couple of pages, but let's now just think of it as "then a miracle occurs." We can write Feynman-like diagrams as illustrated in Figure 4.18.

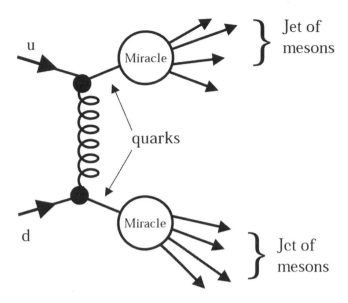

Figure 4.18 Free quarks are never seen in the laboratory. After the scatter, the quarks are converted into "jets" which look like shot-gun blasts of particles, usually mesons.

Jets: The Subatomic Shotgun

We call these "sprays" of particles "jets" and they look like nothing more than a shotgun blast of particles. Another bit of jargon that we use is we say that the quark or gluon "fragments" into a jet. When a quark or gluon fragments, the number of particles in the jet can vary. Sometimes a jet contains only a few particles (say 1–3), sometimes it carries many (30–40). Typically a jet contains 10–20 particles, with the number slightly dependent on the amount of energy carried by the "parent" quark. However, any particular quark will turn into some number of mesons, although we cannot calculate this number for any particular quark in any particular scatter. The best we can do is to calculate what would happen if we allow lots of quarks or gluons to fragment. We can then at least calculate what percentage of the time you get a particular number of particles after fragmentation (e.g. you might get 6 particles 5% of the time, 12 particles 8% of the time, etc.) But we can't calculate what any particular quark will do.

This idea of fragmentation is more than simple theoretical musings. Jets were first observed in 1958 in a balloon-based cosmic ray experiment by the Japanese Emulsion Group, in which plates containing photographic emulsion were carried to great heights and hit by high-energy protons from outer space. Cosmic ray experiments that exhibit peculiar behavior are always a little troubling, since you can't know what particles are involved in the collision. Thus the first time jets were observed in a controlled setting was in 1975 at SLAC (the Stanford Linear Accelerator Center), using the SPEAR accelerator. In Figure 4.19a, I show a real collision that was recorded by the Tasso experiment in 1979. This image is shown (rather than an earlier discovery picture) in order to facilitate comparison with three-jet events which Tasso discovered.

The SPEAR accelerator collided electrons into positrons. Of relevance for this discussion were the particular types of interactions where the electron and positron annihilate into a photon, which then converts into a $q\bar{q}$ pair. According to the above discussion, what one should see are two jets of particles, and further they should be exactly in opposite directions. Figure 4.19a clearly indicates such a signature.

This now raises an interesting question. How does one make a measurement on jets (which is the only thing an experimental physicist can observe) and compare *it* with a calculation involving charged leptons, quarks, gluons, photons, etc. (which is the only thing that a theorist can reliably calculate)? This is the really neat thing. It turns out that the particles in a jet carry an amount of energy very similar to that of the parent quark or gluon. So if you're clever enough, you can devise methods that combine together all of the particles that came from a particular quark (i.e. a jet) and compare what you measure of the jet to the quark and gluon predictions provided by theorists.

Several methods exist to combine the particles, using different rationales. An extremely common technique (and one that is very easy to explain) is one in which you mentally put a cone with a fixed size

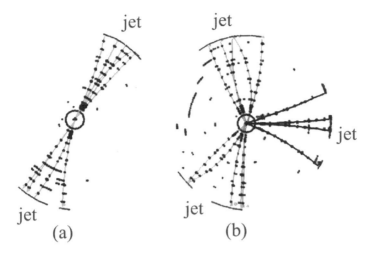

Figure 4.19 Examples of two jet and three jet events as measured by the Tasso experiment. The three jet events were taken as evidence for the existence of the gluon. (Figure courtesy of the TASSO collaboration at DESY.)

around the jet and move the cone around so that you get the maximum amount of energy in the cone. Then you stop. Such an algorithm works well, but is not foolproof. As we see in Figure 4.20, it is possible that a particle might fall outside the cone and you miss it. This happens fairly frequently, but luckily the missed particle usually doesn't carry much energy, so its loss isn't critical. We find the jets in any particular event and we then compare the results to calculations based on Feynman diagrams and, surprisingly enough, it works pretty well.

So now that we have converted quarks to jets and back to something closely resembling the initial quark, one might ask "But how do jets form (i.e. what is the 'miracle' in Figure 4.18)?" To discuss this, we need to dredge up a formula known to one and all: $E = mc^2$. In words, this equation says that matter and energy are the same or, equivalently, that you can convert energy into matter and back again. Keep this in mind in the following discussion.

The easiest way to see how jets form (i.e. how a single color-carrying parton can fragment into many color-neutral hadrons) is to

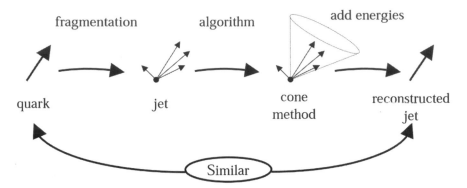

fragmentation   algorithm   add energies

quark   jet   cone   reconstructed
                                     method   jet

Similar

Figure 4.20 The method whereby one compares the quark theory with the reality of jets. A quark undergoes fragmentation and forms a jet. The experimenter puts a cone around the particles and adds their energies back together. The reconstructed jet is similar to the original quark, thereby making theory and experiment comparisons possible.

consider the case where an electron and positron annihilate and recombine into a quark/antiquark pair ($e^+e^- \rightarrow q\bar{q}$). After the collision, the $q\bar{q}$ are moving away from one another along a straight line. Recall that the strong force acts much like a spring or rubber band. When the two particles are separated by a small distance, the force between them is small. However, as the separation increases, the force increases, in the same way as when a rubber band stretches. In the case of a rubber band, what happens is that it stretches so much that the force along the rubber band gets so large that it breaks. The energy stored in the rubber band shows up by making the broken ends move very quickly away from one another. With quarks, the situation is much the same. As the $q\bar{q}$ separates, the color force between them gets stronger and the energy is stored in the tube of color force (the rubber band analog) connecting the quarks. Eventually, the force gets to be so large that space itself cannot allow that much energy stored in that small a volume and the color tube breaks (kind of like a spontaneous spark). But when the tube breaks, this energy is still too large to just pull the ends back like a rubber band. What happens in this case is the energy at the break point is converted into an identical qq̄ pair. Now you have two qq̄ pairs, the original and the newly created one. With a rubber band, we're done at this point, but with quarks we're just getting started. Since the newly created qq̄ pairs aren't moving very fast, the original quark and antiquark are still moving away from them. So the process repeats itself again and again, each time with some of the "moving" energy carried by the original quarks being converted into the mass of qq̄ pairs. Eventually, adjacent quarks and antiquarks are moving with sufficiently similar speeds that they no longer move away from one another. Then the process stops and the qq̄ pairs pair up, each creating a separate meson, with the mesons moving in roughly a straight line, along the direction of the original quarks. So voila! Quarks or gluons have become jets of mesons.

In Figure 4.21, I show the example of an up/anti-up (uū) quark pair separating. At each break point, I randomly chose to make a uū

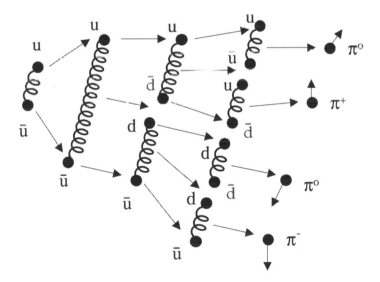

Figure 4.21 A cartoon depicting how jets are formed. As two quarks are separated, they experience an increased strong force (which increases as the quark separation increases). The energy stored in the force field between the two quarks is converted into quark and antiquark pairs. This process continues until the energy is depleted. In the end, one has many particles even though only two quarks began the process.

or d$\bar{\text{d}}$ pair. In this example, we see that an up and anti-up quark convert into a π^+, a π^- and two π^0's. In reality, there are generally many more breaks with a corresponding many more mesons, but the principle is the same.

While we have limited our discussion to the case of $e^+e^- \rightarrow$ q$\bar{\text{q}}$, Figure 4.17 makes it clear that instead of having quarks exit the collision, one could have gluons exit instead. Since gluons carry the color charge, they react much like the quarks we've just finished discussing. Thus both quarks and gluons can make jets.

Proving that gluons could make jets fell to the Tasso experiment at DESY in Hamburg, Germany. Since quarks can radiate gluons, one can imagine viewing events in which three jets were formed. Figure 4.22a shows the Feynman diagram for $e^+e^- \rightarrow$ q$\bar{\text{q}}$g. A second way to represent this particular collision is given in Figure 4.22b, which shows the collision at all three stages: as the e^+e^- are coming together, at the moment of collision, in which a photon is formed and immediately after the collision when the quark, antiquark and gluon are exiting the collision. Because all three final state objects carry the color charge, one could expect to see three jets. Such a collision is shown in Figures 4.19b and 4.22.

Proton Structure: The Miniature Lightning Storm

The knowledge that quarks feel the strong force and manifest this fact by exchanging gluons means that in our discussion of baryons and mesons in Chapter 3, I didn't tell you the complete story. A proton doesn't contain just 3 quarks. In addition, a proton contains the gluons jumping from quark to quark. Since each gluon can split (temporarily) into two gluons or a q$\bar{\text{q}}$ pair (before quickly recombining into a single gluon), at any particular moment, the inner structure of a proton can be very complicated. Figure 4.23 shows a possible proton at a particular time and we see that the reality is very complex. Rather than saying "quarks or gluons" each time we talk about particles in a proton, we coin a new word "parton" which means any

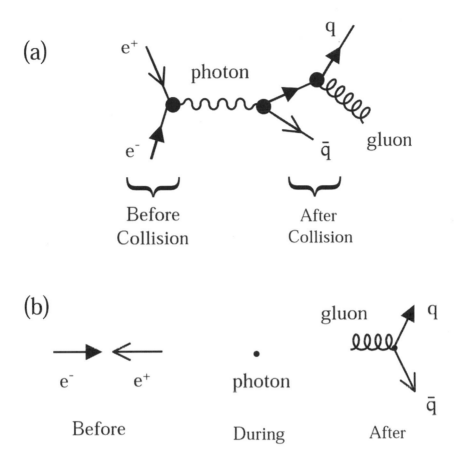

Figure 4.22 A more complicated diagram showing how gluon emission might occur. (a) shows the Feynman diagram, while (b) shows the interaction at three stages. In the third stage, the two quarks and one gluon exiting the collision are shown.

particle that is inside a baryon (e.g. a proton or neutron) or a meson (e.g. a pion).

If the structure of a proton is so complicated, how do we know so much about it? We do experiments. The easiest way to measure the structure of a proton is to fire an electron at it. The electron has no structure and is electrically charged, so a photon from the electron hits a quark in the proton. When the electron interacts with the quark

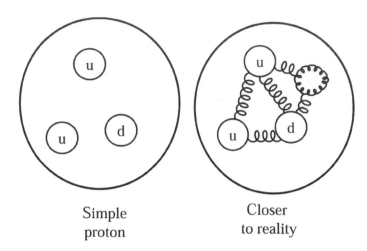

Simple
proton

Closer
to reality

Figure 4.23 A proton is said to contain three quarks, two ups and one down. The reality is more complicated, as the quarks exchange gluons which can pair-produce additional quarks and pairs of gluons. The true nature of a proton is highly complex and always changing.

via photon exchange, the direction and energy of the electron changes. We can measure these changes and use that information to determine what the quark was doing when it was hit. If we scatter many electrons into many protons, we eventually build up an idea of what the structure of the proton is like. You can't look into the proton directly, so you have to see how the particle that interacts with the proton changes and infer something about the innards of the proton from the electron's change. It's similar to a situation in which you want to know something about a room in which you're forbidden to look. One way to do that is to send a person through the room. Even if the person couldn't speak, you would learn something about the room by how long the person takes to get through the room and their condition when they exit. As an example, the room could be empty, packed, on fire, or filled with mean bikers. A person's transit time and condition when exiting depends on the details of what they find in the room; so with sufficient care, you can deduce something about the forbidden room by watching the person.

Even though the situation described above seems complicated, there remains an additional complication. The structure of the proton is not static; it is constantly evolving and changing. This point is better explained when one considers a meson rather than a baryon. For maximal clarity, let's consider a rho meson (ρ^0) containing a u$\bar{\text{u}}$ pair. Because a same-flavor quark/antiquark pair *can* annihilate into a gluon, it does. Gluons can split into q$\bar{\text{q}}$ pairs or pairs of gluons. Also, single quarks can spontaneously radiate and reabsorb a gluon. In Figure 4.24 we follow what is going on inside a neutral rho meson, the ρ^0. In the figure, as we go through time, we move to the right. The vertical lines in the figure show snapshots at particular times. We see the first time we look at the structure of the ρ^0 (which we call t_1) it contains a u$\bar{\text{u}}$ pair. At the second time (t_2), the ρ^0 contains a single

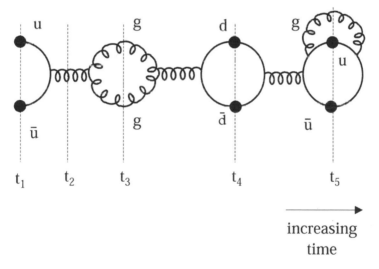

increasing
time

Figure 4.24 The intricate life of a ρ (rho) meson. Because it contains matter and antimatter quarks of like kinds, they can annihilate with one another, converting into gluons, which can in turn convert into entirely different quark pairs. Thus when one looks at a ρ meson, it might have an entirely different internal structure than the previous time at which you looked at it. In the figure, we see pairs of quarks, single gluons, pairs of gluons and quarks and gluons at the same time. In fact, the reality is even more complex.

gluon. At t_3 it now contains a pair of gluons, while at t_4 the ρ^0 contains a pair of d$\bar{\text{d}}$ quarks. Finally, at t_5 the ρ^0 contains a u$\bar{\text{u}}$ pair plus an extra gluon.

There's nothing magical about the pattern given in Figure 4.24 … I just made it up. The pattern I described is not unique, things could happen in a different order. What's really important is the fact that it illustrates some of the important configurations of quarks and gluons that can make up a meson nominally consisting of identical pairs of matter and antimatter. Each "time slice" shows one possible configuration. Basically, what it really shows is that at any particular time, the internal structure of the ρ^0 can be completely different than it was at some other time. So what does it mean to measure the structure of the ρ^0? At any time you look inside a ρ^0, the structure can be extremely complicated and at some later time, the ρ^0 can have a completely different, yet just as complicated structure. Honesty requires that I point out something that I finessed in the above discussion. Although it does not detract from the most important point, the eagle-eyed reader will note something not quite right. Specifically, at time t_1 the particle is a ρ^0 meson, which has no net color (the reader is invited to refresh their memory with the discussion surrounding Figure 3.4 if needed). In Figure 4.24, we see that there are spots where the ρ^0 meson is drawn as a single gluon (e.g. t_2). Since gluons carry a color charge, this would seem to indicate the amount of color can vary. This is not true … color is "conserved" which means no matter at what time you look at the ρ^0 meson, you should always measure zero color. In fact, the minimum number of gluons one needs to make an object carrying no net color (and also reproducing the ρ^0 meson's spin) is three. So at time t_2 there actually should be three gluons, but that just means that the structure of a rho meson is even more complicated and variable than the simple picture of Figure 4.24.

We've been speaking of ρ^0's for clarity, but the discussion is also valid for protons. So when you measure the proton's structure, you are investigating all of the possible complicated configurations. You do this by hitting lots of protons with lots of electrons and eventually you build up a picture of what you are likely to see when you look inside a proton.

We can also take this time to foreshadow some of the interesting and important points discussed in the next chapter. One might wonder from where comes the large mass of the proton. In our simple model, in which the proton consists of only three quarks, you could imagine that the mass of the proton is carried by the three quarks, each carrying 1/3 of the proton's mass. However, we've said that it is possible for the rho meson to consist of only three gluons (which are massless). Thus it is impossible that the mass of the rho is tied up exclusively in the constituent quark and antiquark pair. Similarly with the proton, it is possible for there to exist a great number of gluons within the proton at any one time. Careful measurements and calculations have shown that the masses of the quarks in the proton (up and down) are extremely small, making up only a percent or so of the mass of the proton. So what gives?

We must again turn to Einstein's venerable equation $E = mc^2$. At any particular instant, each quark and gluon will carry a specific amount of energy (although as we've seen in the above discussion that amount changes instant to instant). Because energy and mass are equivalent, the mass we measure of the proton actually reflects primarily the energy of the constituents flying around within. Robert Kunzig, in the July 2000 issue of *Discover* magazine, contributed an article entitled "Gluons" in which he eloquently described the true nature of the proton.

> A proton is made of three quarks, yes, but the quarks are infinitesimal ... just 2 percent or so of the proton's total mass. They're rattling around at near light speed inside the protons, but they're imprisoned in flickering clouds of other particles ... other quarks, which materialize briefly and then vanish and, above all, gluons, which transmit the force that binds the quarks together. Gluons are massless and evanescent, but they carry most of the proton's energy. That is why it is more accurate to say protons are made of gluons rather than quarks. Protons are little blobs of glue ... but even that

picture conveys something too static and substantial. All is flux and crackling energy inside a proton; it is like an unending lightning storm in a bottle, a bottle less than 0.1 trillionths of an inch in diameter. 'It's a very rich, dynamic structure,' says Wilczek. 'And it's very pleasing that we have a theory that can reproduce it.'

In the next chapter, in which we discuss the question of mass, it is important to recall that there we are speaking only of the masses of the quarks, leptons and gauge bosons. The mass of the proton and neutron (and thus consequently the mass of most of the visible matter in the universe) is really a reflection of the subatomic lightning storms contained within the nucleus of atoms.

The complex structure of the proton has consequences when you collide electrons into protons. Because you can't know what the structure of any proton will be at the moment of impact, you can't calculate what sort of quark scattering will occur in that collision. The best that you can do is to combine your knowledge of the most likely types of interactions with the most likely types of configurations of partons within a proton to calculate the most likely interaction that you will observe. Since you can also calculate how often you will get a rare interaction and/or a rare proton structure, you can therefore infer the likelihood of seeing any particular rare type of collision.

Feynman Diagrams and the Weak Force

We now change topics and consider the fourth and last of the known forces. The way we describe how the weak force works in terms of particle exchange is the most complicated of the well-understood forces. Unlike the electromagnetic and strong forces, which are governed by the exchange of a single type of particle, the weak force can be caused by the exchange of one of two quite different particles. These particles are called the W and Z bosons and they differ from the photon and gluon also by the fact that they are massive. The very heavy mass of these particles is the root cause of the relative weakness of the weak force.

The weak force was originally observed in beta decay, where a neutron (which is electrically neutral) changes into a proton (which has positive electrical charge). In order for this behavior to be described in terms of particle exchange, the exchanged particle had to be electrically charged. The particles that were hypothesized were the W particles, of which there were two, one with an electrical charge the same as that of the proton and the other with the same charge as the electron (i.e. the same size as the proton, but opposite sign). We call these particles W^+ and W^-, with the superscript obviously denoting the sign of the electrical charge.

Neutron decay is also called beta decay because of the conversion of a neutron into a proton and a beta particle and can be described in terms of quarks in the neutron emitting a W particle. A down (d) quark in the neutron emits a W^- and turns into an up (u) quark. The W^- decays into an electron (e^-) (a beta particle and hence the name) and an antielectron neutrino ($\bar{\nu}_e$). This behavior is shown in Figure 4.25.

There was a problem with a theory which contained only W's. The theory made really silly predictions. There were Feynman diagrams which were clearly allowed if the W particle existed, but when

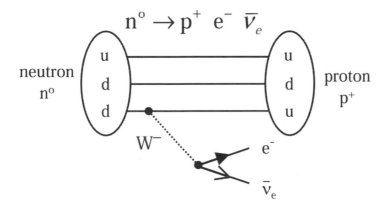

Figure 4.25 Conversion of a neutron into a proton viewed within the context of a conversion of a down quark into an up quark through the emission of a W boson.

appropriate calculations were done, the probability that that particular interaction would occur was greater than the probability of all interactions combined. Since the particular interaction was one of the many allowed, it doesn't make sense that the probability of the one was more likely than it and all of the others added together. So some fixing was needed.

There were two ways in which the theorists knew how to make this problem go away. One was if there existed ultra-heavy electrons that were identical to the conventional electron in *every* way except for their mass. If such a thing were true, the problems with the theory could be resolved. The only problem was that such particles have never been observed. This example is very important because it shows that not all theories are right, even if they make good mathematical sense. Here was a theory that solved a problem but was not correct. The universe didn't cooperate. Physics is ultimately an experimental science and that keeps us all, theorists and experimentalists alike, honest.

Luckily theorists are resourceful people. Just because this theory was wrong didn't stop them. In the late 1960s, a number of theorists also noted that they could fix the theoretical problems if another force carrying particle existed, the Z^0. The Z^0 would be electrically neutral, like the photon, but massive like the W's. Because there is only one kind of Z it is customary to drop the superscript. The Z would couple to the weak charge. Unfortunately there was a problem with this theory as well. Just as a down quark could emit a W^- and change flavor to an up quark, one would expect that a strange quark might emit a Z particle and change into a down quark. The Z would then decay into an electron/positron pair (e^+e^-). These are the so-called Flavor Changing Neutral Current (FCNC) decays. FCNC's have their name because a neutral particle (the Z) changes the flavor but not charge of the quarks. Such a hypothetical decay is shown in Figure 4.26.

The problem was that this decay is also never observed. One could imagine that we might reject this theory too, for the same reasons that we rejected the heavy electron idea. However we are saved by an unlikely avenue. As discussed in Chapter 3, Glashow, Iliopoulos

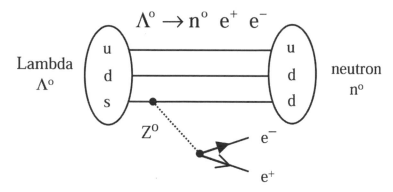

Figure 4.26 Conversion of a Lambda particle into a neutron through the emission of a Z boson. Such a conversion is never seen.

and Maiani in 1970 proposed that the charm quark existed. If the charm quark existed, it would cause the FCNC's to be forbidden. Thus with the observation of the J/ψ in 1974 (which was understood to be a pair of charm and anticharm quarks), the failure to observe FCNC's did not rule out the existence of a Z.

As discussed in Chapter 2, the invention of high flux, high-energy neutrino beams presented a new opportunity for measuring weak interactions. However, the experiments were designed primarily to see "charged current" interactions (i.e. ones involving W's), which involved a neutrino hitting a nucleus and a charged lepton (say $e^{\pm}$ or $\mu^{\pm}$) leaving the collision. Examples of this are shown in Figure 4.27.

The important signature for these types of interactions is that it is initiated by an unobserved particle, with a charged lepton exiting the collision. Since one could calculate typical neutrino energies, it was also apparent that the charged lepton carried lots (but not all!) of the energy of the neutrino. The details of what happened to the proton after it got blasted apart were not as carefully scrutinized.

Now, consider the case of a neutrino emitting a Z, which would hit a proton. This is a hard experiment to get right. The invisible neutrino enters, emits a Z and exits the collision, also invisibly. Thus the only thing one actually observes is the "stuff" associated with the proton

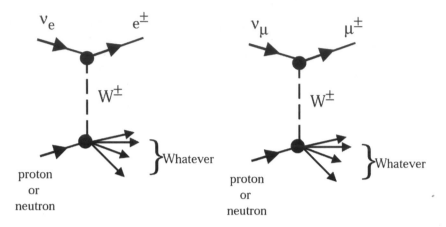

Figure 4.27 Feynman diagram of neutrino scattering in which a *W* boson is emitted. The neutrino is converted into a charged lepton which can be observed. This is the so-called charged current scattering.

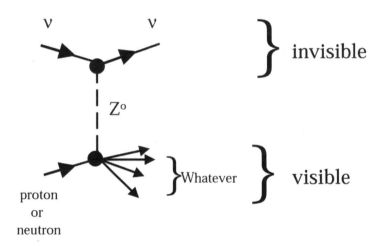

Figure 4.28 Feynman diagram of neutrino scattering in which a *Z* boson is emitted. The neutrino stays a neutrino, thus the outgoing state cannot be observed. This is the so-called neutral current scattering.

getting blasted apart. An example of this type of collision is shown in Figure 4.28.

In 1973, the Gargamelle experiment headed by Andre Lagarrigue, which was a bubble chamber experiment at CERN, observed the occurrence of neutral currents. In the bubble chamber, very clear photos of electrons from hydrogen atoms being hit by invisible particles were recorded. No other experiments observed similar types of events and some very pointed comments were made suggesting that the observation was not real. Nonetheless, the Gargamelle group was confident and after very careful scrutiny and some other events of the same type, they announced their discovery in 1973. Thus neutral currents were established.

Observation of charged current and neutral current events is very encouraging, but doesn't prove the existence of W's and Z's. To be certain, you have to see them directly. Unfortunately, all of the calculations and experimental results pointed towards the W and Z being very heavy, about one hundred times more massive than a proton. No accelerators were available which could generate particles of this mass. Clearly, a new accelerator was in order. In 1976, Carlo Rubbia, David Cline, Peter McIntyre and Simon van der Meer proposed a large accelerator at CERN, which would collide protons and antiprotons together at very high energies. We'll talk more about the accelerators in Chapter 6, but the trick here was to first manufacture antiprotons by smashing protons into a target and converting their energy into the mass of the antiprotons. Because antiprotons will annihilate nearly instantly if they come in contact with a proton, they must not be allowed to hit anything and can be manipulated only by electric and magnetic fields. Finally, the antiprotons must be put into an accelerator, have their energy increased and collided precisely with protons. Quite a feat.

However accelerated particles do not a discovery make. You have to record the collisions too. For this, two large detectors were built, called UA1 and UA2 (UA is the CERN nomenclature for "underground area"). UA1 was the Cadillac experiment, headed by Carlo Rubbia, as driven a man as one could imagine. UA2 was envisioned to be more of a safe-bet, guaranteed to make good and solid measurements in the event that Rubbia's grand ideas didn't work as well as hoped. The fact that the UA1 collaboration was constantly looking over their shoulder points out that UA2 had their own share of sharp and driven people too.

The race for the discovery of the W and/or Z bosons had its own drama, analogous in many ways to the discovery of the top quark described in Chapter 3. This race was interesting enough to warrant its own book, which is listed in the suggested reading. On January 21 & 22, in 1983, UA1 announced to packed houses in the main auditorium at CERN that they had discovered the W. UA2 had similar evidence at the time of announcement, but were corporately a bit more cautious and held back. Luckily for UA1, they were right and, as Wolfgang Pauli once said "only the one who dares can win." A few months later, on April 30, 1983, UA1 first observed the Z, closing the discovery chapter in the story of the electroweak bosons W and Z. Carlo Rubbia and Simon van der Meer shared the 1984 Nobel Prize, for the discovery of the W particle and also for the design and building of the vast accelerator complex that made it possible.

A fair question one might ask is "How do you know that you've discovered a W or a Z?" It turns out that the W is much easier to identify with confidence. To begin with, it is produced ten times more often than a Z. In addition, W's can decay in two ways. The first way is by decaying into a charged lepton and its corresponding neutrino; e.g. $(e\ \nu_e)$, $(\mu\ \nu_\mu)$ or $(\tau\ \nu_\tau)$. In addition, a W can decay into a quark/antiquark pair, say $W^+ \rightarrow u + \bar{d}$. However, the $q\bar{q}$ pair converts quickly into jets. Since jets are manufactured in collisions involving the strong force about 10 million (10^7) times more often than those involving the weak force, the decay of W's are lost in the noise and are extremely difficult to uniquely identify. However, neutrinos cannot be produced by any method other than through the weak force. Even though there are 10^7 times more other, more boring, collisions for each collision in which a W is produced, you can train your detector and electronics to ignore them. We will discuss this aspect of particle discovery in Chapter 6.

Not worrying for the moment about how you actually measure these things, we can draw Feynman-like diagrams that describe what is going on during the creation and decay of a W boson.

Take as a particular example the process drawn in Figure 4.29. We can break up what happens into three phases: before the collision, while the W exists, and after the W decays. To simplify things, we draw only the quarks and antiquarks in the proton and antiproton that are involved in the collision and ignore the rest. Before the collision, we see that the up and antidown quarks are moving in the direction of the proton and antiproton beams; in this figure this is the left-right direction. The W is made and the positron and neutrino come out in a different direction. We say that this new motion is (at least partly) transverse to the direction of the original up and antidown quarks.

Recall that neutrinos and antineutrinos don't interact with ordinary matter, which means that they can escape undetected. So take a

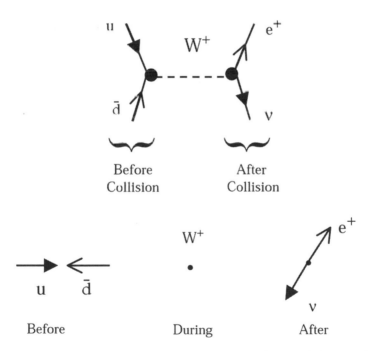

Figure 4.29 The creation of a W boson through the annihilation of an up and antidown quark. The W boson decays into a positron and a corresponding neutrino. The bottom section shows three discrete stages in the creation and decay of the boson.

look at Figure 4.29 and mentally "erase" the neutrino from the picture. This means that what you observe after the W decay is a positron moving sort of upwards and nothing at all moving downwards. So if you see nothing, how do you know that it's there? As I like to tell the students in the college classes I teach, the answer is "Physics!!!" There are certain things that are required to be the same before and after the collision. We see that before the collision the two particles are moving in the left-right direction, with no up-down motion. After the W decay, the positron is moving upwards and because up-down motion must be the same before and after the collision, there must be a particle moving downward to cancel out the fact that the positron is moving upwards. (Recall this is how the neutrino was originally postulated.) Since we see no particle moving downward, this means that an invisible particle (say a neutrino) had to be moving downward. (The reader who has taken an introductory physics class will recognize this reasoning as stemming from the law of conservation of momentum.)

So the way to discover a W most easily is to look for events with a charged lepton moving "sideward" compared to the beams, with nothing on the other side. This type of collision is exceedingly rare, but completely distinct, and it is by looking for these types of collisions that UA1 was able to establish that the W was real.

Finding the Z is similar, although more difficult. To begin with, Z's are created ten times less often than W's. In addition, Z's decay into same-types of fermion pairs; e.g. electron/positron (e^+e^-), muon/antimuon ($\mu^+\mu^-$) or quark/antiquark (say $u\bar{u}$). The earlier discussion we had for when a W decays into quarks holds here as well. They're there, but so rare compared to the more boring collisions involving the strong force that they're hard to identify. So we turn to the case of two leptons in the final state. One thing that's nice about the Z is that you can see both of the leptons into which it eventually decays. The problem is that there are other ways in which two leptons can be made. Figure 4.30 shows two ways an up quark from a proton can annihilate with an antiup quark from an antiproton to form two leptons (which we denote generically as ($\ell^+\ell^-$)).

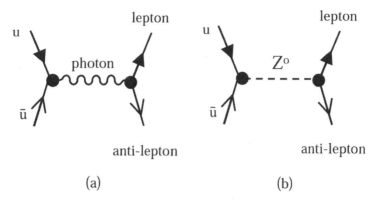

lepton   lepton

u   u

photon   Z⁰

ū   ū

anti-lepton   anti-lepton

(a)   (b)

Figure 4.30 (a) The creation of a lepton/antilepton pair via the creation of an intermediary photon. (b) The creation of a lepton/antilepton pair via the creation of an intermediary Z boson. The two cases are essentially indistinguishable, except for the fact that the Z boson creation becomes more likely when the collision energy is near the Z boson mass.

The production of a lepton ($\ell^+\ell^-$) pair by a photon (Figure 4.30a) is much more likely than through a Z (Figure 4.30b). So how can you establish that a particular set of lepton pairs came from a Z and not a photon? This is done by recalling that a Z is a very heavy particle, with a well-defined and specific mass. When it decays into a ($\ell^+\ell^-$), it gives each particle a specific and unique energy. So you measure the energy of each ($\ell^+\ell^-$) pair and convert that energy back into the mass of the particle from which they decayed. How one does this is described in Appendix D. Since you can calculate how many ($\ell^+\ell^-$) pairs should come from photons at each energy, anywhere you have too many pairs can indicate the discovery of a new particle. Figure 4.31 illustrates this technical point.

In Figure 4.31, we plot the number of lepton pairs that we expect from photons for each energy and compare it with what we see. We see a peak (or excess) of particles at a particular energy, which means that we have too many lepton pairs at that energy. This most likely means that something unexpected is happening there. This often indicates a new particle, and so it was for the Z discovery. UA1 and UA2

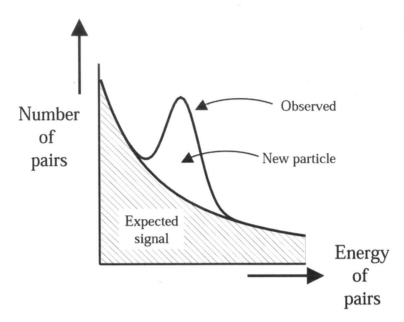

Figure 4.31 Example of how one might observe the creation of a new particle. The energy of the decay products is measured and the frequency of their occurrence is noted. When there is an excess over the expected behavior, this can indicate some new physics process in which a particle is created.

simply had to look for lepton pairs and simply see if they got as many as they expected or more. When they finally measured enough pairs to be sure that the excess wasn't a fluke, they announced.

So in 1983, UA1 announced the discovery of the *W* and the *Z*, with the UA2 collaboration confirming soon thereafter. However, the two experiments only generated a relatively small number of *W* and *Z* bosons between them. To precisely measure the bosons' properties, a new accelerator was needed. Unlike the accelerator in which the *W* and *Z*'s were discovered, which collided protons and antiprotons, the new accelerator would collide electrons and positrons. Because an electron and positron could completely annihilate into a *Z*, it was possible to adjust the beam energy perfectly so that *Z*'s were copiously produced. In 1989, this accelerator, called LEP (for Large Electron-Positron), turned on, with four fancy detectors (called Aleph, Delphi,

L3 and Opal) set to record the collisions. Each experiment recorded about 5 million collisions in which Z's were created and measured the properties of the Z boson with exquisite precision. The detectors and accelerators were so beautifully designed and understood that they were sensitive to the effect of the lunar tides flexing the crust of the Earth and also to when an electric train passed nearby.

In 1996, the energy of the LEP accelerator was increased with the desire to produce pairs of W's. Because of the type of beam, the only way to produce W's was in pairs (e.g. $e^+ + e^- \rightarrow Z$ or $\gamma \rightarrow W^+ + W^-$). Each experiment collected a few thousand events in which W pairs were created and made impressive measurements here as well, although for these kinds of measurements, they had stiff competition from the DØ and CDF detectors at Fermilab. On November 2, 2000, the LEP accelerator was turned off for the last time to make way for a new and vastly more powerful accelerator, the Large Hadron Collider or LHC.

The W and Z bosons are now well characterized and have a mass of nearly one hundred times greater than that of a proton. It is this huge mass that really causes the weak force to be so weak. If the W and Z bosons were massless like the photon, the strength of the weak force would be roughly the same as the electromagnetic force. One can naturally ask what causes the W's and the Z's to be so much more massive than the photon. And in the 1960s, when Weinberg, Glashow and Salam demonstrated that the weak force and the electromagnetic force were just two facets of an underlying electroweak force, physicists naturally asked what was it that broke the symmetry between the electromagnetic and weak forces. This question is very interesting and will be discussed in detail in Chapter 5.

Now that we know of all of the forces and the particles that carry these forces, it would be a good time to sum up what we know. This information is given in Table 4.2.

Since we now have the required tools, I'd like to close the chapter with a discussion of how one "sees" the top quark. In the ensuing discussion, we ignore the fact that we really need detectors to measure

Table 4.2 Important parameters associated with the four known forces.

	Strong	EM	Weak	Gravity
Relative Strength	1	10^{-2}	10^{-5}	10^{-41}
Range (meters)	10^{-15}	Infinite	10^{-18}	Infinite
Force Carrying Particle	Gluon	Photon	W Z	Graviton
Mass (GeV)	0	0	80.3 91.2	0
Year Discovered	1979	Early	1983	N/A
Lifetime (seconds)	Infinite	Infinite	$\sim 3 \times 10^{-25}$	Infinite
Observed	Yes	Yes	Yes	No

things (we'll get back to this in Chapter 6) and I will talk only in terms of Feynman diagrams.

Discovery of the Top Quark

Currently top quarks can only be made at Fermilab, which is (as of 1971 and probably through 2007 or even beyond) the highest energy particle accelerator in the world. Protons and antiprotons are collided with an available energy over 2000 times greater than the energy contained in the mass of a proton and over 5.5 times greater than the minimum energy needed to make pairs of top quarks. In the most common process (illustrated in Figure 4.32), a quark and antiquark come together to annihilate into a gluon, which then splits into a top/antitop quark pair. Both the top and antitop quarks decay essentially 100% of the time into a W boson and a bottom (or antibottom) quark. The bottom quark always forms a jet, but the W boson can decay into quark/antiquark pairs or lepton/neutrino pairs, as described earlier. This fact complicates searches for events in which top quarks are created. Ignoring for the moment that W bosons decay, we draw in Figure 4.32 the typical Feynman diagram in which top quarks are produced.

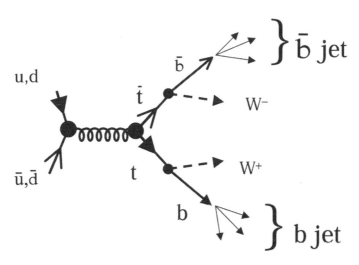

Figure 4.32 An example of how a top/antitop quark pair can be made, with decay modes. The *W* bosons have a number of possible decay modes.

Table 4.3 Different possible decay channels of the W and their probability of occurring.

Decay	Fraction of time
$W^+ \rightarrow e^+ + \nu_e$ $W^- \rightarrow e^+ + \bar{\nu}_e$	1/9
$W^+ \rightarrow \mu^+ + \nu_\mu$ $W^- \rightarrow \mu^+ + \bar{\nu}_\mu$	1/9
$W^+ \rightarrow \tau^+ + \nu_\tau$ $W^- \rightarrow \tau^+ + \bar{\nu}_\tau$	1/9
$W^+ \rightarrow q + \bar{q}$ $W^- \rightarrow q + \bar{q}$	6/9

But we now need to consider the ways in which a *W* boson can decay. This is given in Table 4.3, where the various decay modes are listed, as is the fraction of time that they occur. If both *W*'s decay into leptons (say both electrons, for example), we see the decay chain depicted in Figure 4.33.

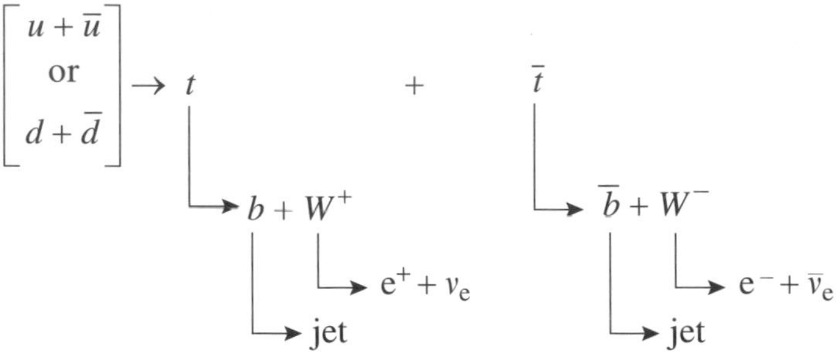

$$\begin{bmatrix} u + \bar{u} \\ \text{or} \\ d + \bar{d} \end{bmatrix} \rightarrow t \qquad\qquad + \qquad\qquad \bar{t}$$

$t \rightarrow b + W^{+}$

$W^{+} \rightarrow e^{+} + \nu_e$

$b \rightarrow$ jet

$\bar{t} \rightarrow \bar{b} + W^{-}$

$W^{-} \rightarrow e^{-} + \bar{\nu}_e$

$\bar{b} \rightarrow$ jet

Figure 4.33 Typical decay chain from a top quark/antiquark pair. In this situation, both W bosons decay into the appropriate electron or positron.

After the decay, we see six objects; 2 jets from b quark decay, 2 charged leptons from W decay and 2 (invisible) neutrinos from the W decay. Thus we see four visible objects; 2 jets and 2 charged leptons, plus missing (invisible) energy.

If only one of the W's decays into leptons and the other into quarks, then we would still see six objects; 2 b quark jets, 2 jets from quarks from one W and a charged lepton and invisible neutrino from the other W. Of course, if both W's decay into $q\bar{q}$ pairs, the final result is 6 jets.

Figure 4.34 is a pie chart that shows how often an event containing a top/antitop quark pair decays in the various ways. We see that the case where both W's decay into quarks is by far the most common, but unfortunately there are many more mundane ways in which 6 jets can be made. So this signature of top quark decay was the last one that was successfully explored because these events are hard to uniquely identify. The case where there were 2 jets and 2 charged leptons, one a muon and the other an electron, is very rare, but it's also a very hard one to fake. For that reason, it was a very attractive configuration for which to search (and before DØ and CDF got into it, the one which was thought to be perhaps the only hope for discovery). The case where one of the W's decayed into leptons, which has the signature of 4 jets, a charged lepton and an invisible neutrino is a reasonable compromise. In the end, the discovery of the top quark was accomplished

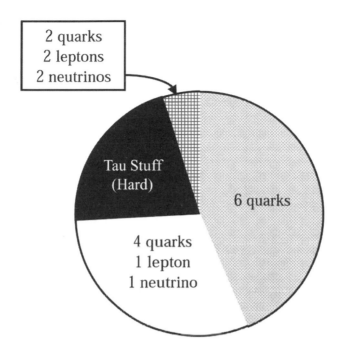

Figure 4.34 Frequency of different decay possibilities in top/antitop events. The number of quarks created is dependent on the W boson decay modes. The most likely situation is that both W bosons decay into quarks, yielding 6 quarks.

using all decay modes of the top quark in which at least one of the W's decayed into a lepton pair.

So now we have become quite expert on all of the particles and forces known to physicists. Taken together, they are called the Standard Model of Particle Physics. It never ceases to amaze me that all of creation can be explained by the 12 quarks and leptons (4 if you don't look inside particle accelerators!) and the 5 force carriers. That so few particles can explain all of the diverse phenomena of the universe (except perhaps for the French fascination for Jerry Lewis and Woody Allen) can only be described as a triumph of scientific achievement. There remain mysteries, of course, like why the W and Z bosons are so much heavier than the photon. We have some (unproven!) ideas about this and the search for confirming experimental signatures is underway (and will be discussed in Chapter 5). There are even deeper mysteries which will be covered in Chapters 7 and 8. Nonetheless, even the incomplete knowledge of which we are certain must be considered to be a success of over two thousand years of scientific inquiry.

chapter 5

Hunting for the Higgs

So Higgs is great. Why, then, hasn't it been universally embraced? Peter Higgs, who loaned his name to the concept (not willingly), works on other things. Veltman, one of the Higgs architects, calls it a rug under which we sweep our ignorance. Glashow is less kind, calling it a toilet in which we flush away the inconsistencies of our present theories. And the other overriding objection is that there isn't a single shred of experimental evidence.

— Leon Lederman, *The God Particle*

With such an uncomplimentary introductory quote, one might wonder why it is that the search for the Higgs boson has engendered such intense interest. Basically, current theory, which the previous two chapters have shown to be quite reliable, strongly suggests the need for the particle's existence. Our journey thus far has covered much of what is known about modern particle physics. I hope that this journey has brought you new and interesting insights. But while the degree to which we know things about the universe is impressive, it can in no way be considered a complete set of knowledge. Believe it or not, the number of questions that continue to perplex modern researchers is really quite large. Of course, this shouldn't surprise you. Knowledge often begets as many questions as answers. Like the maddening two-year old child, who follows each explanation with yet another "Why?", scientists often find that each answer raises yet another question.

While we will discuss interesting questions that are moderately well understood in Chapter 7 and even less understood puzzles in Chapter 8, there exists a particular question that is currently being attacked by nearly 1,000 physicists. Because such effort is being put forth on this topic, we will spend a considerable amount of time trying to understand just what is going on. This subject is the search for an elusive and as yet undiscovered particle, the Higgs boson.

In attempting to discuss this subject, you should realize that we've made the transition from the known to the unknown. You cannot expect pat answers to your questions. That's because there are none available. It's not that I know and I'm just not telling you. I don't know. No one does. But that's what's so exciting. Together we will understand what physicists are looking for and why we think that this is a worthwhile endeavor. This is research in progress. As I guide you through this tale, I will distinguish between those things that we know, i.e. the problems to be solved if the Higgs boson turns out to exist, and those things that are merely informed speculation, such as the nature and properties of the Higgs boson itself.

As we begin our journey to understand the Higgs boson, let us first turn our attention to the modern Periodic Table, given in Figure 5.1. All of the particles listed in this table, from the six quarks and six leptons to the four force carriers, have been observed. To varying degrees, the properties of all of the particles have been well characterized, as was described in Chapters 3 and 4. Two related and critical insights are important to raise here. The first is the fact that there appears to be a pattern in the quarks and leptons. Each vertical column in the figure consists of two quarks and two leptons. We call each column a generation. The intriguing thing is that each generation appears to be a near-clone of the other generations, with

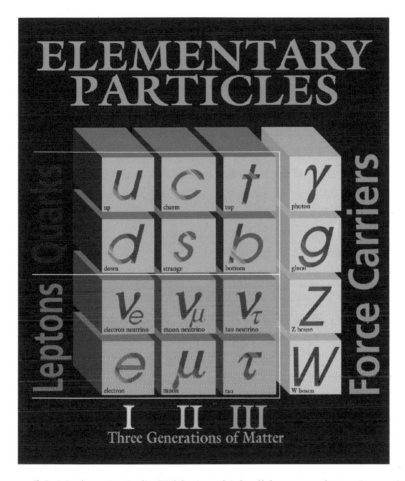

Figure 5.1 Modern Periodic Table in which all known subatomic particles are shown. All of the phenomena that you observe can be explained by these few constituents. (Figure courtesy of Fermilab.)

the notable exception that the mass of the particles involved increases with generation number; e.g. the particles in generation III have a greater mass than the corresponding particles in generation II. Similarly, the mass of the particles in generation II exceeds those in generation I. Of course, an obvious question is "Why should this be?" This points to the study of mass as the obvious variable that needs further investigation.

Mass, Parity and Infinities

So now we have two questions. First, what is mass? And the second question is; why do the different generations have such disparate masses? Let's start with the first question. We all have an intuitive sense of what constitutes mass. While it's wrong, we often equate mass with weight. More weight means more mass. Well, while that can be true, it's not the whole story. Weight requires two objects, for instance you and the Earth. If you were on other planets, your weight would change ... much lower on the Moon and much more on Jupiter. However, your mass remains unchanged. Mass can be thought of how much "stuff" you're made of. I'm much bigger than a baby and thus have a greater mass. (That occasional extra muffin for breakfast probably has something to do with it as well ...)

What is this thing they call mass anyway? Let's think about what the goal for a complete physics theory would be. We would like to reduce the number of particles needed to explain the entire myriad of behaviors that makes up the world. Further, we'd like to have all particles treated equally. Clearly, both of these goals cannot be met with particles that have differing mass. In fact, if you think about it, at large enough energy, mass shouldn't matter. Since the energy of an object can be thought of as a sum of motion energy and particle's mass energy, as the moving energy gets larger and larger, the mass energy becomes an ever less-important factor describing the particle.

It's kind of like an automobile and a feather. Everyone knows that these two things are different. The feather is *much* less massive. It requires considerably less energy to toss around than an automobile which, except for in Hollywood movies, stays firmly on the ground. But now let's take these two objects and put them in the path of a tornado. In a tornado, a car can be tossed around with nearly the same ease as the feather. Thus, we see that in the proper environment, mass doesn't matter all that much. It's only at low energy (or at low wind speed in our example) that objects can be differentiated by how their mass affects their behavior.

In fact, the natural mass for a particle to have is zero. Or perhaps stated more properly, it's easiest to formulate a physics theory with massless particles and this is how it's actually done. In fact, all of the highest energy theories start out with massless particles. This fact sounds like an absurdity or perhaps laziness; after all particles do have mass. People taking introductory physics classes often complain that physicists over-simplify their problems and this is a fair criticism to a point. But simplification is fairly handy. Michael Jordan can accurately predict where a basketball will go, given an initial velocity and direction. (Heck, even I can make that prediction, although for a much lower pay.) One can greatly complicate the question by worrying about the individual molecules in the rubber of the basketball, how the air molecules in the basketball are swirling, the shape of the basketball, whether it's spinning or whatever. To add all of this information greatly complicates the mathematics needed to describe everything that can in principle be known about the motion of the basketball. But if all you're worrying about is whether it goes through a hoop or not, all of this extra detail is an unnecessary complication. You'll get an accurate enough description if you just treat the ball as a simple object with a fixed shape and no internal structures. Similarly with particle physics theories, you can greatly simplify the mathematics and description by treating all particles on an equal footing. Later we can deal with the complication added by mass.

The first time this disconnect between the preferred physics theories, which treated particles as if they had no mass, and the clear observation that particles *do* have mass, occurred in the 1960s. This was the decade when theorists were struggling mightily in an attempt to unify two seemingly dissimilar phenomena: the electromagnetic and the weak forces.

Let's consider the magnitude of the task that these theorists set out for themselves. The electromagnetic force has an infinite range; the weak force has a range of only 10^{-18} meters. The electromagnetic force is about 1000 times stronger than the weak force. Perhaps the most striking difference is a bit tricky to explain. If you take the equations for the electromagnetic force and find all variables that represent length and replace them by the same variable with a negative sign in front of them, you get the same theory with which you started. For the weak force, if you swap all the lengths for their negative, you get the opposite of what you started with. This is kind of hard to say using English and is much easier with the language of mathematics. For those people with math anxiety, you can skip the next paragraph, as it simply says in the math language the same thing I just said. But perhaps you might just take a peek, as it shows the idea extremely clearly.

Suppose you have a quantity that you want to calculate using each of the two forces and that this quantity depends on two lengths, which we call x and y. We want to see what happens when you replace every x by $(-x)$ and every y by $(-y)$. A formula for the electromagnetic force (EM) might be something like $EM(\text{no swap}) = x \cdot y$. If one makes the replacement, we get $EM(\text{swap}) = (-x) \cdot (-y) = (-1) \cdot (-1) \cdot x \cdot y = x \cdot y$. So we find that $EM(\text{swap})$ is the same as $EM(\text{no swap})$. This is to be contrasted with the weak force, which has a different behavior. For the weak force, we have something like $Weak(\text{no swap}) = x + y$. Replacing both x and y by their negative counterparts, we get $Weak(\text{swap}) = (-x) + (-y) = -(x + y)$. So $Weak(\text{swap}) = -Weak(\text{no swap})$. This illustrates how the two forces have very different mathematical properties.

So one of the hardest aspects of trying to unify the electromagnetic and weak forces is the fact that you want to write *one* equation that fully describes both forces, even though when you write the equations for both forces separately, they behave *completely differently*. When you swap all lengths by their negative counterparts, somehow you have to write an equation that manages to explain how both everything changes and nothing changes. This sounds impossible and even a bit Zen-like.

Luckily no one told the young theorists in the 1960s that what they were doing simply didn't make any sense. (Actually, I'm sure that they did know, but they pressed on regardless.) What they tried to do was to take the very successful theory of Quantum ElectroDynamics

(QED), which described the behavior of electrons and photons to eleven digits of accuracy and tried to re-craft the theory of weak interactions using similar equations. You may recall from Chapter 4 that in the early days of QED, theorists tried to calculate properties of the electron, for instance its spin, mass and charge. When the calculations were complete, much to the chagrin of the theorists, the results of the calculations were infinity. Since the measured results were most certainly not infinity, things looked a bit bleak. However, just in the nick of time, Superman (OK, it was really Richard Feynman and friends) stepped in and saved the day. What they did was to show that the theory of QED was *renormalizable*. Renormalizable means some very technical things, but the gist is that through a mathematical slight of hand, one could reorganize the mathematics so that all of the infinities could be hidden in a convenient place in the equation. Poof...with Feynman and friends' spiffy insights, the new calculations were in good agreement with experiment.

When the same approach was tried with the weak force, nothing worked. Infinities were everywhere. There were a lot of smart guys working on the problem. The reasons were simple. First, the idea was very interesting and second, anyone who successfully cracked this nut would gain fame on par with the luminaries who successfully beat quantum mechanics into submission. The main players were Steven Weinberg and Sheldon Glashow of Harvard, Abdus Salam at Imperial College in England, Martinus Veltman and his student Gerard 't Hooft of the University of Utrecht, Netherlands, Peter Higgs of the University of Manchester (England) and Jeffrey Goldstone of CERN. The elder statesmen (having passed their momentous 30th birthday) of the effort were Julian Schwinger, Murray Gell-Mann and Richard Feynman. Eight of the ten names listed here are now very deserving recipients of the Nobel Prize, with the other two waiting for other parts of the theory to be confirmed. With such acclaim, you might expect that they were successful, and you'd be right. The whole story is pretty detailed, so we'll skip a lot of the journey and get straight to the bottom line.

The bottom line is that it turned out to be possible to write equations that included both the electromagnetic force and the weak force. The photon was the carrier of the electromagnetic force and there were three carriers of the weak force, one negatively charged, one positively charged and one electrically neutral. Success!!! Well, sort of. It turns out that the theory required that all four force carriers be massless. Another observation was that it turns out to be impossible to write a theory in which fermion mass particles communicate via boson force-carrying particles, unless everything is massless. (In all fairness, today's impossible is often tomorrow's brilliant discovery, so maybe "impossible" should be redefined to be "nobody knows how to do it.") At any rate, even a massless theory is an improvement, since at least some of the mathematic difficulties have been overcome. The infinities persisted and there was still this mass problem, but progress had been made.

The Higgs Solution

So now we get to the point of the chapter, which is how these remaining issues were resolved. In 1964, a Scottish theorist named Peter Higgs and his colleagues Robert Brout and François Englebert of the Free University in Brussels had a crucial idea. They postulated another field, like a gravity field, which exists throughout the universe. Different particles would interact with it differently. In doing so, we say that the symmetry was broken. Symmetry is physics-ese for saying that things are the same and broken symmetry simply means that they once were the same, but now are different. In the theory, the electromagnetic and weak force-carrying particles are all massless and thus they are the same, which means that they exhibit a symmetry. The field postulated by Higgs and company would interact with these particles differently, thus making them different and breaking the symmetry. Steven Weinberg took Higgs' idea and showed that the aspect of the symmetry that was broken was the mass. From four massless force-carrying bosons came the massless photon and the massive W^+, W^- and Z^0 bosons that governed the weak force. The electromagnetic and weak forces were unified into the electro-weak force (i.e. shown to be the same except for how the particles interacted with the postulated Higgs field). Hurrah!!! Kudos to everyone, except for Glashow, Salam and Weinberg, who would have to be consoled with sharing the 1979 Nobel Prize in physics.

Of course, there was still the question of all of those nasty infinities. Gerard 't Hooft took Higgs' idea and applied it to that question and found out that the Higgs field generated even more evil infinities. (Oh bother, you say... but wait, there's good news.) These infinities had the opposite sign to the original infinities. For his Ph.D. thesis, 't Hooft carefully calculated all of the infinities and then added them up. The result was zero. It's like when my ex-wife's credit card bill and the lottery I won were combined. They neatly cancelled each other out. (OK, well not really, but it's a good story and I hope you repeat it.) But for 't Hooft the cancellation worked. Such a result makes for a heckuva Ph.D. thesis and 't Hooft and his thesis advisor Veltman shared the 1999 Nobel Prize.

So what about poor Peter Higgs? Why has he not been to Stockholm? Possibly luck has something to do with it. But also there is the fact that for all of the spectacular theoretical successes traced to Higgs' idea, we haven't really proven that he was right. For the rest of this chapter, we'll discuss the properties of this field, some of the history of the search to confirm Higgs' idea and the ongoing efforts, in which some of today's most driven and energetic experimental physicists are busting their tails, trying to confirm or refute the theory.

When we discuss the properties of the Higgs field, there are a couple of topics that need special emphasis. We shall begin with the so-called "unitarity crisis," then follow with a longer discussion of the idea of symmetry and how it can be broken. This will lead naturally into a discussion of how interaction with the Higgs field gives differing particles their respective masses.

The unitarity crisis is a fancy name for a simple idea. Particle interaction theory only deals with probability, a consequence of the principles of quantum mechanics. Theorists cannot tell you what will happen when two particles collide. They can only tell you the relative probabilities that various things will occur, e.g. there is a 40% chance of no interaction, 2 particles will come out of the collision 10% of the time, 3 particles 20% of the time, etc. This allows us to know what will happen if we do the experiment many times, for instance in the above example, if we repeat the experiment 200 times (that is, we look at 200 potential collisions), we expect to see about 20 events in which we have exactly two particles coming out of the collision (because 20 is 10% of 200). But in any particular experiment, i.e. in any individual particle collision, all bets are off.

However, one thing that you know must be true is that when you add up all the probabilities of all of the possible things that might happen, they must add to 100%. This means that the particles must do *something*, be it hit or miss each other, or whatever. Without the Higgs field, theorists found out that when they carefully added all of the probabilities, they would get a total probability which was greater than 100%. Taken literally, this means that more particles scatter than were there in the first place. Clearly, this is nonsense, a fact of which the theorists were rather uncomfortably aware. It's always bad when a theory predicts nonsense. But it wasn't really as bad as it sounds. At low enough energies, the theory was OK. Only at higher energies did the probabilities increase until they finally predicted ridiculous values. There, the Higgs field saved the day.

There was ample precedent for such behavior from a theory in the early part of the 20th century. As discussed in Chapters 2 and 4, in 1934 Enrico Fermi proposed his theory of the weak interaction, which was modeled after the earlier theory of the electromagnetic interaction. He knew that whatever particle was exchanged to cause the weak force, it had to be heavy (because the range was short and the force was weak). So if something is very heavy, its mass can be approximated as infinity (like the mass of the Earth compared to the mass of a person). Such an approximation can greatly simplify the calculation and doesn't change the outcome of the calculation enough to matter. Since Fermi's theory was proposed in a time when 100 MeV was a lot of energy, it worked just fine. But any decent theory doesn't make predictions for a single energy; it makes predictions for all energies. Then, if you build a new accelerator with higher energy, you simply change the number for energy in your theory and, presto!, you have a new prediction.

Even in the early days, it was known that Fermi's theory had problems, because eventually it made silly predictions, whereby at an energy of around 300,000 MeV (300 GeV), Fermi's theory's predictions exceed 100%. Bummer. Eventually, it was understood that one could no longer treat the mass of the particle that caused the weak force to be infinitely large; it was simply very large. The relevant particle was the W boson, with its mass of 80,300 MeV (80.3 GeV), far above the 100 MeV for which Fermi's theory was designed. Eventually the approximation of an infinitely massive particle fell apart. Nonetheless, Fermi's idea was pretty impressive, as it was devised to explain phenomena with an energy near 100 MeV and worked quite well even at much higher energies.

Well, even with the correct mass of the W boson, one could calculate the probability of interactions at even higher energies and, depressingly, one finds that at an energy of about 1,000,000 MeV (1 TeV), the theory again predicts the dreaded probability of over 100%. The existence of the Higgs field fixes this problem as well. And this success of the Higgs field, while technical, is crucial in being able to use the theory to make further predictions, because without it, the theory predicts nonsense.

The unitarity crisis stemmed from the fact that something that was very big was considered to be so big as to be indistinguishable from infinity. Eventually, the approximation bit us in the tail. Does this mean that the original theory was wrong? From a purist's point of view, one must say yes, although a less strict person might say no, or at least try to explain why the original theory was "good enough." It's more correct to say that the two theories: "wrong" (which eventually predicted probabilities that exceeded 100%) and "right" (which predicted rational probabilities, at the expense of more complicated calculations) gave indistinguishable answers at low energies and only differed at high energies. To clarify our thinking, one might consider a car and how fast it can go. If you step on the gas with a certain force, you might get the car going 20 mph. Doubling the force on the gas pedal might increase the speed by 40 mph, while tripling the force might make the speed 60 mph. But we know that increasing the 20 mph force by a factor of 10 won't make the car go 200 mph (not unless you drive a much nicer car than mine). We realize that no matter how hard you press the gas pedal, there's a maximum speed that the car won't exceed. This is because things that you initially ignored begin to matter, like air resistance, friction in the engine, etc.

In Figure 5.2, I plot two theories of car speed. The "wrong" theory shows that the maximum speed is proportional to the force on the gas pedal. This theory predicts that any speed is possible in a car, one just needs to press the gas pedal harder. The "right" theory shows

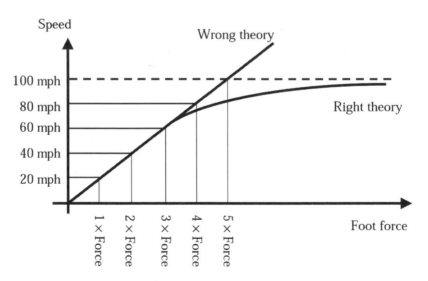

Figure 5.2 Example of how a behavior that seems reasonable at low energy becomes unreasonable at high energy. In this case, how fast one goes in an automobile depends on how hard one steps on the gas pedal. However, above a certain pedal pressure, one cannot go any faster.

a maximum speed of 100 mph, regardless of how hard you press the pedal. But the most important thing to notice is that below tripling the force on the pedal (compared to the 20 mph force), the "wrong" theory works just fine. It's only at high pedal pressures (or particle energy, to return to particle physics) that the two theories differ. So the Higgs theory performs the critical task of keeping the theory from making silly predictions. This fact shows that without the Higgs idea, the theory was incomplete. Since the Higgs field doesn't come naturally from the theory, but is imposed later, this also shows that there must be something more to the theoretical story, because in a complete theory, the Higgs field would emerge naturally. This is a worry for theorists or perhaps an opportunity for a sufficiently clever person to figure out how to integrate the various components of the theory in a more natural way. We'll return to this again in Chapter 8.

Higgs envisioned his field as one that permeates the cosmos and interacts with particles, giving each in turn their mass. This breaks the symmetry of the more "natural" world, in which all particles are massless. Broken symmetry sounds like just a physics buzzword and so I'd like to devote a little discussion to it and show how it relates to the creation of mass.

Higgs by Analogy I

At the entrance to Fermilab, there exists a giant sculpture consisting of three steel beams formed into large arcs that start at the ground and swoop skyward to be joined high overhead. From any angle you view the sculpture, it's an ungainly structure. Each arc is a different height and length and nothing seems to fit together in an aesthetic manner, even the color differs when it is viewed from different directions. However, from a single vantage point accessible to the pedestrian, the three arcs connect together in a way that is pleasing to the eye. Looking from below, the three arcs come together 120° apart, equally spaced, with all arcs looking identical. Robert Wilson, the first director of Fermilab, as well as a passionate artist, built the sculpture

Figure 5.3 The sculpture "Broken Symmetry" located at the west entrance to Fermilab. From all vantage points, the sculpture exhibits no symmetry except for the single vantage point accessible to the pedestrian of viewing it from below. The sculpture represents similar behavior seen in many physical systems. (Figure courtesy of Fermilab.)

and entitled it "Broken Symmetry." The name comes from the fact that the beams only appear identical from a single select vantage point, while appearing dissimilar any other way you look at it. Two views of this sculpture are given in Figure 5.3.

Physical phenomena often exhibit similar features. Things that appear to be very different, under the right circumstances are shown to be one and the same. An obvious example is ice and steam, which have vastly different physical properties, yet are the same substance. In particle physics, a similar thing occurs. In order to unify the electromagnetic and weak forces, it was necessary to postulate four force-carrying bosons, unfortunately massless … a theoretically comfortable idea … but one that is ruled out by observation. As we have said before, the Higgs mechanism (a fancy word for idea), combined these theoretical constructs in technical ways, yielding the four physically-observed electroweak force carrying bosons, the massless photon and the massive W and Z bosons. The symmetry, i.e. the identical nature, of the massless bosons of the simpler original theory was shown to be related, via the Higgs mechanism, into the physically massive (and observed!) bosons. One might think of a similar situation that isn't

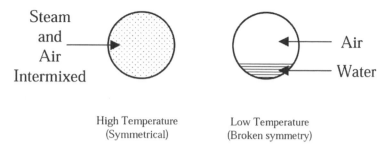

Figure 5.4 The behavior of a particular system under two different energy situations. In the high energy (i.e. high temperature) situation, air and water are mixed uniformly, so that there is no special spot within the flask. At lower energy (temperature), water condenses at the bottom of the flask, with air residing above. At low temperature, the symmetry observed at high energy is not apparent.

quite so abstract. Suppose that you had a hollow metal sphere with a small amount of water in it, a situation depicted in Figure 5.4. If you heat the sphere enough to make the water convert entirely into steam, you would see that within the sphere the environment was every-where identical, with air and water vapor everywhere intermixed in equal proportions. Nowhere in the sphere is there a concentration of water and a lack of air. The situation is everywhere symmetrical. Now consider what occurs if one lowers the temperature below the boiling point, so that water condenses at the bottom of the sphere, while air sits on the top. With the loss of uniformity, we can say that the sym-metry is broken.

The Higgs field has a unique property, as compared to other fields in the universe, e.g. gravitational and electromagnetic fields, etc. This property has to do with how the Higgs field alters the energy in the universe. Generally the existence of a field adds energy to space. The stronger the field, the more energy stored. It's kind of like a slingshot. The further you pull it back, the tighter the slingshot (and one might say, the stronger the "slingshot field.") The further the slingshot is stretched, the more energy stored in the "slingshot field" and farther and faster one can fling a pebble.

The problem is that the universe doesn't like concentrated energy and eventually, be it milliseconds or eons, the universe tries to put everything in the configuration of minimum energy. A boulder on a hillside will eventually make its way to the bottom of the valley... heck, even mountains themselves eventually wear down. If you stretch a spring and hold it, energy is stored in the spring, but eventually the molecules of metal in the spring will move to relieve the tension and the spring will get stretched out. You can try this with a rubber band. Hang it over a hook with a weight that will stretch it, but not break it. After several weeks, the rubber band will be stretched out and will not return to its original size.

Higgs the Way Scientists Think

So the universe doesn't like concentrated energy and force fields (like the electromagnetic and gravitational fields) make energy present, wherever they are. One can reduce the energy at a place by making the field to be zero. If there's no gravity at a spot, that is how one gets the minimum gravitational energy (i.e. none) at that spot. The bottom line is that for most fields, the way to minimize the energy from the field is to have no field at all. Sounds simple, right?

We can solidify our idea by looking at Figure 5.5. On the horizontal axis is the strength of the gravity field and on the vertical axis is the energy stored in the gravity field. The curve shows how much energy is stored in the field for different strengths of the field. You figure out the amount of energy by picking a particular field strength (G_1 in our example) and go straight up until you hit the curve. You then go horizontally until you hit the axis. That tells you how much energy exists for that field strength. The question that one asks is "What is the field strength for which the energy is smallest?" This occurs when you are at the lowest spot on the curve (because that is the lowest energy possible). We see that this occurs at the spot marked G_2, which is when the field is zero. No field means no energy.

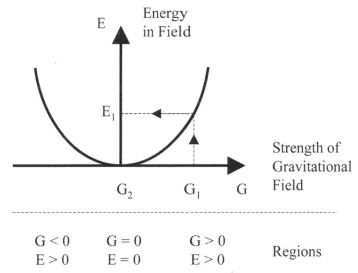

Figure 5.5 Shape of the gravitational potential. As the strength of the gravitational field increases, the energy stored in the field also increases. Only at zero gravitational field strength is there no energy stored in the gravitational field.

The Higgs field also adds energy to the universe everywhere it exists. The question becomes "What is the strength of the Higgs field that adds the minimum energy?" Is it zero too, as in the earlier example? To see this point, we must make a small detour into mathematics. It's a small detour and the gist of the point can be understood by the discussion of the pictures that follow. Our mathematically anxious reader is invited to rejoin us after the next paragraph.

The Higgs field can be inserted into the theory via the usual sorts of mathematical approaches used by my theorist colleagues. They then mercilessly beat on the equations until they arrive at THE ANSWER (this would be a good place for a sound track, because that last sentence really merits some dramatic music). We'll trust our theoretically-inclined comrades and look at the final answer. In the answer, we can write the strength of the Higgs field as H. If H is a big number, the Higgs field is strong. Similarly, if H is a small number, the field is weak. If H is zero, then there is no Higgs field. We will also write the energy caused by the Higgs field using the symbol E. If E is a large number, the Higgs energy is large. We want to find the minimum energy, which is what the universe will do. Then we'll see if the Higgs field is non-zero at that point. So, without further ado, the equation that relates energy and Higgs field strength is (a drum roll please, maestro)

$$E = m^2 H^2 + a H^4$$

See, that wasn't so painful, was it? It's a pretty innocuous equation, not too different from one you see in high school algebra. Let's not worry about the physical significance of a, except to say that a must always be a positive number. It doesn't matter what the actual number is. The variable m is usually taken to be related to the mass of the Higgs boson, the particle that must exist if the Higgs field exists. We'll come back to that particle later. So if m^2 is greater than zero (which makes mathematical sense), one can draw the curve relating the Higgs field strength and energy and one finds that no matter what, the minimum energy occurs when the Higgs field is zero. Such a plot is shown in Figure 5.6a and looks remarkably like Figure 5.5, although it's different in detail. However, if one takes the seemingly silly approach of letting m^2 be negative, then the shape of the curve that relates the Higgs field strength and energy changes. Rather than always increasing from zero, the energy first decreases before increasing. This behavior is shown in Figure 5.6b.

So if the m^2 term is allowed to be negative in the equation, one sees that the minimum energy occurs when the Higgs field is not zero. Literally, this means that if the Higgs field strength actually were zero, there would be more energy in the universe than there is now. Since the universe always eventually settles to the lowest energy state, this means that there must be a non-zero Higgs field that permeates the universe, but only if m^2 is negative ($m^2 < 0$).

So we now enter the final stretch of understanding the Higgs field. You must be asking the question, since we know that if you

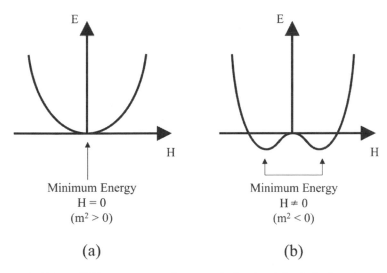

Figure 5.6 Two different types of energy behavior. The left plot shows a situation in which the minimum energy stored in the Higgs field occurs when the Higgs field strength is zero. In the right-hand plot, the minimum energy stored in the Higgs field occurs when the field strength is something other than zero. This is the essence of the Higgs mechanism.

square any number, the result is positive, how can m^2 be negative? At this point, I kind of hope that your Mom is whispering in your ear "Hush child … don't ask the man impertinent questions." The fact is, I don't know. Nobody really does, although there are a good number of theoretical ideas explaining why this should be true. Unfortunately, none of the ideas have been experimentally proven (of course, neither has the Higgs idea). We'll discuss some of the ideas later in Chapter 8. At this point, we should understand that the reason that we write the "m^2" term in the above equation is because of other, similar theories. In its most general form, one could use just an arbitrary variable there, say k. In that case, there is no restriction on whether k should be positive or negative. This is a much safer way to cast the problem, but because much of the literature uses the "m^2" notation, I do so too. But don't take it too seriously and don't let the "how can a 'squared' number be negative?" bother you too much.

The really neat thing about the Higgs idea is the fact that m^2 depends on the energy in the environment (say in a collision). At high enough collision energy, m^2 is positive and thus the minimum energy added by the Higgs field occurs when the Higgs field strength is zero. As the energy of the environment decreases, m^2 gets smaller, eventually becoming zero. As the energy of the environment further decreases, the m^2 term becomes negative and, all of a sudden, the minimum Higgs energy occurs for a non-zero Higgs field strength. At this point the Higgs field "turns on." This behavior is illustrated in Figure 5.7.

It seems kind of odd that there is a "magic energy" at which the rules completely change, but you actually know of a similar phenomenon. Consider a glass of water at room temperature. If you put the glass in a freezer, what happens is you reduce the temperature (i.e. energy) of the water. Nothing much happens until you get to 32°F, at which point the water freezes. Below that temperature, you again simply reduce the temperature of the ice (again by dropping its energy). But something very important occurs at the magic temperature. The water freezes. A material which is a liquid, with all of the properties of a liquid (sloshes around, takes the shape of the container in which it's placed, etc.), becomes solid, with vastly different physical properties. Thus there is a "magic energy" (or temperature) at which the observed behavior of matter dramatically changes, above and below which, the rules only change in a somewhat boring manner. The Higgs mechanism is in many ways analogous.

So the Higgs idea can be summarized by the following. First, you have to assume a Higgs field exists. Secondly, you need to assume that the Higgs field follows the equation listed above. Finally, one needs the m^2 term to be positive for a high collision energy and negative for low ones, implying that the Higgs field is zero for high energy and non-zero for low. Because particles gain their mass through interacting with the Higgs field, this leads us to the inevitable conclusion that at high energies (Higgs field is zero), particles have no mass (as there is no Higgs field with which to interact) and at low energy, the

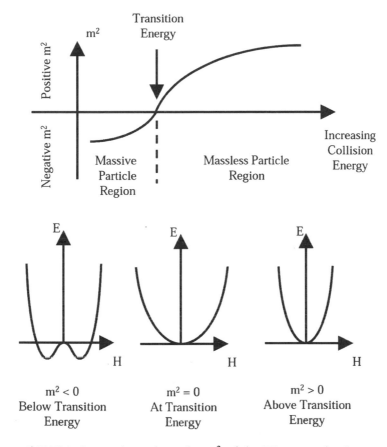

Figure 5.7 This figure shows how the m^2 of the Higgs mechanism can vary as a function of collision energy. If so, the Higgs field strength that corresponds to the minimum energy goes from zero field strength at high energy to not zero at low energy. Because particles gain mass through their interaction with the Higgs field, at high energy all particles are massless, as that is the case when both the minimum energy and the Higgs field is zero.

presence of a non-zero Higgs field gives the particles their mass. That's all there is to it.

We've said many times that the particles gain their mass by interacting with the non-zero Higgs field. Let's talk a bit about what this means. We know that an object gains weight by interacting with the gravitational field. Take away gravity and the weight is gone. With the

Higgs field, it's mass. No Higgs field means no mass. So why do different particles have different mass? This is because they interact with the Higgs field to differing degrees. The super-massive top quark interacts very strongly with the Higgs field, while the massless photon interacts with the Higgs field not at all.

Higgs by Analogy II

One can envision how the particles interact by thinking about water and how objects pass through it. When you go under water, you are immersed in the water, surrounded by it, much like the Higgs field. Different creatures can pass through the water with different degrees of difficulty. A shark, a supremely streamlined fish, can slip through the water with the greatest of ease and can reach very high speeds. In contrast, my Uncle Eddy, a retired sumo wrestler and no stranger to donuts, moves only very slowly through the water. Thus we can say that the shark interacts very weakly with the water, while Uncle Eddy's interaction with the water is very strong.

Another nice analogy is an experiment that you can do while driving your car. Go out driving on a nice open highway, where you can go as fast as our friendly police officers will allow. Make sure there are no other cars around and stick your arm out the window. Open your hand so it's flat and looks like a guy just about to make a karate chop. Rotate your hand so the palm is facing the ground and feel the force of the wind on your hand. We'll call when you have your hand in this position the "down position." Now rotate your hand so your thumb is pointing towards the sky and your palm faces the oncoming wind. We'll call this the "wind position." Note how the force of the wind on your arm has greatly increased. If we take the wind as our analogy to the Higgs field, we see that the down-position hand interacts weakly with the wind, while the wind-position hand interacts strongly.

In 1993, the then British Science Minister, William Waldegrave, announced a contest, with a prize of nothing less than a bottle of good champagne. With a prize of such considerable desirability on the line, physicists crowded around to hear the rules, which were simple. Each person was to submit an essay of no longer than one side of a single piece of paper that would explain the Higgs mechanism in plain English. Many entries were received, each satisfying the criteria of lucidity, clarity and brevity to various degrees. When the contest was finished and the contributions were judged, 5 entries were considered to have been noteworthy essays. The winners of the contest were Mary and Ian Butterworth of Imperial College, London, Doris and Vigdor Teplitz of Southern Methodist University, Roger Cashmore of the University of Oxford, David Miller of University College, London, Tom Kibble of Imperial College, London and Simon Hands of the CERN Theory Division. While all the winning essays were fine examples of clarity, the essay that has garnered the most popularity is the one by Miller, who offered a marvelous analogy of how the Higgs field generated mass and how the Higgs boson comes into existence. The original analogy used Margaret Thatcher, the first female British Prime Minister, as the celebrity in the story, but my colleagues at CERN changed the identity of the celebrity to Albert Einstein, as ol' Al is much more interesting than any prime minister (although Winston Churchill was a pretty colorful guy too). Personally, I think Peter Higgs should get the credit and so I use him in my subsequent discussion.

At any rate, the analogy (depicted in Figure 5.8) goes something like this. Suppose that there's a large room full of physicists at a cocktail party. Just for fun, let's say it's a meeting of all of the collaborators of the DØ and CDF experiments and the four LEP experiments (Aleph, Delphi, L3 and OPAL), which will give us a couple of thousand people. (If you happen to attend the party and notice a devastatingly good-looking guy of obvious brilliance and good taste, you might ask him if he'll autograph your copy of this book. He probably won't, but he'll help you find me ...) This large group of people is dispersed uniformly across the room, which is crowded but still allows the partygoers to move basically freely. The crowd represents the Higgs field.

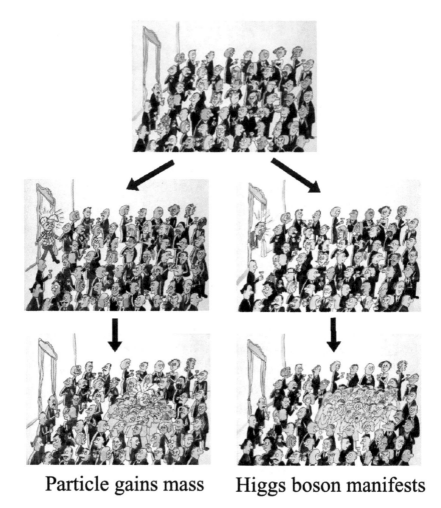

Particle gains mass   Higgs boson manifests

Figure 5.8 An analogy of how the Higgs mechanism can affect the world. In the left-hand side, as the celebrity passes through the room, he interacts with the crowd, thus slowing his passage and effectively giving him a mass. On the right-hand side, no celebrity is present, but the rumor of their immediate arrival causes a clump in the crowd. This clump is analogous to the Higgs boson. (Figure courtesy of CERN.)

At the doorway of the room appears our celebrity physicist who, for obvious reasons, will be played by Peter Higgs. Peter wants to go to the bar and get a drink. He surveys the room, sees the bar and the density of the crowd and estimates how long it will take for him to get from the door to the bar. With a spring in his step, he starts across the room. At this point, the partygoers notice that Higgs has entered the room. These people crowd around Higgs to chat about their most recent attempts to find his boson. Through his interaction with the crowd, Higgs' progress is slowed to a crawl. Meanwhile, another physicist enters the room from the same door as Higgs and, because he isn't famous, crosses the room relatively quickly to get his drink. Both Higgs and our young unknown colleague interact with the crowd to different degrees, just as different particles interact with the Higgs field differently.

But what about the Higgs boson? How does this relate to our analogy? Say Peter has not yet entered the room, but someone finds out that he will soon. They enter the room and tell the people near the door that Peter will be arriving soon. These people will tell their neighbors and soon others will see the knot of excited people and wonder what the hubbub is all about. This clump of people, within the more uniformly distributed people of the crowd, represents the Higgs boson. Further, as some people learn of the rumor and drift away from the clump, others will join. In this way, the clump will work its way across the room, representing the motion of the Higgs boson.

This analogy, while imperfect, really does relate rather well how the Higgs field generates mass and how the Higgs boson comes into existence. The other essays of the contest each conveyed technical points of Higgs physics, but none conveyed the fundamental facts at such a transparently clear level.

We've been talking about the Higgs field as if it were a real phenomenon. If it is real, we should be able to detect it. In order to even begin to verify a theory, a few things must be true. First, the theory must agree with existing observations, in this case, the fact that particles have mass and that the best theory for describing the behavior of these particles requires that the particles be massless. The second thing that must occur is that the theory predicts some new phenomenon or phenomena, not yet observed. When this phenomenon is observed, the theory gains credibility. Failure to observe the predicted phenomenon proves that the theory is flawed and requires modification or, in the worst case, kills the theory entirely. This is the scientific method.

So what is the new phenomenon predicted by Higgs theory? What is predicted is a new particle...the Higgs boson. The Higgs boson can be thought of as a localized vibration in the Higgs field, somewhat like a grain of sand in the beach that represents the Higgs field. Just as the photon is the particle that makes up the electromagnetic field, the Higgs boson is the carrier particle of the Higgs field.

The Higgs boson has some very specific predicted properties. The particle is fundamental, which means it is not made of even smaller particles. It is electrically neutral, i.e. it has no electric charge. It has an as yet undetermined (but non-zero) mass. It is a scalar particle, which means that it has no quantum mechanical spin. This is a unique property, as no other fundamental particle thus far discovered has zero quantum mechanical spin. In contrast, the electron can be thought of as a little spinning top. The only particle of which we have spoken that has a spin of zero is the pion, which is a meson and has a structure and thus isn't fundamental. The axis of rotation defines a direction. While the vagaries of quantum mechanics greatly complicate the issue, we can measure the direction of the spin axis, and so for an electron, one direction is "special" or different from all other directions. This property is true of all other fundamental particles except for the Higgs boson. Because the Higgs is a scalar (i.e. spinless) particle, all directions look the same. This property is one that a number of physicists find to be troubling.

Finally, the general interaction of the Higgs boson is known. It interacts and gives particles their mass. It does this by interacting with different particles to different degrees. One might even say that the Higgs boson is buddies with the massive top quark, while its interactions with the very light electron are distinctly cool. The Higgs boson doesn't interact with the massless photon at all. (I'm told that they had a falling out when they were young...something about a cute little fermion...)

Desperately Seeking Higgs

So how do we expect to find this elusive particle? We'll get to the modern search techniques in a while, but first I'd like to discuss a bit of the history of the search for the still-undiscovered (and possibly non-existent) Higgs boson. Since there was little guidance as to the mass of the Higgs boson, some early experiments looked for particles with masses of some few tens of an MeV (in contrast, we now know that the mass of the Higgs boson exceeds 100 GeV, fully 10,000 times more massive than those early searches). [Note: Since we're going to talk about such a wide range of energy, you may want to quickly review Appendix B.] We know that since the Higgs boson generates mass, it will couple primarily to the heaviest particle that it can. Since the Higgs boson is unstable, it quickly decays into pairs of objects. The pairs can be pairs of charged leptons (say an e^+e^- or $\mu^+\mu^-$), pairs of quarks (say $u\bar{u}$, $d\bar{d}$, etc.) or even pairs of massive bosons (say ZZ or W^+W^-). Exactly which particles into which the Higgs boson can decay is determined by their mass. A Higgs boson cannot decay into pairs of any particle whose mass is more than 1/2 the mass of a Higgs boson.

Let's illustrate this point with a simple example. Let's say that the Higgs boson has a mass of 10 (we aren't worried about the units here). This Higgs can decay into two particles, each of mass 5. If the Higgs boson decayed into two particles of mass 4.5, that's OK too. The two particles have a combined mass of 9, with 1 energy unit left over. Since energy and mass are interchangeable (remember $E = mc^2$?), the remaining unit manifests itself as energy. A common form of energy is motion energy, so these two 4.5 mass objects must be moving to make the books come out right. The trick is to remember that the energy before the decay must be the same as the energy after. Let's now consider the possibility that this hypothetical Higgs decays into two objects, each with a mass of 5.5. Since the combined mass of these two objects is 11, which exceeds the initial Higgs mass of 10, such a decay is not allowed. In our example, the Higgs boson cannot decay into any object with a mass greater than $10/2 = 5$ units.

If we look at Tables 3.3 and 4.2, we see the mass of the various particles into which the Higgs can decay. Ignoring the massless photon and neutrinos (more on that in Chapter 7), which do not interact with the Higgs boson, and restricting our discussion to the charged leptons, the least massive particle is the electron, with its mass of 0.511 MeV, followed by the muon at 106 MeV. In order to decay into an electron-positron pair (e^+e^-), the Higgs boson must have a mass of at least ($2 \times 0.511 = 1.022$ MeV). Similarly, in order for decays into muons to be allowed, the Higgs boson must have a mass exceeding ($2 \times 106 = 212$ MeV). Above 212 MeV, the Higgs boson is allowed to decay into both pairs of electrons or muons, but it will generally decay into muons, because the Higgs will prefer to decay into the most massive particles allowed. Because we know that the Higgs boson has a mass exceeding 100 GeV (100,000 MeV), we can see that decays into pairs of bottom quarks is preferred (although if the mass exceeds 160 GeV, pairs of W bosons will be preferred).

But in the early 1980s, the question of the mass of the Higgs boson was still one of great conjecture. Experiments were done in which K-mesons (K) were allowed to decay and people searched for a pi-meson (π) and a Higgs boson (H) in the debris (K $\rightarrow$ πH). Studies of this type ruled out the possibility that the Higgs boson was less than 212 MeV or so. With the turn-on of the LEP accelerator in 1989, additional results increased the lowest allowed mass of the Higgs boson to 24 GeV (24,000 MeV) and later results increased the lower limit to 65 GeV.

In 1995, the LEP accelerator at CERN undertook a significant upgrade. Originally built to carefully study the Z boson, which required the collision energy be carefully tuned to 91 GeV, the experimenters decided to increase the energy of the accelerator in a series of steps until the collision energy reached 209 GeV. With this much higher energy, a whole new series of measurements were possible. Each measurement in principle could generate ever-heavier Higgs bosons, but at the worst they would raise the limit of the lowest possible Higgs boson mass. The whole process, while interesting, was business as usual until the summer of 2000. Then things started to heat up.

The LEP accelerator was scheduled to be turned off at the end of September 2000. The accelerator was to be dismantled and replaced by an entirely new accelerator, the Large Hadron Collider or LHC. The decision was made to increase the energy to beyond the maximum design energy. (I'm somehow reminded of Kirk telling Engineering that he needs 110% power to the warp drive. The CERN accelerator scientists and engineers, like Scotty, managed to deliver.) The idea was that, given the fact that the LEP accelerator was scheduled to be turned off soon anyway, they might as well run it into the ground. With the maximum energy and maximum number of particles in the beams, they just might be able to discover some new physics (like the Higgs boson, for example). If this effort destroyed the accelerator, then nothing much was lost.

In April 2000, the LEP accelerator turned on with a maximum energy of 209 GeV, more than twice its original design energy. Four superb detectors were spaced around the ring, with the names: Aleph, Delphi, L3 and Opal. During the months of June-September, the Aleph collaboration observed four collisions that had characteristics that were consistent with being the creation of a Higgs boson with a mass of 115 GeV. The other experiments searched as well, with little success, although the Delphi collaboration did observe a collision that looked a little like a Higgs boson, but they weren't sure that it wasn't just an expected and more mundane collision type. As the end of September neared, the experiments sifted and re-sifted their data, looking for the "smoking gun." Finally, they petitioned the CERN laboratory leadership for an extension of the run. The CERN directorate was nervous, as any extension of the LEP run meant a delay in the beginning of the construction of a new accelerator, the Large Hadron Collider or LHC. Changing construction contracts can have significant fiscal consequences and so the leadership compromised and offered a one-month extension, not much but something. The first weeks of the month of October yielded no events with the hoped-for characteristics, until October 16, when the L3 experiment observed an event every bit as good as the earlier Aleph events. The observation of a good event by a different experiment gave physicists confidence that they might actually be seeing the first signature of the Higgs boson.

Naturally, with such a potential discovery on the line and the end of October looming ever closer, the LEP scientists tried very hard to make their case for yet another extension. With a fervor normally associated with the National Rifle Association, they organized a "grassroots" effort among the world's particle physicists, put together a petition and did everything they could think of to convince the CERN directorate to extend the run. After hearing the sides, reviewing the evidence and considering the consequences, the CERN Director General, Luciano Maiani decided to *not* extend the LEP run and, on November 2 at 8 AM, the LEP accelerator was turned off, nominally for good. Undaunted, the LEP experimenters realized that the final decision to dismantle the LEP accelerator and experiments would be undertaken by the CERN Council (basically a board of overseers) on December 15, 2000. Until the accelerators and experiments were dismantled, there remained hope. The LEP experimenters vowed to have final results available in one month, far faster than is usual. Maybe the CERN Council would overturn the Director's decision. However, in the end, the CERN Council ratified the Director's decision. The Council weighed the probability that the data really indicated the discovery of the Higgs boson against the very real fiscal penalty of several million dollars that would be incurred if the LHC construction were delayed. Money won.

So now, in the fullness of time, did the Director and the Council make the right decision? Luckily, the answer appears to be yes. By December 2001, the LEP experimenters announced that they had made an error in their earlier estimate of the background. Background are those things that look like what you're looking for, but aren't. It's as if someone took a handful of cubic zirconia and tossed them on the floor, interspersed with a few diamonds. The cubic zirconia look like the real thing but aren't. The LEP experimenters had underestimated the number of other more mundane collisions that looked a lot like Higgs. When they did their calculations more carefully, they realized that the data wasn't as suggestive as they had originally thought. So the Director made the right choice. No one should fault the experimenters for their error, as they were under an inhuman time pressure and they did all that could reasonably be expected (and more!) But it does underscore the need for double-checking and reflection when a big discovery is on the line. It's also important to note that the error was uncovered by the experiments themselves. When they realized the error, they announced that as well. This is a nice example of the self-correcting nature of honest scientific research.

So if the CERN experiments didn't discover the Higgs boson, what did they accomplish? We can now say with confidence that the Higgs boson, if it does exist, must have a mass exceeding 115 GeV. Such a result is very useful, as telling others where *not* to look saves everyone time. We also know other things about the Higgs boson. If it were much more massive, a problem would occur. Since Higgs bosons interact with particles with a large mass, Higgs bosons would also interact with themselves (as they have mass too). This "self interaction" puts an upper limit on the mass of the Higgs boson as less than about 500–1000 GeV. Above that, the theory would fall apart, or at least the idea of the Higgs boson being structureless would be suspicious. So the allowed range of Higgs boson masses is in the range of 115–500 GeV. If a theoretical idea called supersymmetry is true, this puts a constraint on the Higgs boson and makes the maximum allowed Higgs boson mass much less, say about 200 GeV; in fact it predicts the most likely Higgs boson mass as between about 100 and 130 GeV. Supersymmetry is an unproven idea of which we will speak a great deal more in Chapter 8. A prudent reader will take these predictions with a grain of salt. The best prediction we have for the Higgs boson mass (without invoking new theoretical ideas) is in the range of 115–150 GeV (once all measurements are taken into account). But time will tell.

The Current Search Story

So with the LEP accelerator dismantled and the LHC accelerator many years from completion, what is happening in the meantime? In March of 2001, the Fermilab Tevatron turned on for its attempt to capture the elusive Higgs boson. Two large experiments, DØ and CDF, are currently on the hunt, each hoping to find it first. Like the earlier race for the top quark, the competition is intense and the stakes large. In the next few pages, I'll give you a flavor of what is needed for an experiment to discover the Higgs boson and how we'll go about it.

If the Higgs boson has a mass between 115 and 160 GeV, the heaviest particle into which it can decay is the bottom quark, with its mass of 4.5 GeV. Above 160 GeV, the Higgs boson could decay into two W bosons (each with a mass of 80 GeV). Since the difficulty in detecting a particle goes up with the particle's mass, physicists first must explore the 115–160 GeV range, in order to find the Higgs boson or show that its mass is even greater. Such a choice helps when one designs an experiment. If the most likely decay mode of the Higgs boson is into a bottom and antibottom quark pair (b$\bar{\text{b}}$), it is clearly crucial that the detector have the capability to detect b-quark pairs with good efficiency and accuracy. To this end, both DØ and CDF have built sophisticated silicon vertex detectors (discussed in Chapter 6). The idea is that both bottom quarks can only decay via the weak force and thus they live a long time. Consequently, the b-quark pairs up with another more mundane quark (say an up or down) and forms a B-meson. The B-meson can travel great distances, perhaps a millimeter or even more. (Yes, a millimeter is very small, but you can actually see something of that size. It's huge in comparison to the more characteristic size of particle interactions; say a few quadrillionths (10^{-15}) of a meter.) The silicon vertex detectors can identify events in which a B-meson decays, far from the place where the original collision occurred. We see a comparison of a jet formed from b-quarks and other more ordinary quarks in Figure 5.9. b-quarks, like all quarks, form jets; but with other quarks, all particles

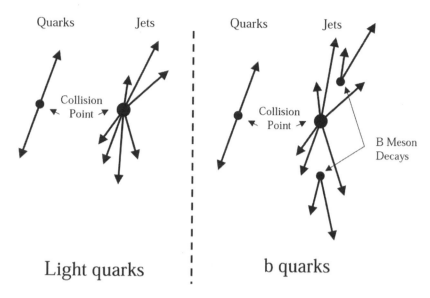

Figure 5.9 This figure illustrates one method whereby one may find the b-quark jets that signify the existence of a Higgs boson. Because b-quarks can decay individually only through the weak force, they live a long time. Consequently, mesons carrying b-quarks travel great distances (i.e. greater than a millimeter) before decaying. Seeing a particle decay well separated from the primary interaction is one way to identify a b-quark.

appear as if they come from a common point, the collision point. In contrast, a B-meson travels a distance from the collision point before it decays. We look for that signature as a strong indication that a b-quark was present in the event.

So our strategy is clear, right? (Of course, you know that when I ask such a question, there must be a catch.) A gluon in the proton combines with a gluon in the antiproton to make a Higgs boson, which decays in turn into a $b\bar{b}$ pair ($gg \to H \to b\bar{b}$, shown in Figure 5.10a). You then look for events in which two b-quarks are produced and look for something that clearly indicates a Higgs boson. At this point, a small diversion is in order. The reader who is really on the ball should be screaming right about now. We know that the Higgs boson only interacts with massive particles and we recall from Chapter 4 that gluons are massless. Thus two gluons creating a Higgs boson should be impossible. This is strictly true, but the explanation is complicated. Rather than muddying our discussion, we pass over this point. The interested reader should peruse the more detailed explanation in Appendix E.

While we see that one can try to find the Higgs boson by looking for events in which a b and anti-b quark were created, the real situation is complicated by the fact that one can create pairs of b quarks far more frequently via a much less interesting process. Instead of the two gluons from the beam particles combining to form a Higgs boson, they form an intermediary gluon, which then decays into a b$\bar{\text{b}}$ pair. This is a background event, one that looks like a Higgs boson event, but isn't. Such an event is shown in Figure 5.10b.

The existence of such a background isn't necessarily the kiss of death, especially if the background is rare enough. Unfortunately, the physics background (i.e. ordinary collisions containing b quarks) is thousands of times more likely than the desired production of Higgs bosons, even when detector performance is not considered. Even worse, the detector can mistake the much more prevalent, lower energy and common collisions (i.e. ones which the detector incorrectly

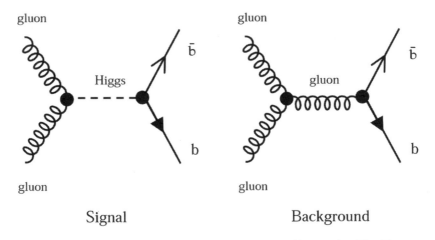

Figure 5.10 There are several ways to make pairs of b-quarks. The Feynman diagram on the right exceeds the desired one on the left by a factor of some few thousands. Thus, finding Higgs bosons in this way is extremely challenging and will not be pursued for the foreseeable future.

concludes contains b quarks, even though they don't). Because of this, the background becomes tens of thousands of times greater than the signal. Even nastier, the background is basically indistinguishable from the signal. So we're out of luck. Well not entirely. Bloodied, but unbeaten, physicists returned to the drawing board and rethought their options. Eventually, it became clear that they could look for an even rarer type of collision, one in which quarks from the proton and antiproton combined to form a peculiar type of matter (called a virtual particle), which would then decay either into a *W* boson and a Higgs boson or a *Z* boson and a Higgs boson. This peculiar state of matter looks much like a conventional *W* or *Z* boson, except that it is too massive. Such an outlandish idea is allowed by the bizarre rules of quantum mechanics, which allows for a particle to exist with entirely the wrong mass, as long as it lasts for only a short period (see Appendix D for a refresher). We indicate one of these weird *W* or *Z* bosons by giving them a "*" for a superscript. These possibilities for making Higgs bosons are shown in Figure 5.11.

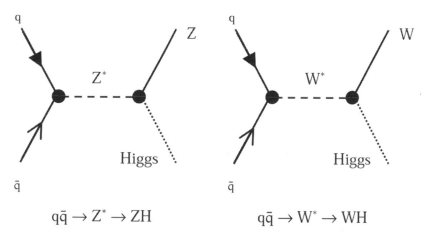

$$q\bar{q} \rightarrow Z^* \rightarrow ZH \qquad q\bar{q} \rightarrow W^* \rightarrow WH$$

Figure 5.11 If a Higgs boson is created in association with a weak boson, it has a distinct experimental signature. Such an experimental signature will only rarely be made by more common physics processes. Associated production is the primary method whereby we intend to search for Higgs bosons because of the small amount of background.

This "associated" production (so-called because the Higgs boson is produced in association with a massive weak boson) is more rare than the more common possibility (gg → H → b$\overline{\text{b}}$), but it has one noteworthy advantage. The events that resemble this process, but come from more mundane types of interactions (i.e. background) are extremely rare.

Using associated production means that in any particular event of interest, we must find both a Higgs boson and either a W or Z boson. Let's use for an example the case of a Higgs boson produced at the same time as a Z boson. We recall that finding a Z boson is easier if we look for cases in which it decays either into electron-positron or muon-antimuon pairs. The preferred Higgs boson decay is still into b$\overline{\text{b}}$ pairs. Thus events in which we find two b-quark jets and two muons or two electrons (for instance b$\overline{\text{b}}$e$^+$e$^-$ or b$\overline{\text{b}}\mu^+\mu^-$) will be given special scrutiny, especially if it is determined that the charged leptons come from the decay of a Z boson.

So just how tough will it be to find the Higgs boson at the Fermilab Tevatron? The probability of creating the Higgs boson is estimated to be about 10–20 times less than the probability of creating top quarks (which were discovered in 1995). However, this new set of experiments will generate 10–20 times as much data as last time. So it's a wash [(1/10 the probability) × (10 times the data) ~ 1]. Except the background in a Higgs boson search is somewhat larger than was present in the searches for the top quarks. A tremendous irony is presented by the fact that the top quarks that were so difficult to find during the last data-taking period contribute significantly to the background for Higgs boson searches. So the search will surely be challenging.

So what are our best predictions? If the LEP evidence turns out to be a fluke, as it now appears, DØ and CDF can rule out the existence of the Higgs boson with a mass of 115 GeV by about 2004. On the other hand, if LEP was actually beginning to see something, the Tevatron expects to have data with similarly suggestive qualities by about 2004–2005. It will take until about 2007 for enough data to be accumulated to announce the discovery, uncomfortably close in time to the turn on of CERN's Large Hadron Collider (LHC). When the LHC begins its collisions, with an energy fully 7 times higher than the Fermilab Tevatron, DØ and CDF will rapidly become obsolete for this particular particle search. So the two experiments are under considerable time pressure. We have just a few scant years to find the Higgs or, in the words of Maxwell Smart, "Missed it by *that* much, Chief."

Finally, what if the Higgs boson exists, but is much heavier than the data from LEP would suggest? The answer to that is technical, but DØ and CDF would eventually have something to say about the existence of the Higgs boson up to a mass of about 190 GeV. Beyond that is the rightful domain of the LHC, which could easily extend the search to 600 GeV. If even the LHC fails to find the Higgs boson, then something is very wrong with our theory. While that possibility is very real, our best knowledge suggests that Fermilab has a good chance to find the Higgs boson (or something very much like it) in the latter half of this decade. Keep an eye on the news or Fermilab's web page, which will always contain the latest information.

Of course, while we've spent a lot of time discussing the Higgs field and the search for the resultant Higgs boson, one might ask just what direct evidence has been observed to prove the correctness of the Higgs theory. The answer, as of late 2003, is none, nada, zilch, nothing. There has been *no* direct evidence for the existence of the Higgs boson. So why all the fuss? Well, we do know some things. The electroweak theory works and has been tested to exquisite precision. The theory works so well that for the last few years, presentations of results from the LEP experiments grew somewhat boring, as calculations and experimental results agreed so well. While this is a triumph for the Standard Model of particle physics, there was no inconsistency between data and theory to baffle physicists and cause us to rethink our theories.

We recall that the electroweak theory predicts massless force-carrying bosons which, through their interactions with the Higgs field, gain their observed masses. Since our experiments and theory agree so extremely well, isn't that proof that the Higgs field exists? The answer is, most emphatically, no. What appears to be true is that the electroweak theory is an accurate theory and that *something* breaks the symmetry between the behavior of electromagnetism and the weak force. But it doesn't have to be the Higgs field. What appears to be true is that there is a mechanism of electroweak symmetry breaking and that whatever it turns out to be, it will have to act much like Higgs' original idea. But it could be different.

Of course, the question is, "How different?" The Higgs boson, as currently predicted is an "electrically neutral, massive fundamental scalar" particle. Neutral and massive mean what you think. Fundamental means that it contains no internal structure and scalar means that it has no quantum mechanical spin. The problem is that while we know of other scalar particles, they aren't fundamental and the fundamental particles of which we know aren't scalars. So the Higgs boson would have unique properties. This absolutely could be true, but it's a little fishy.

So, if not the Higgs boson, then what? There are a number of other theories that explain the electroweak symmetry breaking, each with their own theoretical problems. The most popular of the competitor theories is called Technicolor. Technicolor postulates the existence of techniquarks, yet another new particle type … a new layer in the cosmic onion. In analogy with quarks, which can combine into mesons, a techniquark and an anti-techniquark can combine to make techni-mesons, one of which will play the role of the Higgs boson. Another theory, supersymmetry, which we will discuss in Chapter 8, even predicts many types of Higgs bosons, including ones with electrical charge.

So what is the answer going to be? I have no idea. Currently physicists are looking very hard for whatever it is that breaks the symmetry between electromagnetism and the weak force. The Higgs mechanism is simply the leading contender among several competing theories. Since all of the theories make similar predictions (the precision data taken by the LEP experiments tightly constrain the options),

I do not list all of the contenders here. There must be something there, but exactly what it is, no one can say. Physicists can be broken into believers, unbelievers and skeptics (others have called them Higgs believers, atheists and agnostics). The believers expect that we'll find the Higgs boson as described here, while the unbelievers think fundamental scalars are impossible and one of the other theories must be correct. The skeptics, with whom I must include myself, take a wait and see attitude. A few, deeply disturbed (but possibly correct!), individuals predict that we'll not find anything that serves the same function as the Higgs boson and we'll have to revamp the entire theory. The other day, a colleague of mine, Eric Myers, currently a professor at Vassar College, predicted that we'll discover the Higgs, only to find that it's quite different than we expect. Maybe in a few years the Tevatron's Run II will sort it out, or perhaps CERN's LHC. No matter. The journey will be great fun.

chapter 6

Accelerators and Detectors:
Tools of the Trade

Take interest in these sacred dwellings which we call laboratories. There it is that humanity grows greater, stronger, better.

— Louis Pasteur, 1822–1895

It's said that university administrators prefer theoretical physicists to experimental ones. Experimentalists need large and expensive equipment, while theoretical physicists need only paper and wastebaskets. (I'm told that they like philosophers even more, as they don't even need wastebaskets.) Nonetheless, it is in these intricate experimental apparatuses that discoveries are made. One of the weaknesses in books for the non-expert is that they are often written by theorists and the question of how one actually discovers and measures new phenomena is frequently neglected, or at least glossed over. This is a shame, as the techniques whereby one makes discoveries are nearly as interesting as the discoveries themselves.

In the preceding five chapters, we have discussed one of the great triumphs of mankind, an extraordinary understanding of the universe at a deep and fundamental level. While it's true that our understanding is incomplete, the fact that we have a detailed understanding of the universe at sizes, speeds, energies and temperatures so far removed from the realm of common human experience is a testament to the thousands of focused men and women who have dedicated their lives to asking the hard questions. However, whenever I give a public lecture on particle physics, I usually have some member of the audience who transfixes me with a gimlet eye and says, "You expect me to believe all of this stuff? Go ahead ... show me a quark." And, irritating though it sometimes may be, it's a good question. The physics that we discuss in this book is so far removed from ordinary human experience that you shouldn't believe what you read just because I say so. The fact is that quarks or electrons are simply too small to see. What my critically thinking audience member is really asking is, "How do you see the unseeable?"

While the approach is easier to see if you're doing a simple experiment like dropping a ball and asking how long it takes to fall, the short answer is the same for any science experiment. You spend a while arranging the environment to make the conditions suitable for the measurement. For example, you might be curious about how fast you could run a mile. You'd first make sure that the track was dry and the weather was nice. Simultaneously you prepare your measuring device, say a stopwatch. This chapter first details how one accelerates particles to very fast speeds. This is critical, as in order to routinely measure the most energetic (and hottest) conditions accessible to modern science; you need to smash two or more particles together. However, even if one smashes extremely energetic particles together, the act is pointless unless the collision can be recorded. To this end, giant detectors are built, weighing thousands of tons. It is these two tasks that dominate this chapter. Towards the end of the chapter, we will briefly mention other experimental techniques in which we dispense with an accelerator. Before we move to a technical discussion, let's take a moment to get a flavor of the environs of America's premier particle physics laboratory.

A Drive through Fermilab

As my colleagues and I go to work each morning, our day begins in the usual way. A final cup of coffee, hugs to the kids, a kiss for the spouse and off to our cars we troop. Graduate students, being younger, usually forgo the kids and often the spouse, replacing those rituals with cold pizza or possibly a morning jog. The Fermilab staff calls many places home. Some, especially my European colleagues, live in Chicago, trading a long commute each day for the energy and excitement of one of America's largest cities. Others, often our technicians, live to the west of the lab, where one can still buy land in parcels larger than an acre. Their commute is about as far as the city dwellers', although the view of stop and go traffic is replaced with the open road and expansive farms.

However, most Fermilab workers live in one of the surrounding communities. Affluent and conservative, in many of these towns the local ballots list only one candidate per position, as the area votes Republican with nearly the same dependability as Chicago votes Democratic. The only community that deviates from the norm is Aurora, just southwest of the lab, with its more diverse and interesting demographics. Each of these towns was once truly distinct, but suburban sprawl has been a homogenizing force. Growing around the perimeter of the towns, the sprawl has interlinked them into a vast Chicago metroplex, with the uniform blandness that so many decry as the tragic death of regional diversity. Each town has resisted the creep with varying degrees of success. Most towns retain at least some of their original character in their downtowns with many cute shops.

While the approach to Fermilab could be a trip through so many other places, upon entering the site, everything changes. Although Fermilab is currently home to the world's highest energy particle accelerator, you wouldn't know it as you entered. The lab looks like a park, a vast expanse of undeveloped land, surrounded by the urban sprawl. Comprised of some 6800 acres, the terrain retains much of the flavor of northern Illinois of a thousand years ago. Consisting of mostly open grassland, interspersed with small stands of woods, Fermilab's scientific efforts include an attempt to reseed the flora and some of the fauna of the original prairie. Fermilab even owns a small herd of 50–100 American bison. Many hundreds of white-tail deer, ground hogs, coyotes, pheasants, herons, egrets, red-tail hawks, as well as thousands of transient Canada geese, make their home here, living essentially as they have for hundreds of years. Dotted across the open expanse, one sees a few lone trees, majestic sentinels who watch the coming and going of people and particles and indeed the seasons with equal equanimity.

While one's initial impression of the lab is rather reminiscent of a nature preserve, the lab's main purpose is to conduct world-class particle physics research. The heart of this research program is the Fermilab Tevatron, a large particle accelerator in the shape of a ring about four miles in circumference. Spaced around the ring are six "interaction regions" at which two counter-rotating beams of protons and antiprotons can be made to collide. At two of these interaction regions sit a building, each housing a particle collision detector. Both of these detectors are similar in design and functionality, although technically quite different. They each weigh about 5000 tons and require about 500 physicists and a considerable number of Fermilab's support staff to keep them operating.

These two detectors sit at their interaction regions and record a select subset of the particle collisions that occur at their center. Of the millions of collisions that occur every second, each detector can only record about 50 for future analysis. From the data that is recorded, physicists try to unlock the secrets of the universe.

While these two large detectors are currently the lab's dominant focus, there are other experiments unobtrusively located at other parts of the lab, using particles from one of the other support accelerators. As you cast your gaze across the prairie, the frenzied efforts involved in physics data taking are hidden. Mostly you only see the prairie, although one architectural feature grabs your eye ... a fifteen-story building that is the administrative center of the lab.

Wilson Hall, named after Robert Rathbun Wilson, the first director of Fermilab, locally known simply as "the high-rise," was supposedly inspired by the Beauvais Cathedral in Beauvais, France. Having seen both, I'm not totally convinced, but some resemblance is evident. Viewed from the front, the shape of praying hands is easy to see and there is no questioning Wilson Hall's identity as a cathedral of science.

While the high-rise is the easiest building to spot, driving across the lab, one spies many other interesting buildings. The 15′ bubble chamber is surrounded by a large geodesic dome made, local lore insists…perhaps even truthfully…from 120,000 recycled aluminum cans, donated from nearby communities. The original beam line control building, no longer in use, is a pagoda, rising above the prairie, while its replacement (also now unused) looks like a sort of post-apocalyptic bunker straight out of "Mad Max." The New Muon Lab reminds me very much of the Hamamatsu-shi Budokan (the municipal gym and martial arts center in Hamamatsu City, Japan), while the old Meson Lab mixes a WWII airplane hanger heritage with what can only be described as an industrial-chic roof. A pumping station on the north side of the lab is surrounded by a wall described by a mathematical spiral. The Fermilab village, situated at the east side of the lab, provides housing for visiting scientists and their families. Several original farmhouses were moved from other places on the site and turned into dorms and apartments. One of them, called Aspen East, is especially impressive. The remaining housing is provided by what was once the village of Westin. Finally, spread across the lab are the original barns, once supporting the farms that dotted the area, still useful in the particle physics era. A very nice feature of the Fermilab site is the pioneer cemetery, still respectfully maintained by Fermilab and housing many original settlers and one new resident, Bob Wilson, who died in 2000 and wanted to spend eternity at the heart of what was ultimately his greatest scientific legacy.

I've described Fermilab and its environs, partly because it is the lab with which I am most familiar and also because it really reflects Bob Wilson's deep-seated conviction that world-class science and a firm commitment to the environment are not at odds. I've not even mentioned the art, some of it Bob's own sculptures, sprinkled around the lab. Other labs each have their own character, with Brookhaven National Laboratory nestled in the coastal forest of Long Island and the urban-sited DESY, with its ring located under the city of Hamburg, Germany. The CERN laboratory has far less of the nature-friendly flavor, being a highly developed site on the outskirts of Geneva, Switzerland. But Geneva has its own appeal, with a beautiful lakefront, a large working clock, built into the side of a hill, the face of which is made of flowers. It also has La Chocolaterie du Mont Blanc, the very finest chocolate shop I have ever encountered. If you ever go there, try the green and brown chocolates, as well as the ones with the walnuts on top, but bring lots of money.

The Fermilab described here is the one that all visitors immediately perceive. (Note that Fermilab welcomes visitors, although since 9/11, access is occasionally restricted... call ahead for current information. We at Fermilab all look forward to the day when we can return to the days of unimpeded access for our neighbors.) The façade that I've described conveys the fact that cutting-edge research and a commitment to the environment, a respect for the area's unique heritage and a deep appreciation for the artistic aspects of mankind's spirit are not at all in conflict. However, I've communicated nothing of the true purpose of the laboratory. Fermilab is fundamentally a particle physics laboratory. At its heart is the huge accelerator, the Tevatron, and its cutting-edge particle collision detectors. It is this type of equipment that is ultimately responsible for Fermilab and other laboratories' discoveries about which you read in your newspapers and popular science magazines. When I was younger, I wondered how physicists made the measurements they did. Let's spend the rest of this chapter learning how.

Not All Accelerators Are in Cars

So we begin our journey with the question of why it is necessary to accelerate particles at all. There are two primary reasons. The first reason is just a matter of knocking the quark or electron hard enough that nothing else matters. Let's consider the case of the late great Chicago Bear running back, Walter Payton, running with a football. Suppose I want to knock the ball out of his hands. I might first try the low energy attempt of having my young nephew run and tackle Payton. All through this impact, Payton's grip on the ball would be essentially unchanged. So let's up the energy a bit and have me (who's a fairly large, but not very athletic guy) run at Payton at full speed and tackle him. In this collision, Payton might (although I wouldn't bet on it) even fall down and have to use two hands to keep control of the football, but he wouldn't drop it. But at least the impact had some effect on the ball. Now let's up the collision energy another notch. Let's suppose that Walter Payton gets hit by that other great football player, Mike Singletary, the Chicago Bear middle linebacker (let's say it was an inter-squad practice). When these two guys collide, there would be a god-awful impact and at least occasionally the collision would be so great that the football would get jarred out of Payton's hands and fly off without any chance for him to adjust his grip.

Particle collisions are similar. Suppose, as an example, you wanted to collide an electron with a proton with the intent of liberating a quark (neglecting, for the moment, the jet phenomenon discussed in Chapter 4). If the collision were at low enough energy, the electron would bounce off the proton, perhaps jiggling the quarks a bit, but certainly not liberating any. Even a much more violent collision might hit the proton hard enough to knock a quark just a little way outside the proton, but the residual force between that quark and the rest of the quarks in the proton would be strong enough to pull the quark back into the proton, although now the proton would be substantially disturbed. Finally one could have the electron hit the proton so hard so as to knock the quark clear out of the proton with no chance of the quark returning (and so a jet forms). So a hard collision can liberate a tightly held object, which a lesser collision would never reveal.

The second reason why extremely energetic collisions are interesting requires our old friend $E = mc^2$. A more extensive discussion of this important equation is given in Appendix D, but we can learn a lot about the equation "just by looking" (with apologies to Yogi Berra). It has properties common to all equations…it has a left side and a right side. The left side just has an "E" which stands for energy. The right side has an "mc^2," with the "m" meaning mass and the "c^2" meaning "the speed of light squared." The actual number for the speed of light depends on what units you choose (just like the number that tells you your weight depends on whether you use pounds, ounces or tons, even though your weight is the same). A very common number for the speed of light is 3×10^8 (three hundred million) meters per second (although 186,000 miles per second is the same thing). Squaring a big number gives a *very* big number. So c^2 is just a very big number.

Thus the famous equation says, "Energy is equivalent to a certain amount of mass, multiplied by a huge number." A little mass means a lot of energy, while more mass means even more energy. To give some necessary perspective, if one takes a paperclip, with its mass of about one gram, and converts it into energy, the amount of energy would be basically the same as the energy released in each of the atomic bombs that devastated Hiroshima and Nagasaki.

The atom bomb example shows that a little mass can be converted into a lot of energy (because c^2 is so big). But the equation works both ways. One can equally well convert energy into matter. And this is where particle accelerators come in. If one accelerates particles to great energy, so that they are moving very fast, and collides them so they are no longer moving, one might ask where their energy went. Prior to the collision, there was a lot of moving energy and after the collision there was none. Since energy can neither be created nor destroyed, the moving energy can be converted into mass, because energy *is* mass and vice versa. However, just like we saw that a little mass can make a lot of energy, in order to make a very light particle, you need a huge amount of energy.

So to make new and heavy particles, one needs to accelerate more traditional particles to huge energies and collide them together.

This is why it is only Fermilab that can make top quarks. It's the only accelerator with enough energy to make these heavy particles. This is also why particle physicists are always trying to build bigger particle accelerators. More energy means that we can create heavier particles. If the particles exist, we can discover them. And discovery is what scientists live for (well, that and a conference in Paris).

So how does one accelerate particles? In general, if one wants to accelerate something, then that object needs to feel a force. If you want a large acceleration (and consequently a high speed and energy), you need a strong force and the largest force over which we have significant control is the electric force. The only problem is that electric fields only affect electrically charged particles. Further, in order to accelerate a particle for a long time, you can only use stable particles (i.e. ones that don't decay). The naturally occurring charged particles that don't decay are the electron (e^-) and the proton (p). With a little cleverness, we can also accelerate their corresponding antiparticles, the positron (e^+) and the antiproton ($\bar{p}$). We'll get to that later. And, of course, naturally occurring neutral particles, like the photon (γ) and the neutron (n) cannot be accelerated by this method.

To begin our discussion, let's talk only about accelerating a proton. If we put the proton in an electric field, it will feel a force in the same direction that the electric field is pointing. It's like gravity in a way. The gravity field always points downwards and so when you drop an object, it falls in that direction. The nice thing about an electric field is that we can orient it any way we want and consequently we can choose the direction that we want a proton to move, a point illustrated in Figure 6.1.

There are lots of ways to make an electric field. The simplest way is to attach two metal plates to a battery. The metal plates can be any shape, although we traditionally draw them as squares. Also the plates can be any size and separated by any distance, although the math and explanation gets *much* easier if you make the separation much smaller than the length and width of the plates. You take a wire from one end of the battery and hook it to one plate and use another wire to

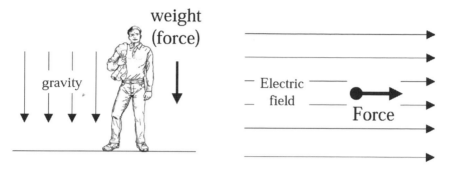

Figure 6.1 Electric fields are conceptually similar to gravity fields, except that electric charge, not mass, is the important quantity. In addition, electric fields can be directed as desired. (Drawing by Dan Claes.)

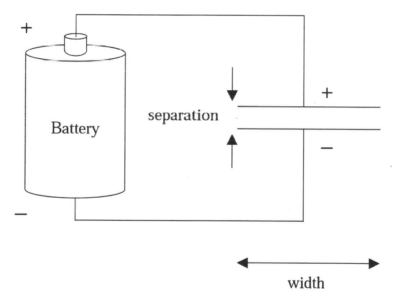

Figure 6.2 Electric fields are easily made between parallel plates of metal, connected to a battery. Real world batteries are more powerful than this simple D-cell, but the principle is the same.

connect the other end of the battery with the other plate. Figure 6.2 shows what such a contraption looks like, with the two parallel lines representing the parallel plates.

The little "+" and "−" in the figure shows which end of the battery is hooked to which plate. The actual battery that one uses is typically much more powerful than the simple D-cell that I drew here, so in drawing Figure 6.3 we drop the battery and just draw the plates. In order to remind us that the plates are attached to some kind of battery, we label the plates with the same "+" and "−" that tells us to which side of the battery they are connected. Finally, we draw the plates in such a way that the separation between them is much greater than the width (especially in Figures 6.3 and following). This is just for artistic and clarity reasons. The plate separation is really much smaller than the width. With all of these things in mind, we can draw the most primitive particle accelerator.

In Figure 6.3, I drew two plates and the electric field between them. In the center of the right-hand plate, a small hole is drilled in order to let the proton escape. A proton with an initial speed of zero is released at the left plate. It feels a force towards the right and so it

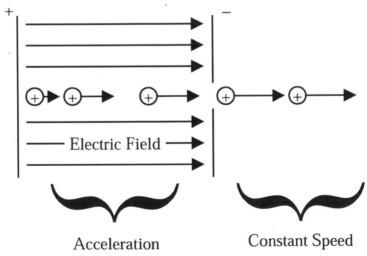

Acceleration   Constant Speed

Figure 6.3 Accelerators work by putting a charged particle in a region carrying an electric field. In that region, they experience constant acceleration. When they exit the region containing the electric field, they move with constant velocity.

accelerates, moving faster and faster while between the plates and finally passing through the hole in the right-hand plate. After leaving the space between the plates, the proton moves with constant velocity. Voila! You've just accelerated your first proton.

The energy that the proton carries when it leaves the space between the plates is related to the strength of the battery. In order to get more energy, you need to use a stronger battery. Of course, there's a practical limit on just how strong one can make a battery, so one has to think up clever ways to get around this limitation. We'll address this question in a little while. But before we do this, we need to define some useful concepts. The most important one is to have a useful measurement of energy. You may have heard of some units: BTU's (British Thermal Unit), ergs or joules. While joules are relatively useful, there is a unit that particle physicists much prefer; the electron volt or eV (each letter is pronounced, i.e. "E-V"). The electron volt is extremely useful. The reason is that the strength of batteries is given in volts (like the 1.5 Volt D-cell battery or a 12 Volt car battery) and the most likely particles to be accelerated are the electron or proton, which have the same size electrical charge (although opposite sign). We would need to adjust the following discussion for particles with electrical charge of a different size than that of a proton (like an α particle, with the charge of twice that of a proton), but we'll ignore that here. With a proton or an electron however, the calculation is simple. If a proton is accelerated from one plate to the other and the battery has a strength of one volt, the proton will have an energy of 1 eV. If the strength of the battery is 1000 volts, then the energy would be 1000 eV (or equivalently 1 kilo-electron Volt or 1 keV, pronounced "K-E-V"). For stronger batteries, the pattern continues. Table 6.1 shows this continuing pattern.

Clearly this unit of energy is very convenient and allows for quick conversion between energy and battery strength. Now some sense of scale is in order. The energy involved in holding electrons in orbit around the nucleus of an atom is some few tens to thousands of an eV. Standard dental x-rays use photons of approximately 10 keV.

Table 6.1 Prefixes that are used to describe big numbers.

Voltage (Volts)	Word	Prefix	Symbol	Energy
1	One	—	—	1 eV
1,000	Thousand	kilo	k	1 keV
1,000,000	Million	Mega	M ·	1 MeV
1,000,000,000	Billion	Giga	G	1 GeV
1,000,000,000,000	Trillion	Tera	T	1 TeV
1,000,000,000,000,000	Quadrillion	Peta	P	1 PeV

The particle accelerator that you call a television also accelerates electrons with an energy of about 10 keV. Radioactivity in naturally occurring uranium ore is approximately 1 MeV. The highest energy particle accelerator in the world has a maximum energy of nearly 1 TeV.

Because 1 eV is a fairly small unit of energy, Einstein's Theory of Special Relativity doesn't change things much. At 1 eV, an electron is traveling about 370 miles per second, while the much more massive proton has the still impressive speed of 8.5 miles per second. On the other hand, a paperclip with a mass of one gram, dropped from a distance of 1 meter (3.2 feet) has a kinetic (i.e. moving) energy of 6×10^{16} eV (6×10^4 TeV), fully 60,000 times more than the highest energy to which one can currently accelerate a sub-atomic particle. So an electron volt is a very small unit of energy. It's just very convenient to use it in the subatomic world.

The simplest particle accelerator with which you are familiar is your TV or computer monitor, shown in Figure 6.4. Electrons are accelerated to an energy of approximately 35 keV and made to hit the screen. Quickly changing electric fields cause the electron beam to scan across the screen. Without these rapidly changing electric fields, the electron beam would just go in a straight line and make a white dot in the center of your television screen. If you recall seeing this when you turned off the television 30 years and more ago, then you're as old as I am. If you weren't around, ask your parents or grandparents.

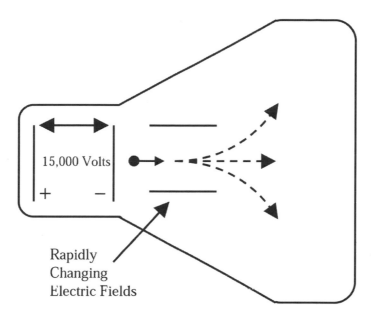

Figure 6.4 A television screen is the most familiar particle accelerator. The voltage between two plates accelerates the particle. Subsequent plates deflect the particle transverse to the particle's original direction of motion. The polarity of the plates is given for a hypothetical positively charged particle rather than the electron of a regular television.

However, 35,000 eV is quite a bit smaller than the 10^{12} (1,000,000,000,000) eV that a world-class accelerator can do. It turns out that a 1 trillion volt battery is not possible, so we have to be clever. Well, the easiest thing that anyone could think of was simply to put a bunch of pairs of plates in a row. Each set of plates adds the same amount of energy. So in Figure 6.5a, with its three sets of plates, a particle would be accelerated to $3 \times 35{,}000 = 105{,}000$ eV (using the two plates of a TV as an example). The set of plates labeled 1 are exactly as described above; a particle starts at rest at one plate and leaves the other with some speed. But with the set of plates labeled 2, the particle enters with an initial speed and leaves with an even greater speed, a pattern that repeats for plates #3. If that's a little hard to see, the situation is exactly analogous to the situation where you drop a ball. If you drop a ball through a distance of one foot, the ball hits the

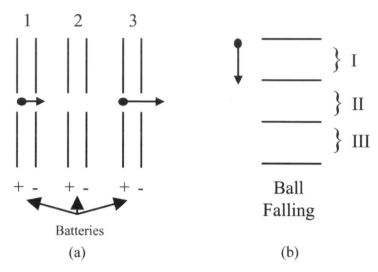

Figure 6.5 A linear accelerator works via the simple principle that one can use more than one set of plates. Each set of plates is separated by an electric field that will further accelerate the particle. Each set of plates can be thought of as being similar to dropping a particle through an additional distance. The particle will go faster if you drop it from a greater height. Similarly, additional sets of plates will make the charged particle go faster.

bottom of the one-foot distance with a particular speed. In order to increase the final speed (or equivalently energy), you simply drop the ball through a larger distance, say two or three feet, as shown in Figure 6.5b. The type of accelerator that one can make using this technique is called a LINAC (for Linear Accelerator), because it accelerates along a line.

This method for making an accelerator works quite well, but requires that one make multiple copies of the same basic accelerator unit and that's expensive, especially for a large number of copies. It would be nice if one could somehow reuse the single accelerator. Once this became apparent, many solutions for this problem were offered and built. This isn't a book on the history of accelerators so I won't detail them all here; the interested reader can refer to the bibliography for suggested reading. But there is one technique that is very frequently used in modern accelerators and this one will get some additional explanation.

Not All Rings Are for Engagements

Remember when you were a kid and you went to the playground? There's a piece of play equipment, which we used to call the "spinny thing." (Some people tell me that they called it a Merry-Go-Round.) It was like a wheel set on its side and it could rotate freely. Children would sit on it and someone would grab a handle and push it so that the whole ride would spin like an old record turntable. The children would spin in a circle and squeal (and throw up, in my experience). One way to get the spinny thing going very fast was to have an adult stand near it, grab a handle and throw it so the wheel would spin. When the handle came around again, the adult grabbed it and threw it again. This pattern repeats itself with the wheel getting faster after every throw. So with just one point of acceleration (the adult), the spinny thing could go very fast, simply by having the handle pass the adult many times.

There is a type of accelerator that works by a very similar principal, the synchrotron. In a synchrotron, a particle is made to move in a circle back to a powerful, but relatively short LINAC, where the particle is accelerated in the direction of motion. The basic idea is shown in Figure 6.6.

To give a sense of scale, the Fermilab Tevatron, the largest "ring" in the United States, has a circumference of about 4 miles, but the acceleration region is about 50 feet. The much larger (and higher energy) LHC accelerator in Europe (with its circumference of 18 miles) has an acceleration region of about only 10 feet. A curiousity is the fact that when the LEP accelerator occupied the space now owned by the LHC accelerator, its acceleration region was about 2000 feet. This was because the LEP machine accelerated electrons, which are intrinsically more difficult to keep at high energy than the heavier protons.

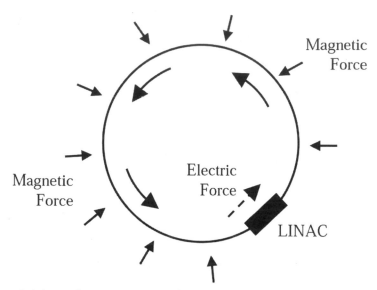

Figure 6.6 A synchrotron is an accelerator that reuses the single acceleration region. Electric fields placed at a single spot on the ring accelerate the particles. Magnetic fields placed around the ring will return the particle back to the acceleration region, where its speed is increased further.

Exactly how one makes the beam move in its circular orbit is extremely complicated, taking years of intense study to understand in detail. However, the basics are pretty easy to understand. While technically one could use electric fields to provide the force to make the particle move in a circle, it turns out to be easier and cheaper to use magnetic fields to do the job. While you could in principle use common magnets, like the one that holds your kid's art to the refrigerator, even the most powerful magnets of this type aren't nearly powerful enough to make an accelerator of a reasonable size. Luckily, we know of a way to make even stronger magnets. Recall in Chapter 1 how we said that Oersted found that by putting current through a wire, he could make a magnetic field? Well, in modern accelerators, we exploit this same phenomenon. Large "electromagnets" are made (so named because they use electricity to make a magnet) by taking coils of wire and putting current through them. One uses coils because each loop in the coil adds equally to the magnetic force, e.g. a 100-loop coil will generate 100 times the force of a single loop. So physicists use as many loops as is practical to get more magnetic field for the same amount of current. They also push as much electrical current through the wires as possible. The problem is that the longer the wire, the harder it is to push current through it and more loops means a longer wire. So compromises must be made.

If one could make the resistance of the wire lower, electricity would flow more easily through the wire. With a larger current, one can make a larger magnetic force, with a reduced electricity bill. Serendipitously, it turns out that nature exhibits a peculiar and useful behavior. When the temperature of certain metals is made to be very low, the resistance of the wire not only gets small... it goes to zero!!! So physicists now make some magnets so the current carrying wires are about −450°F (or about −270°C). Very cold. Any magnet that is cooled to such frigid temperatures is said to be superconducting. The first large accelerator complex that came online using superconducting magnets was the Fermilab Tevatron, which commenced operations in 1983.

At this point, I'd like to take a little detour to mention a useful spin-off of this research. While particle physicists use these large, superconducting, multi-coil magnets to guide particles in their orbits, the same basic design is behind the MRI (Magnetic Resonance Imaging) magnets one sees in hospitals. In hospitals, huge superconducting magnets form the core of the same MRI devices that are used to take pictures inside the body, with less damage to the body than caused by the more traditional x-rays. While particle physicists cannot claim credit for discovering superconductivity, they can claim responsibility for "industrializing" it, that is in pushing the envelope in making large numbers of big magnets (recall that the Fermilab Tevatron is 4 miles in circumference and it takes 1000 magnets to cover that distance). So anyone you know who's had their life saved by a MRI scan, should know that they are alive (in part) because of a spin-off of particle physics research.

With the idea of the LINAC and the synchrotron, we can now understand how a modern high-energy particle accelerator works. There are many accelerators currently operating or under construction, so I will discuss the one with which I am the most familiar, the Fermilab Tevatron. The Tevatron is the highest energy accelerator in the world and acts as the crown jewel in the diadem of American accelerators. Situated at Fermi National Accelerator Laboratory (a.k.a. Fermilab), about 30 miles west of Chicago, the Tevatron is capable of simultaneously accelerating both protons and antiprotons to nearly 1 TeV. Let's neglect the antiprotons for a moment and concentrate on how one can accelerate a proton to such high energies.

It turns out that one can't just plop a proton in the Tevatron and turn it on. While one could in principle do just that, it's just not efficient. So the Tevatron (and all modern accelerator complexes) actually consist of a series of accelerators, called an accelerator chain, each tuned to work most efficiently in a particular energy range. The proton is steered from accelerator to accelerator, gaining energy at each step, culminating finally in the 1 TeV energy of the Tevatron. A good analogy is an old manual transmission car. While you can, in principle, put the car in 5th gear and start from rest, one can accelerate much more efficiently by using the lower gears and gradually working the car up to top speed. The Fermilab accelerator consists of 5 "gears." They are, in order of increasing energy, the Cockroft-Walton, the LINAC, the Booster, the Main Injector and the Tevatron.

If one wishes to accelerate a proton, one must first somehow obtain a bare proton, a moderately difficult task. Luckily, there exist atoms which contain only protons in their nucleus; the hydrogen atom. The problem with hydrogen is that each proton is associated with an unwanted electron. Obviously, you'd want to strip the electron off and throw it away, and while that's not incredibly hard to do, there are technical reasons why we make another choice. So, rather surprisingly, we instead add an electron. While this seems like exactly the wrong thing to do, it does an obviously good thing; it gives the hydrogen a net electric charge. Hydrogen atoms, consisting ordinarily of one positive proton and one negative electron, are electrically neutral (i.e. have no electric charge). (Hydrogen molecules, which consist of two hydrogen atoms, are similarly neutral.) By adding an electron, the hydrogen now has a negative electrical charge and can be accelerated by electric fields. In fact, protons are not accelerated one at a time, but rather in great bunches of 10^{12} or more. These protons are not all in the same place, rather are spread out forming what we call "beams." In fact, the way the protons travel reminds me somewhat of the cars on a circular Nascar track. (When we get back to antiprotons, it will be worthwhile to to think of them as a similar beam of cars, this time driving in the opposite direction.)

The Cockroft-Walton, the first accelerator in the Fermilab chain, conceptually consists of two plates, with an exceptionally fancy "battery." The negatively charged hydrogen enters the region of electric field and is accelerated from zero energy to 750 keV. The hydrogen ion is then injected into "gear two," the LINAC. The LINAC consists of a large series of repeating plates, roughly as shown in Figure 6.5a. The hydrogen enters the LINAC with an energy of 750 keV and leaves with an energy of 401 MeV (and a velocity about 75% that of the speed of light). As the hydrogen leaves the LINAC, it passes through a thin foil of carbon. As the atoms pass through the foil, they interact with the electrons in the foil and the electrons are stripped off the hydrogen. The electrons are discarded and bare protons can now be formed into a beam. The third accelerator is the first synchrotron in the chain and is called the Booster. The Booster accepts the 401 MeV protons and accelerates them to 8 GeV in 0.033 seconds. The proton beam then passes into the newly constructed (1999) Main Injector. This "fourth gear" in the process is also a synchrotron and accelerates the protons from 8 GeV to 150 GeV. The final gear, the Tevatron, is a synchrotron with superconducting magnets. It accepts the 150 GeV proton beam and accelerates the protons to 1000 GeV, or 1 TeV. (In fact, when they're feeling a bit catty, our competitors correctly point out that the Tevatron can reliably operate only at 980 GeV, or 0.98 TeV and therefore we're a bit presumptuous to call it a Tevatron. What they say is true, but my response is "Picky, picky, picky…" And besides, we've tested the Tevatron up to 1.01 TeV, but we run at 0.98 TeV for a 3% safety margin.) Figure 6.7a shows a photo of the Fermilab accelerator complex and Figure 6.7b provides a diagram to understand the photo.

(a)

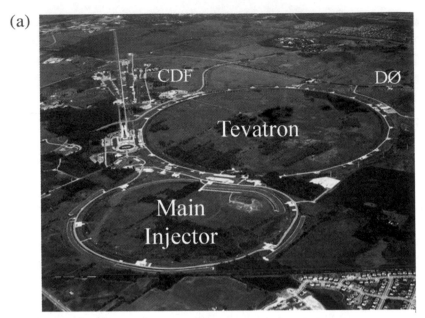

FERMILAB'S ACCELERATOR CHAIN

(b)

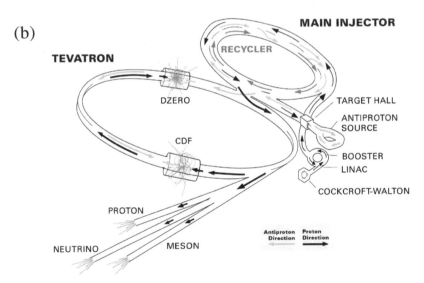

Figure 6.7 The Fermilab accelerator complex. (a) is an aerial view, while (b) shows the various components. (Figure courtesy of Fermilab.)

Table 6.2 Basic parameters of the Fermilab accelerators.

Name	Year Turned On	Initial Energy	Final Energy	Maximum Velocity (% speed of light)	Acceleration Time (seconds)
Cockroft-Walton	1971	0	750 keV	4	1.6×10^{-7}
LINAC	1971	750 keV	401 MeV	71	8×10^{-7}
Booster	1971	401 MeV	8 GeV	99.45	0.033
Main Injector	1999	8 GeV	150 GeV	99.998	1
Tevatron	1983	150 GeV	1 TeV	99.99996	20

Table 6.2 details the important parameters of the 5 stages of the Fermilab accelerators, while Figure 6.8 shows photographs of the various accelerator components.

Targets and Beam Types

When one finally has a beam of particles (protons in our example), the question becomes "What next?" We know that in order to do our experiment, we need to collide the protons into something, but how do you do it? There are two basic techniques. The first one used was simply to take the beam and aim it at a stationary target (we call this a "Fixed Target" experiment). The target can be anything and is often a canister of hydrogen chilled so that it is a liquid, but a small block of solid material works just as well. In the past, I've done experiments where the target is beryllium, copper, carbon, lead and others. The most important requirement is that you have a chemically very pure target (or at least precisely know its composition). Otherwise comparing your experimental results with calculations can be a nightmare.

When a beam particle (say a proton) hits a stationary target particle (say another proton), all sorts of possible interactions occur.

Figure 6.8 From top center, going in a clockwise direction: the hydrogen bottle from which protons are generated, the Cockroft-Walton accelerator, the LINAC, the Booster, the Main Injector and Antiproton Recycler and finally the Tevatron and the Main Ring (the predecessor of the Main Injector). In the image containing the Tevatron, the top series of magnets is the currently-unused Main Ring, while the smaller ring of magnets is the Tevatron. (Figure courtesy of Fermilab.)

But, whatever the interaction, the particles that come out of the collision all go generally in one direction, e.g. if the beam comes from the left moving towards the right, after the collision the debris tends to

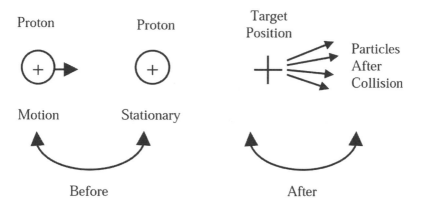

Figure 6.9 Typical fixed-target collision topology. Initially a moving particle hits a particle at rest. After the collision, many particles are created and travel generally in the direction of the initial moving particle.

move towards the right, as shown in Figure 6.9. For this reason, in a collision of this sort, the detectors tend to be all on one side of the collision. In many respects, a collision of this type is like shooting at a watermelon. After the impact, the spent bullet and all of the watermelon guts shoot out one side.

One very nice consequence of experimenting in this way is you don't have to steer the beam with meticulous precision; after all, you can simply make a wider target. A second nice feature is that you can increase your interaction rate by making a thicker target. Recall that collisions are incredibly rare. If you had a target that was one inch thick, most of the beam would pass through the target without interacting. If you made the target two inches thick, those majority of particles which didn't interact in the first inch could have another chance in the second inch. A third inch helps even more. The practical maximum length of a target depends very heavily on beam type, target type and experimental setup, but a target of liquid hydrogen, say 20″ thick, is not at all uncommon. Even with such a thick target, if the beam were protons, only about 4% of the protons would interact in the target and most of those interactions would not be very violent and thus would be generally uninteresting collisions.

The last thing to discuss about fixed target experiments is the fact that it allows us to get around one of the first points that we made about the sorts of particles one can accelerate successfully; the stable ones. The particles that come out of the collisions are all sorts, most commonly pions, but in any event, they are generally unstable. If one can collect the particles extremely quickly, keep the desired ones and dump the ones not wanted, then one can focus the particles with the use of magnets and, presto!, you have a beam of unstable particles. Only particles that decay via the weak force live long enough to use this technique, but beams of pions, kaons, muons and neutrinos have been made, to name a few.

While fixed target experiments have a long and very respectable past, they suffer from one major limitation. They don't supply as much energy to the "interesting part" of the collision as you might imagine. Both before and after the collision in Figure 6.9, the parti- cles are moving towards the right. This means that only lower mass particles can be created. Contrast this to the case where two identical particles, carrying an identical amount of energy (one can use the analogy of identical cars traveling at the same speed, but traveling in opposite directions), hit head-on. After the collision, both particles (or cars!) are stopped dead in their tracks. Recall that energy both before and after the collision must be the same and recall further that the total energy a particle has is the sum of the effects of its own mass and its energy of motion. In the case of a head-on collision, there is no energy of motion after the collision and thus *the entire energy of motion before the collision goes into mass energy after the collision.* (Purists will note that what I just said applies more to the situation where an electron collides with a positron, rather than when two pro- tons collide. They're right. Let's ignore them.) The effect is much bigger than you'd imagine. Let's consider what happens when you collide a 1000 GeV electron with a 1000 GeV positron going in exactly the opposite direction. (We're keeping the purists happy here.) Such a collision can create a 2000 GeV particle. In contrast, let's con- sider the case when we collide a 1000 GeV electron with a stationary positron. Intuition says that we should be able to make a particle with half the energy of the first case, or 1000 GeV. However, intuition is wrong. In this situation, only 1 GeV is available for making new particles. This means a colliding beam experiment with electrons and positrons can make (and discover!) particles 2000 times more massive than a similar fixed target experiment. For protons hitting antiprotons, the difference isn't quite as large, but a colliding experiment can create particles nearly 50 times more massive than the corresponding fixed target experiment. With such an enormous possible gain, it is obvious that physicists would eventually figure out how to make a colliding beam machine.

The first collider of which I am aware was a small electron-electron collider that ran in 1965. This collider was built at the Budker Institute of Nuclear Physics (BINP) in Novosibirsk, Russia, although a similar accelerator was also being built at the time at the Stanford Linear Accelerator Center (SLAC). I've actually seen the first BINP collider, called VEPP-1. The whole thing is about the size of a pool table. There were two rings, each half the size of the table, looking like two wedding rings pushed together so they touch. This first colliding beam accelerator had a maximum energy in each beam of 160 MeV. From such a humble (yet impressive) beginning came the modern leviathan accelerators with circumferences as large as 18 miles.

Because the purpose of the Fermilab Tevatron is to seek out new and massive particles, it was clear that it would have to be made to run in a colliding beam mode. When you decide to build a collider, you first need to decide on what particles you want to collide. There are many options. You can collide positrons and electrons (e^+e^-), like at LEP or electrons and protons (ep) like at HERA. (LEP and HERA are other big accelerators, both in Europe.) At Fermilab, as has been indicated earlier, we collide protons and antiprotons ($p\bar{p}$). While which configuration you choose is usually driven by what measurements you want to make, there are pros and cons to each of the choices. For instance an e^+e^- machine is very nice because the electron and positron have no internal structure, so all of the beam energy can go into making new particles. Plus you know the collision energy of each collision with extraordinary precision because each collision is the same. Further, positrons are relatively easy to make (more on that later), so you can make very dense beams and thus have many collisions. The down side is that electrons are very light and thus lose their energy very quickly as they move in a circle. Further, the fact that you can set the beam energy so precisely can be a problem. If the particle you're looking for has a mass of $100 \, GeV$ and you set the beam energy to $98 \, GeV$, you might just miss it entirely. There are ample examples of this in the past. I've met many physicists who bemoan their membership in the "I *just* missed discovering the J/ψ" club. (See Chapter 4 for a refresher if needed.)

Proton-antiproton machines on the other hand have their own features. Protons are heavy, so they don't lose much energy as they revolve around the ring. On the other hand, antiprotons are hard to create and work with, which makes things tricky. The real difference, however, is related to the structure of the proton, which we recall from Chapter 4 is an extended object, containing partons. Each parton carries a random fraction of the energy of the proton or antiproton. Let's concentrate on the three "valence" quarks or antiquarks contained within the beam particles. While the energy of the proton and antiproton are known to good precision, one must recall that the interesting collision is between the partons (quarks in our example), not the whole proton or antiproton. I like to compare a p$\bar{\text{p}}$ collider as like two swarms of bees passing through one another. Most of the time, the swarms pass through one another with little interaction, but every so often two bees hit head on. But it is never true that all of the bees in one swarm simultaneously collide with all of the bees in the other swarm. So it is with quarks.

Because each quark carries typically 10–30% of the proton's energy, that means the interesting collision involves much lower energy than the total beam energy. This negates much of the advantage one gets from using a p$\bar{\text{p}}$ collider. In fact, the typical "interesting" collision energy in the Tevatron is about the same as the highest energy collisions accessible by LEP, the world's highest e$^+$e$^-$ machine. So, you say, why do it?

Why use a p$\bar{\text{p}}$ machine, with the irritating complication that every collision has a different energy? Well, in addition to the obvious answer that we want to know what happens when you collide quarks, there is an even more pressing reason. While the valence quarks within the proton typically have about 20% of the energy of the proton, occasionally they have more … say 50–80%. Thus very rarely you can get really violent and highly energetic collisions, far above the energy accessible to existing e$^+$e$^-$ machines. And for particle physicists, energy is life. Energy is discovery potential. Energy is good.

But for all that, p$\bar{\text{p}}$ collisions are very messy. They have been called garbage can collisions, with the interesting interaction being between a broken alarm clock in one can and a worn out shoe in the other. You need to find the interesting collision signature buried in the rest of the glop that comes out of the collision. e$^+$e$^-$ machines are much cleaner, with particles after the collision only coming from the interaction itself. In e$^+$e$^-$ collisions, there are no spectators. It's fair to say that p$\bar{\text{p}}$ colliders are for discovery, but it's hard to beat an e$^+$e$^-$ machine for precise measurements, once you know where to look.

Enter Antimatter

One topic that we have glossed over thus far is the antimatter beams that are used in colliders. Both the Tevatron (p$\bar{\text{p}}$) and LEP (e$^+$e$^-$) consist of two counter-rotating beams of matter and antimatter. This is also true of CERN's much earlier (and lower energy) SPS (p$\bar{\text{p}}$) and earlier accelerators at SLAC (e$^+$e$^-$). This choice is not unique, as the earlier CERN accelerator, the ISR (for Intersecting Storage Rings) and the behemoth LHC (or Large Hadron Collider), currently under construction, both were (or will be) proton-proton (pp) colliders. As always with any design, there are pros and cons associated with either choice. The most notable advantage to using the complementary matter and antimatter beams is that like kinds of matter and antimatter can completely annihilate into energy, which can then in turn convert into a new undiscovered particle with a high mass.

The problem with antimatter is that it is very hard to store. If a positron comes into contact with an electron, it will convert immediately into energy. The same is also true for a proton and antiproton. Thus the first thing one needs to worry about is if you *did* have antimatter, how do you isolate it from ordinary matter? The second question is "Where do you find antimatter?" If you look around the Earth and indeed the universe, for all of the billions of light-years we can see, we observe no large deposits of antimatter. Sure, we occasionally see a particle of antimatter in a cosmic ray, but that doesn't really count. In all of the incomprehensibly large amount of matter we see in all of the billions and billions of galaxies (with apologies to Carl), nowhere do we see so much as a thimbleful of concentrated antimatter. (Note: Why this is true is a huge mystery, currently under intense study. In Chapter 7, we discuss the issue in detail.)

But the bottom line is that there is no place in the universe where we can go and buy or mine antimatter. We need to first somehow manufacture it and then store the antimatter without ever touching it. It's really a daunting challenge and yet we are now capable of doing just that. At any one time, Fermilab has the world's largest supply of antimatter, a truly huge amount. Of course, huge is relative. If one took all of the antiprotons ever manufactured in the 20 or so years Fermilab has been making them and put them in one place, the whole lot would weigh far less than a dust mote. And even though I've emphasized how much energy could be released by combining matter and antimatter together, if we took all of the antiprotons that we've ever made and let them simultaneously mix with an identical number of protons, the resultant energy would raise the temperature of a 20 ounce cup of black coffee (none of this sissy double decaf nonfat raspberry mocha soy latté nonsense...) by 90 degrees Fahrenheit (50°C)... just about enough to re-warm your room temperature beverage to something approximating a drinkable temperature. So our neighbors are quite safe.

So the real question is how does a lab like Fermilab make such a "vast" quantity of an exotic substance like antiprotons? Again we go back to that famous equation: $E = mc^2$. The one source of antimatter that we can always exploit is to create it from pure energy. As we mentioned in Chapter 2, antiprotons were first observed at the Bevatron at Lawrence Berkeley Laboratory in 1955. A beam of protons with an energy of a little over 6.5 GeV (or about seven times greater than the mass of a proton) was steered onto a proton (i.e. hydrogen) target. In all of the millions and millions of collisions that were inspected, a few antiprotons were made. So few antiprotons were made because (a) it's hard to create exactly the right three antiquarks and have them "happen" to form an antiproton and (b) the accelerator energy was just barely enough to make antiprotons. The really astute reader will wonder why we needed a beam energy of seven times the mass of a proton to make an antiproton (which has a mass the same as that of a proton). The reason is that we started with two protons and we need to have those two protons in the final state (since they can't annihilate each other). In addition, matter and antimatter are made in pairs, so for every antiproton we make, we also have to create a proton. Thus at a minimum we needed to have 3 protons and one antiproton after each collision. Finally, the experiment was of the fixed target variety and thus the particles had to move all in one general direction after the collision. When all of these effects are added up, the minimum beam energy needed to make an antiproton is about 6.5 GeV.

Antiprotons are made at Fermilab in much the same way. Protons are made to hit a target and lots of particles come out. The antiprotons are culled out and the rest of the particles are dumped. So how is this done? The antiprotons are culled out by simply taking all heavy, negatively-charged particles and put them in a storage ring. All particles except antiprotons quickly decay.

We run in collider mode, that is, we have protons and antiprotons in the Tevatron, with two beams sitting there counter-rotating, colliding head on. We can keep the two beams in the ring for a full day or so, which is really nice. You fill the ring with the two beams and then you have your detectors measure collisions for about a day. Then, when the beams have deteriorated in quality, you dump them (e.g. smash the beams into the walls in a specially prepared place), reload the accelerator with two fresh beams and repeat.

While the beam is circulating in the Tevatron for those 24 hours, we could have the four other accelerators sitting idle. But instead of not using them, we spend the time making antiprotons. Every 2.5 seconds, we use the four accelerators, up to and including the Main Injector, to accelerate protons to an energy of 120 GeV. We then aim these protons into a nickel target and collect the antiprotons that come out. The yield is very poor. For instance, each pulse of protons contains 5×10^{12} protons and yields on average 5×10^7 antiprotons. Because we need about 3×10^{11} antiprotons to do interesting physics research, we repeat the antiproton creation step over and over, each time storing the antiprotons in a special particle accelerator called, for obvious reasons, the "accumulator." It takes about 7 hours to accumulate enough antiprotons to do a useful experiment. Over the course of the 24 hours, we can accumulate (we say "stack") about 10^{12} antiprotons. The technical details of how this is done are challenging enough to warrant a piece of the 1984 Nobel Prize, shared by Simon van der Meer and Carlo Rubbia.

The positrons used in an e^+e^- collider are generated in a similar way. While the biggest ring has electrons and positrons colliding, one of the smaller accelerators is plugging along, happily accumulating positrons. By the time the beam quality in the big ring has degraded, they have gathered enough antimatter that they can refresh the beam in the big ring and continue the experiments.

Accelerators of the World

While I have concentrated on the Fermilab complex, this reflects my own familiarity and should not make you think that Fermilab is the only place that can build impressive accelerators. Long before Fermilab was wrested bare-handed from the Illinois prairie by Bob Wilson and his cronies, truly world-class accelerators were functioning at Argonne National Laboratory, Brookhaven National Laboratory, the Stanford Linear Accelerator Center, Los Alamos National Laboratory and the

Lawrence Berkeley Laboratory. Overseas, there was the international laboratory CERN (Conseil Européen pour la Recherche Nucléaire, the European Council for Nuclear Research, a.k.a. the European Organization for Nuclear Research) in Switzerland, the DESY laboratory (Deutsche Elektronen Synchrotron, the German Electron Synchrotron) in Hamburg, Germany and the KEK laboratory (Kou Enerugi Kasokuki Kenkyu Kikou, the High Energy Accelerator Research Organization) in Tsukuba, Japan. The Soviet Union had sharp scientists and they built some impressive accelerators of their own. While Fermilab is now the highest energy accelerator ever built, CERN built an amazing and competitive (but complementary) accelerator called LEP (for Large Electron Positron), which housed four extraordinary experiments and made Fermilab scientists hustle to compete. And while Fermilab still reigns supreme, in a few years (say about 2007 or 2008), the torch will pass once again to CERN when the LHC (Large Hadron Collider) will turn on with fully seven times the energy and ten times the number of collisions per second that Fermilab can currently generate. Had the SSC (Superconducting Super Collider) in Waxahachie, Texas been built, America would have retained its title as the energy leader, but no longer. You may recall that the SSC was supposed to be a competitor facility to the LHC, colliding beams of protons and antiprotons with an energy of 20 times Fermilab's Tevatron. It was cancelled by Congress in the fall of 1993. (Can you tell I'm jealous? Luckily, CERN, like Fermilab, is a truly international facility and so Americans will be well represented.) However, even now a consensus is arising that the next accelerator should be an enormous linear accelerator, called the NLC (Next Linear Collider), which will be a very high energy linear accelerator with electron and positron beams. Design work is ongoing and site selection is still years away. (Although it seems to me that 30 miles west of Chicago would be an ideal choice ...)

The current suite of accelerators allows for a very rich physics research program. It's rare for a modern particle based accelerator to have only one purpose, but typically each has a primary focus. The

Tevatron, originally built to discover the top quark, now pursues the Higgs boson. LEP was built to measure the properties of the Z boson with great precision and subsequent upgrades allowed them to characterize W bosons and also to search for the Higgs. The HERA accelerator (an electron-proton collider) was designed to look deeply inside the proton and to investigate physics processes in which each collision always includes an electron and a quark. The LHC collider will either search for the Higgs (if Fermilab fails to find anything) and/or it will either find or kill the theory of Supersymmetry (discussed in Chapter 8). RHIC (the Relativistic Heavy Ion Collider) at Brookhaven National Laboratory can collide bare gold nuclei together and is looking for an entirely new state of matter, the so-called quark-gluon plasma, in which the energies will be so great over such a large volume that the quarks and gluons are thought to be able to break out of the protons and neutrons and intermix freely. In addition, there are much lower energy, but very high beam rate, accelerators designed to look for rare physics, specifically CP violation (discussed in Chapter 7). These accelerators serve the BaBar detector at the Stanford Linear Accelerator Center and the Belle detector at KEK in Japan. These competing accelerators have their energy tuned to copiously produce $b\bar{b}$ pairs (pairs of b quarks and b antiquarks). See Table 6.3 for details.

All of these myriad accelerators, while they differ in many details, have a lot in common. They all collide counter-rotating beams of charged particles (except CEBAF). They all accelerate particles with electric fields and steer the beams with magnets. And all accelerators are a crucial component of the experiments that ask and answer the interesting questions posed by frontier physics.

While the ability to produce such technically complicated accelerator complexes and the resultant beams is amazing, one must remember that the accelerators are just tools. The reason we accelerate particles is to better understand the behavior of matter under extreme conditions. Accelerators provide us matter under these extreme conditions, but unless we record the collisions, we are just wasting time.

Table 6.3 Summary of some of the world's accelerators.

Accelerator	Lab	Turn On Date	Beams	Energy GeV	Ring Diam.	Major Goal
Tevatron	FNAL	1983	p$\bar{\text{p}}$	1960	2 km	Top, Higgs
LEP	CERN	1989	e$^+$e$^-$	90	8.6 km	*W, Z*
RHIC	BNL	1999	Many*	200**	1.2 km	qg plasma
HERA	DESY	1992	ep	310	2 km	Proton structure
LHC	CERN	2008?	pp	14,000	8.6 km	Higgs, SUSY
KEK B	KEK	2000	e$^+$e$^-$	10.6	0.5 km	b CP physics
PEP II	SLAC	1999	e$^+$e$^-$	11.1	0.7 km	b CP physics
CEBAF	TJNAF	1994	ep	4.8	***	Nuclear
SSC	SSCL	****	p$\bar{\text{p}}$	20,000	19 km	Higgs, SUSY

*can accelerate several types of nuclei from hydrogen to gold
**indicates energy per nucleon, but in gold there are many nucleons
***does not have circular structure
****cancelled 1993

What we need is some sort of camera that can record each collision for later study. Ideally, we'd like to have a camera that can take a clear picture of the "interesting" part of the collision. But all of the "interesting" stuff occurs in a space about the same size as a proton (or even much smaller). You might imagine that the desire to take such small pictures would result in a tiny camera, but in fact modern particle detectors can weigh thousands of tons. So the real question one should ask is "How do physicists record data from these colossal collisions and how do they make sense of what they see?"

Particle Detectors: The World's Biggest Cameras

When you first see a modern particle physics detector, you're struck by how large it is. Smallish detectors weigh about a thousand tons, while modern top-end detectors weigh in at about 5000 tons. Currently under construction are two huge detectors that dwarf even these. So there are some interesting questions. Why do they have to be so big? Also, how do you measure the particles? Let's think about what we would like to measure in the ideal world. In a 2 TeV proton-antiproton collision, if one requires that the collision have some moderate violence (more on that later), one can have 100–300 "final state" particles (that is, particles after the interaction). In order to completely measure the event, there are certain things that you want to know. First, you'd like to know the identity of each particle (i.e. is it a photon, electron, pi meson, muon, etc.) You'd like to know each particle's energy and momentum; which is to say the direction that the particle is going and how energetic it is. In the non-relativistic world with which you're familiar, this means how fast the particle is going. While the rules are different in the relativistic world of particle physics, the idea of speed being related to energy is useful enough to help you get the idea. See Appendix D if you want more detail.

The final information you'd like for each particle is the exact point in space where the particle came into existence. If you know these four things (position at creation, particle type, energy and momentum) you know all that one can reasonably hope to know about the collision. If you were being greedy, you'd also like to know as much about each particle's history as possible. For instance, a pion might be made in a jet as part of fragmentation or it could come from the decay of a heavier particle, which itself was created in the jet. The bottom line is that you want as much information as possible about each particle.

So while there are hundreds of final state particles in a typical collision, let's concentrate on just one. For reasons that will become apparent in a while, let's concentrate on a muon. A muon is a useful first particle to consider because it's relatively stable (i.e. a low energy muon can travel thousands of feet before it decays and a high energy muon can travel further). The muon has an electric charge, so it can be manipulated by electric fields and most importantly, this electric charge allows the muon to ionize matter. Ionization is the most-used technique to measure the energy of high energy particles.

Ionization and Tracking

When you think about it, one detects anything by how it interacts with its surroundings. You detect a white object by the fact that it reflects photons into your eye. You detect a black object because it reflects nothing. You know that there's a baby in the next room, because when it cries it causes the air to vibrate, which in turn vibrates your eardrum, which stimulates nerves and so on. Something is detectable if it interacts with its surroundings.

So too is it with elementary particles. Because a charged particle carries an electric field, as it passes through matter, the atoms in the material "see" (i.e. feel the effects of) the charged particle. Both the atomic nucleus and the atomic electrons see the charged particle, but because the nuclei are so massive, they don't move very much. The electrons however, have such a small mass that they can be knocked completely off the atom. Generally atoms have the same number of electrons and protons, but if they don't (as in the case when one knocks off an electron), we call this atom an ion. The process of knocking electrons off atoms (and thus converting atoms to ions) is called ionization, and is illustrated in Figure 6.10.

Once this point is understood, you're 90% of the way towards understanding particle detection. If a particle passes through matter, it ionizes the atoms within the matter. We collect the electrons knocked off the atoms (details on how in a short while) and infer the passage of the charged particle by the existence of the liberated electrons. Further, since in each liberation of an electron, the charged particle loses energy, we can determine the energy of the particle by how deeply it penetrates. For instance, if each time an atom is ionized, the charged particle loses 1% of its energy (which is much too large a number in reality), then after the particle passed 100 atoms, it would lose all of its energy and stop. It's something like a skidding car.

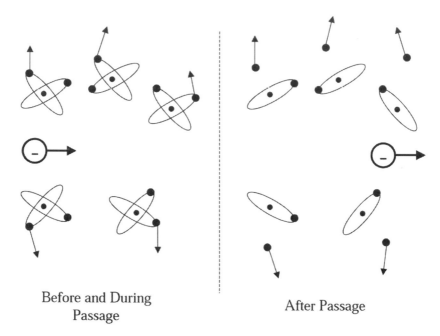

Before and During
Passage

After Passage

Figure 6.10 A cartoon of how ionization works. A moving particle, carrying electric charge, moves through matter, knocking off electrons as it goes. These electrons are collected and detected.

A fast-moving (i.e. more energetic) car will leave a longer skid mark than a slower one.

While how far the particle can travel through matter is a good measure of its energy, the number of electrons that it knocks off is even more useful. The reason is that once you have electrons, you can put them into specialized electronics and manipulate the electrical signal in useful ways. We see in Figure 6.10 how electrons can be knocked off the atoms. If we put an electric field in the area through which the charged particles pass, the electrons will feel a force and get pushed to one side. If you set up the electric field as we described when we were discussing accelerators, the electrons move towards the positive plate. Once they hit the plate (which, as we recall, is made of metal), the electrons flow through wires to electronics which essentially count the number of electrons. So before we get back to a more

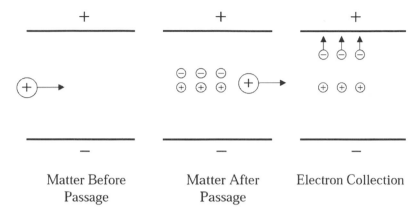

Matter Before   Matter After   Electron Collection
   Passage   Passage

Figure 6.11 The essential steps in ionization electron collection. A particle moves through matter in the presence of a transverse electric field. The particle knocks electrons off atoms, leaving electrons and positively charged ions. The light electrons are then collected on one of the plates and directed to electronics.

general discussion of particle detection, let's recap what we know about ionization. A charged particle passes through matter. The electric field of the particle knocks the negatively charged electrons from the atoms in the matter, leaving positively charged ions. An electric field that we impose pulls the electrons and ions even further apart (so they can't recombine, which they would ordinarily do). The electric field causes the electrons to move very quickly (because they have a very low mass), but the positively charged ions move very little (because they are very heavy, although they eventually do move). This sequence of events is shown in Figure 6.11.

There is one final important point. Ionization is the process whereby electrons are knocked off the atoms in the matter. This process slows down the particle crossing the matter and it occurs whether or not we collect the electrons. So electron collection is irrelevant to ionization. We only collect the electrons to *measure* the ionization.

It's kind of like having a car lose its brakes and crash into a tree farm (with little tiny trees). The car will hit many trees, breaking them (i.e. ionizing atoms). Each breaking of a tree slows the car a little bit. Eventually, the car comes to rest. Whether or not you look at them, trees are broken (ionized), which causes the car (particle) to stop. By counting the number of trees (i.e. collecting the electrons), you can measure how fast the car was going, which is related to the car's original energy.

While we talk about counting electrons to determine a particle's energy, we still need to know how far a particle can penetrate before it stops. That's because we need to know how much material to put in front of the charged particle. If it turns out that a particle will stop in one foot of material, and we only put up one inch, the particle will slow down (really lose energy) through its passage through the material, but it won't stop. We'll find this property useful presently.

As you'd imagine, how far a particle can penetrate through matter is related to the density of the matter through which it passes. A dense material means more atoms. More atoms mean more electrons to ionize in the same amount of space. So a particle stops more quickly in a dense material than a non-dense one. Taking as an example three materials; a gas (like air), plastic and solid iron, one can calculate the amount of material needed to stop a particle. For example, let's consider a muon carrying 10 GeV of energy (this is a respectable, but not impressive, amount of energy). We would need 56 kilometers (33 miles) of air, 50 meters (160 feet) of plastic, and 8.5 meters (27 feet) of iron to stop our example muon. Twenty-seven feet of iron is a lot (and recall that this wasn't really a very impressive muon). So we'll need to be cleverer somehow. But recall that ionization is at the heart of everything that we'll discuss from now on.

There's another important technique that we can exploit and that we've already discussed. This is the behavior of a charged particle in a magnetic field. If a charged particle is moving in a magnetic field, it gets deflected to the side. The path the particle travels is along the circumference of a circle. You will recall that this was the reason that we used magnets to make the great circular accelerators. For a fixed and constant magnetic field, the radius of the circle over which the particle travels is related to the energy (really the momentum) of the charged particle. Higher energy particles travel along the circumference of bigger circles. Finally, whether the particle travels clockwise or counterclockwise around the circle depends on the sign of the charge of the particle. Positive and negative particles travel in opposite directions. Which goes which way is arbitrary (as we can change it by changing the direction of the magnetic field or by changing where we stand to look at the particles). But negative and positive particles always travel in opposite directions. These points are illustrated in Figure 6.12. We see in Figure 6.12a that in the absence of a magnetic field, the charged particle travels in a straight line. In Figure 6.12b, we see how particles with negative and positive charge travel in opposite directions. Finally, Figure 6.12c illustrates the fact that lower energy particles are affected much more by magnetic fields than high energy ones. Low energy particles travel obviously curved paths,

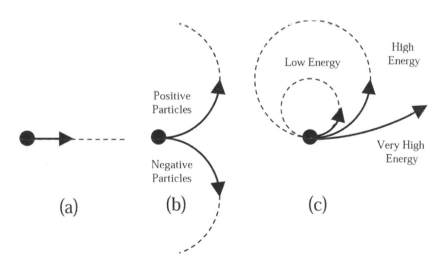

Figure 6.12 A magnetic field causes a charged particle to move in a circular orbit. A positively charged particle will be deflected in the opposite direction as a negatively charged particle, although which direction each is deflected is arbitrary. In addition, the radius of the circle in which the particle moves is related to the particle's momentum. Higher momentum particles travel in larger circles.

while very high energy particles travel paths that look nearly straight to the casual observer. Now that we know about the two crucial techniques of ionization and magnetic fields, we're ready to tackle the interesting question of particle detection.

Probably one of the neatest images in particle physics is given in Figure 6.13. In it, we see dark lines against a lighter background. Each dark line is an actual photo of the path of a charged particle through matter. The fact that the paths are curved indicates that the photo was taken in a magnetic field, which caused the particle to deviate from the more natural straight path. More modern techniques are better in many ways, but it's hard to beat a bubble chamber photograph for clarity, so we will spend some time discussing this photo to solidify our knowledge of ionization and magnetic fields.

The story of the invention of the bubble chamber is really quite interesting. It goes something like this. It was a dark and stormy night. Well, actually it wasn't but that's always a good way to start a

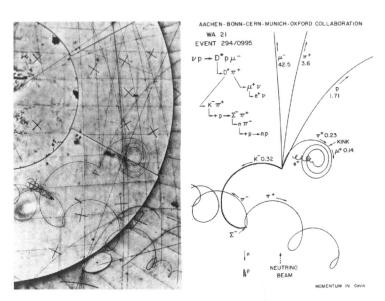

Figure 6.13 A bubble chamber photograph showing an interaction within the detector. The diagram on the right identifies the particles involved in this particular interaction. (Figure courtesy of CERN.)

story. The accepted legend has it that in 1952, Donald Glaser was in a bar in Ann Arbor, Michigan, where the University of Michigan is situated. He was staring moodily into his beer, watching the bubbles rise to the top. One of his fellow patrons remarked that the bubble made a nice track and Glaser was struck by the idea of how he could use radioactive particles to form bubbles in a properly prepared liquid. Glaser put a liquid (diethyl ether in fact) in a glass vessel and heated it under pressure so that it was just below the boiling point. Just before the particles hit the chamber, he'd reduce the pressure, leaving the liquid in a "superheated" state. A liquid in a superheated state is hotter than the boiling temperature, but without boiling. All that is required is a slight nudge to induce boiling. This nudge was the transit of a charged particle. As the particle crosses the liquid, ionizing as it goes, it leaves a little trail, something like a jet contrail, which can be photographed. You can see something like this phenomenon by putting salt in a glass of soda. (Rumor has it that Glaser used beer, but you should definitely use soda, as beer is far too valuable to be used for such experiments.) At the 1953 meeting of the American Physical Society, Glaser announced his work. In 1960, he was awarded the Nobel Prize at the unusually young age of 34. This tale is really quite marvelous. Any story that starts with beer and interesting physics and ends with a Nobel Prize is OK in my book. It also provided a great example to my fellow graduate students and me after a long day smashing atoms. If Glaser could use beer for inspiration, that was good enough for us. We were even willing to forgive Glaser's later career change to biology research.

While Glaser invented the bubble chamber, it was Luis Alvarez who turned it into a real detector technique. Glaser's first example chamber was only a few inches in diameter and made of glass, because Glaser believed that only glass was smooth enough to not introduce unwanted bubbles from imperfections in the vessel's surface. Alvarez quickly increased the size of the bubble chamber by using metal containers with glass windows in order to photograph the tracks. In addition, Alvarez switched the liquid from diethyl ether to liquid hydrogen.

Further improvements culminated in Fermilab's huge 15-foot diameter bubble chamber, which commenced operation on September 29, 1973 and was retired from service on February 1, 1988, after an astonishing 2.35 million photographs. For Alvarez's work using the bubble chamber, through which he discovered numerous particles, he was awarded the 1968 Nobel Prize. Alvarez was a true Renaissance man, dabbling in many things besides physics. Some of his more interesting endeavors were the use of cosmic rays to search for hidden chambers in Egyptian pyramids, his analysis of the famous "Zapruder" film, which depicts John F. Kennedy's assassination and perhaps his most extraordinary accomplishment; the explanation of the extinction of the dinosaurs by a meteor or comet impact. Alvarez was an eclectic guy.

So looking again at Figure 6.13, we see the tracks caused by particles as they passed through the liquid hydrogen, causing it to boil along the particle's path. A photo is taken of the string of bubbles caused by the particle's transit. And finally, the presence of the magnetic field ensures that the particle's path is bent, with the degree of bending related to the particle's energy; the greater the bending, the lower the energy.

The bubble chamber, while the simplest to visualize detector technology, suffers from a number of limitations. It's relatively slow, with a maximum repetition rate of a few times per second. It's also a little difficult to ensure that the bubble chamber records the 'right' interactions (more on that later). Also, analysis of bubble chamber data requires actual people to look at film … a very tedious and error-prone method. Clearly a new approach was desired.

The crucial idea of how to overcome the limitations of a bubble chamber came from a Frenchman, Georges Charpak. A member of the French Resistance and a survivor of a Nazi concentration camp, Charpak became one of the most prolific of the dabblers in the technology of particle detection, a predilection that garnered him the 1992 Nobel Prize. Probably his most notable work was the invention of the wire chamber. A wire chamber is pretty easy to understand. Taken to its simplest element, a wire chamber can be visualized as a very thin wire under tension and surrounded by a gas (often a mixture of gases with a significant fraction of argon). When a charged particle crosses the argon gas, the argon is ionized. Through a clever arrangement of electric fields, the electrons from the ionized argon gas are made to move towards the wire, where they "hop on" and are directed along the wire and out to waiting electronics. Thus the passage of a charged particle is announced by a signal in your readout electronics. It's kind of like a kid pulling a prank on you by ringing your doorbell and running away. While you never actually see the kid, the pulse of electricity that rings the doorbell is proof of the kid's passage across your porch.

While a single wire can illustrate the principle of a wire chamber, what is usually done is many wires are arranged in a plane of parallel wires, much like a harp. When a particle crosses the plane, a signal is generated in the wire which is nearest to the particle's passage. If you think about it, this gives you position information (because you know which wire the particle was closest to and you know where that wire was located). You can then measure the particle's path by positioning many planes of wires and noting which wires were hit. This idea is demonstrated in Figure 6.14. I have drawn the planes of wires so that

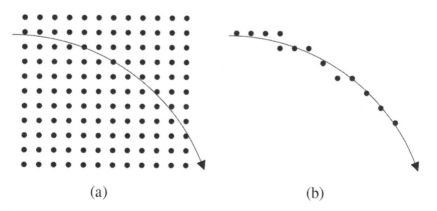

(a)   (b)

Figure 6.14 The essential aspects of a wire chamber. Each dot represents a wire, viewed end on. A particle, deflected by a magnetic field, hits a series of wires. By seeing which wires are hit, one can reconstruct the particle's path.

we are looking along them, thus each dot signifies the end of a wire. In Figure 6.14a, I show all wires and the particle passing through them. In Figure 6.14b, I only show the wires that were near the particle's path. Using the crucial skills you picked up in kindergarten, you can connect the dots and reveal the path of the charged particle. The fact that the particle's path is curved reveals, of course, the presence of a magnetic field.

By using this technique, one can make as large a detector as desired. We see that we have achieved many of the goals that we set out when we listed what we'd like to see in our detector. We know the particle's position and path to great precision, as well as its energy (by the amount of bending seen in the particle's path). Further, we know that the particle carries an electric charge and, since we know the direction of the magnetic field, we know if the charge is either positive or negative (by the clockwise or counterclockwise motion of the particle). We've really made great progress.

While the "harp" detector (really called a wire chamber) works well and has a long and storied past, it too has its limitations. One can only put the wires so close together before the chamber becomes hard to operate for technical reasons. So even with clever electronics, one can only measure the position of particles to a precision of a couple of tenths of a millimeter. While impressive, even better precision is needed for many measurements. Another important consideration stems from the fact that high energy collisions can have several hundred particles coming out after the collision. If two particles are near one another, they may both cross near the same wire and the two particles will leave only one signal. The only solution to these problems is to decrease the spacing between adjacent wires; something we already said was difficult beyond a certain point. So what do we do?

It turns out that one can replace the wire and gas technology with little strips of silicon. These strips can be extremely small, thanks to the extraordinary efforts of the computer industry to make ever-smaller chips. In a modern detector, the width of a strip can be 0.05 millimeter. Think about it. If you hold up a meter stick, the smallest separation on the stick is one millimeter. These silicon strips are so thin that you can place 20 of them side-by-side in the space of a millimeter. Perhaps just as impressive is the total number of silicon strips that comprise a modern detector, which approaches one million. New detectors are being built that will dwarf even these amazing numbers. These new detectors will come online in about 2008. The net result is that we can put silicon based tracking devices near the center of the detector and get very precise information of the behavior of the collision at distances very close to the spot at which the interaction occurred.

Of course, we don't yet know how to observe a neutral particle (i.e. one that carries zero electric charge, say a photon or a neutron). Further, we can't easily distinguish between an electron (e^-) and a negative pion or muon (π^- or μ^-) or, conversely, a positron (e^+) and a positive pion or muon (π^+ or μ^+). Further, a little thoughtful reflection will reveal another limitation. As we know, particles carrying a great deal of energy are not bent much by magnetic fields and travel through straighter and straighter paths.

As we see in Figure 6.15, eventually one gets particles of such energy that they hit a single row of wires. Higher energy particles will travel even straighter and still hit just the same set of wires. Thus once

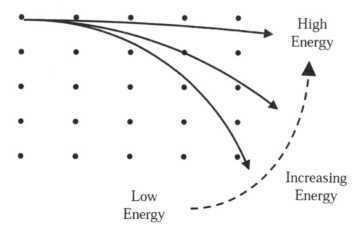

Figure 6.15 Charged particles traveling through a wire chamber, each with greater energy. Above a certain energy, all particles will hit a single set of wires. This sets a limit on how high an energy a particle can carry and be characterized by this detector technique.

the particle gets above a certain energy, you cannot distinguish between it and those of higher energy. Since all particles of that energy and higher hit the same group of wires, they all look the same and thus they all have the same measured energy, which we denote "lots." So yet again, we need to invent additional techniques.

Earlier, we talked about ionization and we figured out that a 10 GeV muon would take 27 feet of iron to stop. That's a heckuva lot of iron and so we'd like to figure out how we could shorten that. If ionization were the whole story, we'd be out of luck. However, there is a phenomenon that we've neglected. This is called showering. Muons don't exhibit this showering behavior; so let's talk about an electron.

Calorimeters: Measuring Energy

When an electron passes through matter, it causes ionization like any other charged particle. However, it can also do something else. When it gets near the nucleus of the atom, it is affected by the electric field of the nucleus and it gets deflected. As a consequence of the deflection process, it emits a photon. So while before the collision there was a single electron, with a particular energy (say 100 energy units), after the collision there is an electron and photon, each carrying (say) half of the energy of the original electron (so they each have 50). The thing that happens next is what seems so miraculous. The electron with 50 units of energy travels a short distance and hits another atomic nucleus and emits another photon, each carrying half of its energy (and so each have 25 units). The question becomes what happens to the photon. Well photons can convert into electron and positron pairs (e^+e^-) and they do this easily near an atomic nucleus. Each of the electrons and positrons carry half of the photon's original energy (so they each have 25 units). But now the electron and positron can hit an atomic nucleus and radiate more photons. Thus what we have is a cascade, or shower, of electrons, positrons and photons that grow rapidly in number, with each increase in number

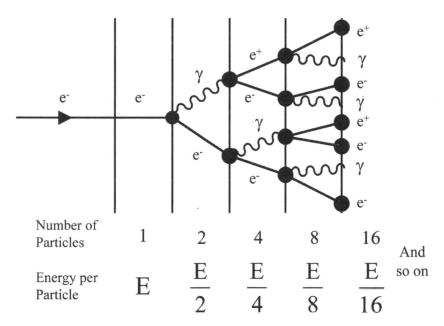

Number of Particles	1	2	4	8	16	And so on
Energy per Particle	E	$\dfrac{E}{2}$	$\dfrac{E}{4}$	$\dfrac{E}{8}$	$\dfrac{E}{16}$	

Figure 6.16 The essential features of calorimetry for an electron. A single electron interacts with an atom in the detector, giving off a photon. The electron continues on and interacts again. The photons eventually pair-produce electron-positron pairs. The number of particles grows geometrically, with each daughter particle carrying a fraction of the energy of its parent. In the end, a single high energy particle is transformed into many low energy particles.

yielding a corresponding decrease in energy. This effect is seen in Figure 6.16.

The distance between subsequent creations of particles is called a radiation length. Each particle travels on average one radiation length before splitting into pairs ($e \rightarrow e\gamma$ or $\gamma \rightarrow e^+e^-$). The distance that makes up a radiation length depends heavily on the material, but typically is a low number in metals (e.g. in lead, it's 6 mm, or 0.25 in, while for iron it's 18 mm, or 0.71 in).

The real importance of showering is the fact that we convert a single high energy particle into many low energy ones. These low energy particles ionize the material and eventually stop. Since the distance a particle can travel through a material, losing energy to ionization, is proportional to the energy of the particle, by increasing the number of particles (and decreasing their energy), you greatly decrease the thickness of the material needed to absorb the original particle. For instance, if you increased the number of particles by a factor of 100 (and thus reduced the energy to 1/100), the amount of material needed to stop the particle would be 100 times less.

The actual depth of iron needed to contain all of the energy from the electron must include the length necessary to let the shower grow. As we see in Figure 6.16, after the first few interactions, there are only a few particles. It takes several radiation lengths to build up to the maximum number of particles (which can reach several thousand). When all effects are included, we find that a 10 GeV electron can be entirely contained in 15–20 inches of iron. This thickness is needed to catch the few odd particles that randomly penetrate unusually deeply. The bulk of the energy is stopped in less than 6–7 inches of iron. The distances in lead are about 1/3 those seen in iron.

Actually from what I've said so far, there's no way to justify the estimate of 15 inches of iron, as the shower could continue indefinitely until the energy of each particle was infinitesimally small. This is because the picture I've described to you is a little too simplified. In reality, if the energy of a particle gets below a threshold, it no longer converts into pairs of particles and instead just passes through the material, ionizing as it goes. It's kind of like an airplane. Above a certain speed, it can fly. It may fly clumsily or in an agile manner, depending on speed, but it flies. However, once the speed goes below a threshold, the plane simply no longer flies. Below a particular energy, electrons, positrons and photons no longer make pairs.

I like to compare particle showers to fireworks. A particle enters the material, only ionizing. It then interacts with an atomic nucleus, making more particles, which in turn make more particles, with each subsequent particle having a lower energy, until the low energy particles march through the material, only ionizing. Similarly, when a firework is launched, it often shows a little glowing trail as it moves upwards, reminiscent of ionization. The firework has an initial explosion (interaction) and throws off little "bomblets," which then undergo additional explosions. The cascade of explosions continues, until the bomblets run out and the dying embers fall, glowing and winking out one by one, just like the final stages of ionization of the last particles in a shower.

The discussion of showering has thus far only covered electrons, positrons and photons. What about the hadronic particles, say protons and pions? Well, they shower too, but in a more complicated way. In order to shower, the hadrons need to actually hit an atomic nucleus, rather than just getting near it. This is more rare and thus the particle must travel further before it interacts. Further, each interaction is more complex than with electrons. Rather than just creating two particles after each collision, some random number, perhaps 3–8, of particles (usually pions) come out. These pions travel through the material and eventually interact with other nuclei. So the shower grows like it did in the case of electrons, but when all effects are taken into account, these showers tend to be quite a bit longer than ones initiated by an electron or photon, but still of manageable size (approximately 4–5 feet, if we continue with our 10 GeV example, this time of a pion). Figure 6.17 shows the moderately more complex shower of a hadron.

The examples of particle showers have been given in metallic materials, lead, iron and so on. And the behavior I've been describing occurs. However, there's a problem. I don't know how to easily measure the ionization of particles in metal. The typical method of measuring ionization (described earlier) consists of a special gas and electric fields in order to gather the ionization electrons. But gas is much less dense than metal and so a shower in air could take literally miles of air to complete. So what do we do? The answer is actually rather clever. What you do is to mix the two ideas. You take a slab of metal and follow it with a space filled with a low density material that can measure ionization. You then follow with another slab of metal and space with easily ionizable material and repeat this pattern many

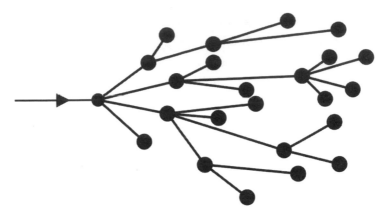

Figure 6.17 A cartoon of a hadron interacting with matter. For each collision, many particles come out, each with lower energy. While more complicated in detail, this is similar to how an electron builds an equivalent shower.

times. The metal lets the shower progress rapidly, while the low density material allows you to "peek" at the shower at various spots along its development. Since the less dense material we've described in the past was gas, you could imagine building a detector consisting of alternating slabs of metal and gas and this would work, but it turns out that there is another, preferred, option for measuring the ionization in the gaps between the metal plates. This new material is called scintillator.

Scintillator is typically made of plastic with some special chemicals mixed in. It looks very much like sheets of slightly-purple Plexiglas. Unlike a gas, in which the ionization is made evident by the presence of electrons which you can measure if you're clever enough, when scintillator is ionized, it gives off a very fast blink of light, typically violet or a very light blue. The light bounces in the plastic until it hits a specialized bit of equipment that can convert light into an electrical signal. Once electricity is generated, it can be fed into specialized electronics and computers and thus be analyzed.

The specialized bit of electronics referred to in the last paragraph can be one of several different technologies. Recently, there has been a push towards solid-state devices, but there is a much older technology that is easier to explain and is just plain cool. Further, this "older technology" is still used by many modern experiments. A photomultiplier tube (also called a PMT (each letter is pronounced) or just a phototube) converts light into electricity extremely quickly. You need to know two things to understand PMTs. The first is that even a low energy photon (i.e. visible light) can knock an electron from an atom. An understanding of this phenomenon is one of the things that garnered Einstein his 1921 Nobel Prize (not, as many people think, his theories of relativity). Aha you say ... we're there. But unfortunately, that's not true. One photon can liberate (up to) one electron, and one electron is an extremely tiny amount of electricity. So we need to somehow amplify this single electron, liberated by a single photon, into enough electrons to make a big enough electrical signal that modern electronics can register it. This brings us to our second bit of knowledge, the fact that when an electron is moving fairly quickly and hits certain metals, it can knock several electrons out. In fact, typically one can knock out 4 electrons for one electron in.

So we now can make a phototube. We take a hollow tube of glass and paint one face of the cylinder with a mixture of metals that can easily convert a photon into an electron. We then add an electric field which accelerates the electron to a plate or mesh of metal, where it knocks out 4 electrons. Now comes the clever part. We take these four electrons and accelerate them to another plate, where each of them knocks out 4 more electrons, giving us 16 ($= 4 \times 4$). Another plate will give us 64, and so on, with each additional plate contributing another factor of 4. A cartoon of how this works is shown in Figure 6.18. Typically, a phototube will have 8–14 plates. A 12 plate phototube will convert one photon into 4^{12}, which is about 17 million, electrons. Thus, a phototube can convert a single photon into enough electrons with which modern electronics can work. The phototube is even more amazing, because it does this multiplication in about 10^{-8} seconds, and in a very small package. A typical phototube is a cylinder with a diameter of about 1–2″ and about 4–6″ long.

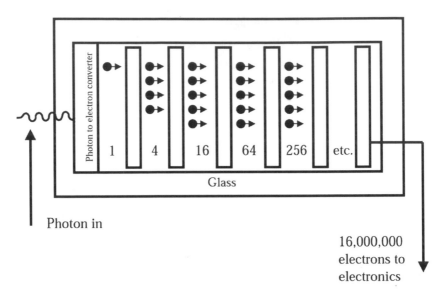

Figure 6.18 Principles of photomultiplier operation. A photon enters from the left and dislodges an electron. An electric field directs the electron to another plate, where it dislodges four more electrons. A series of electric fields directs the output electrons, with each electron dislodging four more electrons. In this way, a single photon can generate millions of electrons and be detected in specialized electronics.

So why do you combine such outlandish technologies together to form giant detectors, consisting of alternating plates of metal and either plastic scintillator or gas? It's because it provides a nice way to measure a particle's energy. The number of particles generated in a shower is roughly proportional to the incident particle's energy. By sampling the number of particles in the shower, you can measure the energy of the incident particle. Such a detector is called a calorimeter (or energy measuring device). Calorimeters are integral components of modern particle detectors, as we shall see.

Thus we now know a lot of useful tricks that can aid us in identifying the kinds of particles that we get in our collisions. You may recall that this was one of our requirements when we listed the features we'd like to see in our particle detectors. Each type of particle interacts with matter differently. Taking 10 GeV to be a representative energy, a muon, which does not shower, penetrates about 27 feet of iron. An electron, which showers electromagnetically (or the short way), is absorbed in about 1–1.5 feet of iron. Finally, a hadron (pion or proton for example), which showers via nuclear processes (or the long way) takes about 4–5 feet of iron to absorb. Further, a photon showers very much like an electron. Neutrinos, which don't interact very strongly with matter, can travel through light-years of iron. Finally, we should recall that the neutral particles do not ionize and thus can't be seen unless they initiate a shower.

One Detector: Many Technologies

Knowing these characteristics, we can draw a cartoon of a detector and see how the various particles (muons, electrons, hadrons, photons, neutrinos) interact. We see that each has a distinct behavior and thus signature, enabling us to identify what the particle's identity was, by how it interacts in our detector.

In Figure 6.19, we see roughly the signatures we expect to see for various particles. On the left is a region consisting of only an easily ionizable gas, while on the right is a calorimeter consisting of alternating metal plates and sheets of plastic scintillator. A muon leaves a single track in both the gas and the calorimeter. An electron leaves a single track in the gas, but generates a shower that penetrates only a little way into the calorimeter. A pion, on the other hand, has a similar signature in the gas, but has a much deeper shower. A photon, being electrically neutral, leaves no track in the gas, but generates a shower very similar in character to that of an electron. Finally, the non-interacting neutrino leaves no signal in either detector.

With this knowledge, we are able to get a handle on the essentials of a modern particle detector. We recall that there are two kinds of beam-related particle physics experiments, fixed target and collider. While fixed target experiments can be very sophisticated (and I learned the trade on one such experiment), we will concentrate on

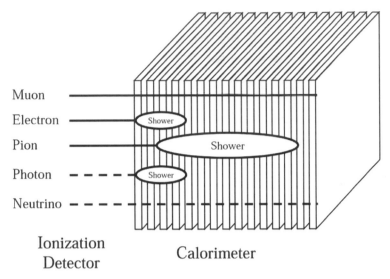

Muon

Electron

Pion

Photon

Neutrino

Ionization
Detector

Calorimeter

Figure 6.19 A cartoon of how different particles interact with a large block of matter. A muon will travel through the matter, undergoing only ionization. An electron will interact in a short length, depositing its energy. In contrast, a pion looks similar, but with a greater degree of penetration. Photons look like electrons, except, being electrically neutral, they leave no signal prior to undergoing a shower. Neutrinos do not interact in the detector and escape undetected. In this manner, one can identify the various particles that hit your apparatus.

collider experiments, as these are a more modern method for measuring the behavior of matter under extreme conditions.

Because most (but not all) collider experiments consist of two beams of equal energy, counter-rotating and colliding head on, the basic geometry is such that particles can come out of the collision in all directions. This is because the beams are of equal energy. If one were of much higher energy, the particles would tend to come out more in the direction in which the higher energy beam was going. But because we're discussing the case of equal energy beams, this sets the basic geometry of the detector. In the perfect world, you'd like your detector to be a sphere, centered on the collision point, with two little holes to let the beams enter and exit. Imagine a basketball with

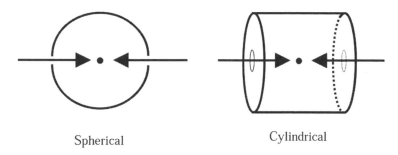

Spherical   Cylindrical

Figure 6.20 Conceptual diagram of a collider beam detector. In both cases, the detector nearly surrounds the collision point. The cylindrical geometry shown here is more often used.

the center of the basketball at the location where the beams collide and you've got the basic idea. The most important point is that you should fully surround the collision so as to collect all of the particles created in the interaction. If you lose particles, you lose information.

Unfortunately, engineering concerns, e.g.. questions of mechanical support, portability of components, ease of construction, etc., make a different geometry a preferred choice. The more practical geometry is a cylinder, with the flat ends perpendicular to the beams and the center of the cylinder located at the collision point. Figure 6.20 shows the two geometries. From now on, we will treat collider detectors as cylinders.

While the design of a large particle detector is an art and thus one can make numerous choices, there are some basics, shown in Figure 6.21. Typically the center of the detector consists of a silicon detector, in order to precisely measure the characteristics of the collision near the origin. This is followed by a larger volume, generally filled with a tracking chamber; often using the gas and wires principle discussed earlier, although other options are possible. This entire volume is usually filled with a magnetic field, in order to deflect the particles so as to measure their energy. This central tracking volume is usually followed by two layers of calorimeters; a thin one designed to make the showers from electrons and photons as short as possible and a second one that is much deeper in order to contain the hadron (e.g. pion and proton) initiated showers. These calorimeters consist of

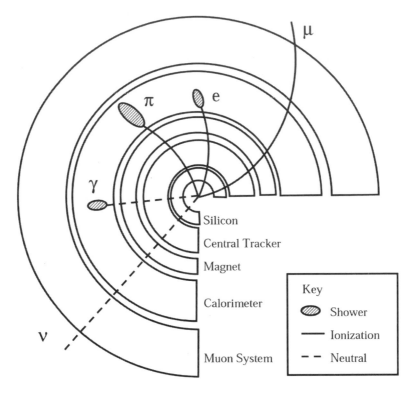

Figure 6.21 Cartoon of an end view of a typical detector. In this figure, the beams would be going into and out of the page. The missing part of the detector in the lower right quadrant is to leave place to label the parts, the actual detector would be circular in this view (and cylindrical in a three dimensional rendering). Overlaid are representative tracks and signatures of typical particles in which a modern experiment would show interest. Dashed lines indicate undetectable (in that equipment) while solid lines indicate the detector would observe the transit of that particle.

metal plates separated by regions of easy-to-readout material. The types of metal can vary, with lead and iron being traditional choices, although uranium (^{238}U, the kind that doesn't blow up) is a very attractive choice too.

After the calorimeters come the muon detectors. Muon detectors are somewhat misleadingly named, as the central tracker and calorimeters can measure muons too. However, all particles except muons and the invisible neutrinos are stopped by the thick calorimeters. Thus any particle that makes it through the calorimeters and is visible in the "muon detectors" must be a muon. One quarter of a modern detector is shown in Figure 6.22.

While the detector configuration shown in Figure 6.22 can be taken as typical, each detector can vary significantly from the base design shown. This is because each detector is a hugely complex piece of equipment and, given the limited budget, compromises must be made and each group designing a detector needs to make decisions on which capability is the most important and for which one the group is willing to accept less than optimum performance. Such decisions are taken most carefully, using the best scientific judgment available. But at the frontier, the choices are not always black and white and thus intuition comes into play. History rewards those who make the right choices and forgets those who don't.

At a particular laboratory, one often has two large groups building competing detectors. The reasons for the duplication are many: redundancy, in the event that a disaster seriously damages a detector; competition, in order to encourage the research groups to work hard (as if that were needed); and as a cross-check, in case one group makes an error, the other will likely catch it. At Fermilab, the two huge detectors are DØ (pronounced D-Zero) and CDF (an acronym for Collider Detector at Fermilab, a throwback to the time when it was the sole detector). Photographs of these two detectors are shown in Figure 6.23. At the HERA accelerator, ZEUS and H1 grace the ring, while at the LEP accelerator, fully four experiments were present: ALEPH, DELPHI, L3 and OPAL. The huge LHC accelerator, scheduled to turn on in the second half of this decade, will be supplying beam to two truly huge detectors, ATLAS and CMS, as well as a couple of other more specialized and smaller detectors.

Just for fun, I show in Figure 6.24 two examples of events as recorded by the DØ detector. Both DØ and CDF each recorded over 60 million collisions over the course of the last data-taking period (1992–1996) so, as you might imagine, I picked two especially

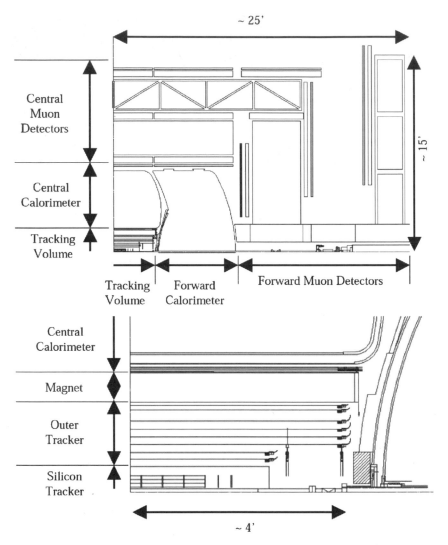

Figure 6.22 A quarter view of a typical particle detector. The detector clos-est to the beam is made of small strips of silicon. The surrounding detector is often made of parallel wires, surrounded by gas. Two calorimeter detectors follow, one for measuring electromagnetic energy, while the next one is designed to measure hadronic energy. The entire detector is surrounded by a muon detector. The bottom figure is a close-up of the central tracking volume. In both figures, the beams travel in the horizontal direction and collide in the lower left-hand corner of the figure (i.e. the center of the detector). (Figure courtesy of Fermilab.)

Figure 6.23 The Fermilab DØ detector (left) and CDF (right). (Figure courtesy of Fermilab.)

photogenic collisions. Figure 6.24a is an event in which a top and anti-top quark were created. This image views the detector from the end, with the beams going into and out of the page. As discussed in Chapter 4, four jets, a muon and a neutrino are clearly seen (as well as an additional muon in a jet, which usually indicates a b-quark jet). The second image, Figure 6.24b is my personal favorite. This collision is (as of this writing) the most violent, yet well understood (i.e. from the 1992–1996 data-taking period) collision ever recorded (and fully understood) between two subatomic particles, with fully 900 GeV of energy involved, probing distances about 10,000 times smaller than a proton. That's a lot of energy and a tiny distance, even for modern particle physicists. In addition, when I show this event to experts who have not seen it before, they usually say something like "Is it real?" This is because the event is extremely clean, with none of the confounding icky signals that often contaminate real data. This event was recorded on December 25, 1994 ... one of the best Christmas presents an experiment could hope for. Unlike Figure 6.24a, this event is displayed by viewing the detector from the side, with the beams entering the detector from the left and the right. In the current data-taking run, we might hope to get about 10–20 more events like it

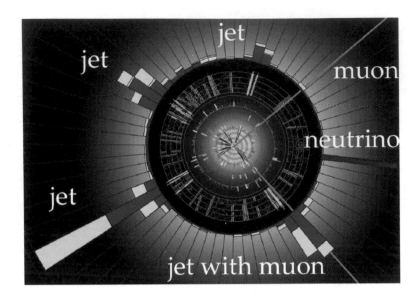

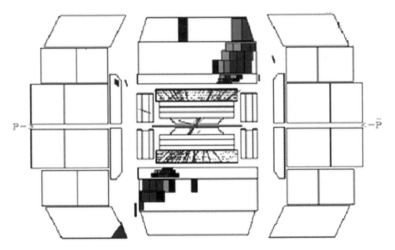

Figure 6.24 Two representative "photographs" of exceptional collisions, recorded by the DØ detector. The top photograph is one in which a top/antitop quark pair is believed to have been made. The bottom photograph is the single most violent, yet well understood, collision recorded as of this writing. (Figure courtesy of Fermilab.)

(in fact, we've already seen a few, but our current understanding of our apparatus suggests we should be cautious). This suggests that our new "most violent" collision will involve even greater energy. I can't wait!

While we now have a good idea of how a modern accelerator and detector work, we have neglected a very important question. When the Fermilab accelerator is running at full luminosity (i.e. flat out, with a maximum set of beams), approximately ten million collisions occur each second in each of the two detectors. However, most of the collisions will be of a boring type, which have been thoroughly studied before. Because each experiment can only record to computer tape about fifty collisions each second, one must somehow tell the detector which collisions are most interesting and which ones are boring. In a rough manner of speaking, each detector must look at and discard 100,000 collisions for every one that is recorded; and they must record the "right" ones. If you record the wrong ones, you will never make any interesting discoveries, a concern that keeps researchers up at night. So how is this done?

Well clearly in order to make a decision at such a breakneck pace, the whole process needs to be automated. If a particle physics detector can be thought of as a camera, it must also be a smart camera that decides what pictures to take. This capability is provided by sophisticated and fast electronics. These electronics are taught what sorts of signals in the detector are likely to indicate interesting physics and to keep those particular types of collisions. The process of deciding which collisions to record is called a "trigger."

Typically, there are many levels of triggers, often numbered 0, 1, 2 … with different experiments having a different number of levels. Level 0 is usually something simple like requiring that two groups of particles crossed in the center of the detector and that at least one pair of beam particles hit each other. Because of the intensity of modern beams, this only reduces the number of collisions that later electronics need to inspect by a few factors of 10. The next level (Level 1) might require that the detector had a lot of energy somewhere in it. This indicates that the collision had some degree of violence and that perhaps it might be interesting. Subsequent levels make more and more strict requirements on the event, in order to see if it is interesting enough to be recorded. Finally at the highest level of trigger (which could be Level 3 or so), one can ask for quite sophisticated requirements, for example that there are two highly energetic jets and one lower energy one, and further the event has an electron and a muon. In order to make such a sophisticated evaluation of the event takes a long time, even for computers (well 1/1000 of a second or so … long is relative, as anyone who's been on an airplane with a fussy baby can tell you). The reason that one must have so many levels of triggers is that each level takes ever longer to perform. If an event doesn't have at least a couple blobs of energy, somewhere in the detector, asking for jets, electrons and other complex requirements to be satisfied would be silly. Thus Level 1 rejects the event and Level 2 and Level 3 never see it. The fewer events that the highest level trigger needs to evaluate, the longer that the trigger can spend on each event and thus the more sophisticated the algorithm that you can use. Thus we now see why properly designing a trigger is one of the most critical tasks of an experiment. If any level of your trigger rejects that Nobel Prize winning event, it will never be recorded and the competition just might take that trip to Stockholm instead of you. And you'll just have to applaud politely.

Lifestyles of the Rich and Famous

When one considers just how large a modern particle physics experiment can be, naturally one wonders in part about the people who build them and just how long it takes to do an experiment. As you might imagine, such an endeavor takes many complementary skills to successfully complete. From what we've talked about so far, it is clear that you need physicists, mechanical and electrical engineers, computer scientists and civil engineers, as well as the often-overlooked technical support: technicians, administrative support, infrastructure (heat, lights, janitorial services, etc.) While the physicists write the books, take the credit and, to be fair, provide direction, all of these people play crucial roles in our discoveries.

It's interesting to understand just what sort of human scientific component is involved in a modern experiment. While it's hard to unequivocally state the number of people on a modern experiment, it's hard to imagine an experiment with fewer than 200 physicists. Both the DØ and CDF experiments at Fermilab have about 500 physicists and the monster experiments ATLAS and CMS at the Large Hadron Collider (LHC) have nearly 1500 each.

Taking the DØ experiment, with which I'm most familiar, we can learn a little more. There are 500 physicists working on the experiment, of which approximately 100 are graduate students who, in a different era, would be apprentices, learning the trade while working with more experienced scientists. While they're called students, by the time they start to work directly on the experiment, they have finished their class work and they spend their time not only doing research, but more importantly learning *how* to do research; learning how to think critically and how to extract meaningful information from imperfect data. After they graduate, they have learned skills that allow them to work independently and to initiate their own experiments.

Those who choose to pursue particle physics as a vocation have a long journey in front of them. After 4 years of college, they enter graduate school, where they work towards their doctoral, or Ph.D., degree. Depending on circumstances, personal drive, effective mentoring and, to an unsettling degree, luck, this can take about 4–8 years. After graduation, they usually take a postdoctoral research position and this lasts for 3–8 years, although the norm is 5–6. After this, they can take a junior faculty or national laboratory position, which is still provisional. It is during this time when they are expected to take a leadership role and show that they have a proper vision of what can be considered to be important research topics. If they succeed in this phase, which takes approximately 6 years, they are considered to be leaders in the field and they are granted tenure. The full journey from neophyte to acknowledged expert could take about 20 years.

The numbers given here are for a person following an experimental physics path in a purely American setting. Theorists and people in some other countries can take less time and, in a few countries, it can take even longer. Regardless of details, a journey along this path is a significant commitment. But as one who has made the trip, I can say that the rewards are great, especially for the insatiably curious.

In addition to the 500 physicists on an experiment, there are approximately a like number of technical support; engineers, computer professionals and technicians, although the number and ratio of these people vary over the lifetime of an experiment. The efforts of these dedicated professionals are critical to our research success.

While we've spoken of the life cycle of a physicist, experiments have a lifetime of their own. Experiments take about 5–10 years to plan and a similar amount of time to build. To do the experiment takes about 5 years, during which we try to collide beams for every second of every day, or as close to that ideal as is remotely possible. No physicist worth his or her salt would advocate anything less than continuous operation, for time without beam is data not recorded. And data is why we're here.

After about 5 years, the detector usually needs refurbishment or an upgrade to keep up with inevitable improvements in accelerator performance and also to address questions unasked when the experiment began. Eventually, new accelerators and detectors come online and the torch is passed to a new generation. But a properly designed collider experiment can exist in various incarnations for 20–30 years.

Water Is for More Than Just Drinking

While so far we have concentrated on accelerators and the detectors used for colliding beam experiments, knowledge in particle physics and its cousin field cosmology can also be advanced by experiments that don't involve accelerators. In Chapter 7, we discuss the interesting field of neutrino astrophysics, in which a detector might consist of a huge underground tank consisting of 50,000 tons of water. Because there are great numbers of large underground water detectors, I'll spend just a few moments to explain here the principals whereby they operate.

In 1934, Pavel Alekseyevich Cerenkov was working as a research assistant to Sergei Ivanovich Vavilov. Cerenkov was assigned to understand what happens when the radiation from a radium source penetrates into and is absorbed in different fluids. Previous researchers had noted the blue glow when radium was placed near water, but prevailing wisdom suggested that the mechanism was fluorescence, which had been observed earlier. Cerenkov investigated more closely and showed, by changing the fluids, that fluorescence could not be the explanation. He further showed that the real source of the effect was the electrons from the radium hitting the water. He published these experimental results in the period of 1934–1937. What was lacking was an explanation of the question: "If not fluorescence, then what?"

In 1937, Igor Yevgenyevich Tamm and Il´ja Mikhailovich Frank were able to mathematically explain the origin of what has been called the Cerenkov Effect. The mathematics are tricky as usual, but the basic idea isn't so hard. As a charged particle travels through a material, it can agitate the atoms of the material. The agitation of the atoms can be relieved by the atom emitting a bit of light. So far, this doesn't really constitute Cerenkov light. In order to properly be called Cerenkov light, the charged particle has to move faster in the medium than light travels in the same medium. (You may have heard that nothing can travel faster than light, but this is only true in a vacuum. In a material, particles can travel faster than light.) Because of the fact that the particle is traveling faster than light does, the light emitted at different points along the path adds together and one can use this as a signature of the passage of a charged particle in a transparent material (and, very usefully, water). If a charged particle travels a relatively short distance in the water, it emits light in a cone, which when it hits a flat wall, looks like a circle. This behavior is illustrated in Figure 6.25. By analyzing the size of the circle and the exact moment that the light hits the various spots on the wall, you can infer a lot of information about the particle that was in the water. As you will see in Chapter 7, this technique is critical in being able to understand

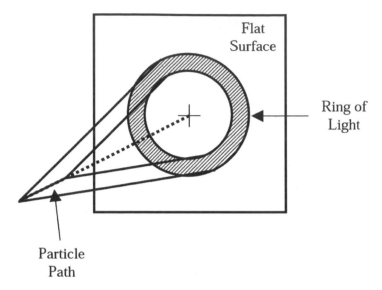

Figure 6.25 The essential features of Cerenkov radiation. A charged particle traverses a transparent medium and gives off light in the shape of a cone. The particle is detected by the circular ring of light observed on the detector walls.

a constant flux of neutrinos from outer space. For their efforts, Cerenkov, Tamm and Frank shared the 1958 Nobel Prize.

The technologies and techniques that are described in this chapter make up only a tiny fraction of the possible approaches that modern physicists use to try to unravel the tiniest mysteries (or perhaps the biggest, depending on how you think about it). There's no way that a single chapter in such a book can cover all of the intricate ideas that we use. The interested reader is invited to peruse the suggested reading in the back of the book for additional information. Nonetheless, we've learned the really important ideas. The extremely energetic particles made in modern particle physics experiments are viewed through their relatively mundane interactions with ordinary matter. Researchers are always looking for new ways to apply these well-known ideas in innovative ways, but if you're comfortable with your understanding of the methods we've discussed in these pages, you're likely to have a pretty good grasp on the accelerators and detectors used for the foreseeable future.

c h a p t e r 7

Near Term Mysteries

Round about the accredited and orderly facts of every science there ever floats a sort of dust-cloud of exceptional observations, of occurrences minute and irregular and seldom met with, which it always proves more easy to ignore than to attend to ... Anyone will renovate his science who will steadily look after the irregular phenomena, and when science is renewed, its new formulas often have more of the voice of the exceptions in them than of what were supposed to be the rules.

— William James

While Chapters 3–5 describe much of what we know or think we know about particle physics, there nonetheless remain intriguing questions. The next two chapters concentrate on those questions which are shrouded in mystery. These mysteries can be separated into two classes. The first class are those mysteries that are partially understood and one can do experiments now with some hope of shedding light on the question on the time scale of the next few years. The second type are those mysteries which are clearly present, but for which there exists little (if any!) experimental guidance and no real hope for any immediate insights (although they're very neat problems and people think about them a lot). Chapter 7 deals with the first type, while Chapter 8 concentrates on the second. Thus, as you read through the next two chapters, you should keep in mind the fact that as you do, you are approaching closer and closer to the frontier, the very boundaries of knowledge. If the descriptions of the science seem to becoming fuzzier, that reflects our ignorance. For all of the topics that will be discussed in the next two chapters, I don't know what the final answer will be. No one does. Further, I judge this to be the most difficult chapter in the book, with the remaining ones quite a bit easier. This is because there is a lot of information about these topics, but incomplete understanding. Thus I'll tell you what I know, but because the understanding is still not as solid as we'd like, I can't always give you the bottom line, as I do not fully understand what the answer will be. At any rate, the next chapters are easier so don't get discouraged. And besides, this chapter isn't really *that* hard … it just gives a flavor of the confusing life at the experimental frontier. The next chapter gives theory's frontier.

Although the final answers remain obscure, there remain still considerable and interesting bits of knowledge to be discussed. After all, in order to know that there is a mystery at all, you need to know *something* … just not understand it. In this chapter, we will discuss two very interesting questions. The first is the question of neutrino mass. While neutrinos were postulated to have a very small mass and perhaps be massless, some relatively new evidence suggests that perhaps they have a (very low) mass after all. Unlike quarks or charged leptons, where the masses are directly measured, the mass of neutrinos manifests itself by the peculiar behavior of neutrino oscillations. This is the phenomenon whereby the different flavors of neutrinos (ν_e, ν_μ and ν_τ, or electron, muon or tau neutrinos) "morph" into one another. Fairly solid evidence for this phenomenon has recently been observed. The second question addressed in this chapter is the problem of why there appears to be only matter in the universe, when all particle experiments show that matter and antimatter are made in equal quantities. The experimental knowledge that we have on this topic is called CP violation. The theory of CP violation closely mirrors that of neutrino oscillations, although its physical manifestation is different.

Mystery #1: Neutrinos from the Sun

With the conclusive proof of the existence of the neutrino in 1956 (discussed in Chapter 2), naturally physicists were interested in investigating neutrinos from other sources. While the initial neutrino discovery was accomplished by looking at the neutrino flux from a man-made nuclear reactor, there exists a much more powerful nuclear reactor, not too far from you. This nuclear reactor is the Sun, which puts out 2×10^{38} electron-type neutrinos every second. Not all neutrinos from the Sun carry the same amount of energy, as various fusion reactions occur in the Sun. While the most common reaction in the Sun is: $p^+ + p^+ \rightarrow d^+ + e^+ + \nu_e$ (two protons fuse together to form a deuteron, a positron and an electron neutrino), the energy of the created neutrino is very low. When the neutrino's energy gets low enough, it becomes even more difficult to detect. Thus other (and higher energy) neutrino sources within the Sun would make the search easier. Luckily, there are many subsequent fusion processes, e.g. a proton and a deuteron can fuse into a helium-3 nucleus ($p + d \rightarrow {}^3He$). In fact, all stable isotopes of hydrogen, helium, lithium and so on are present in the Sun and are available for fusion. While the details of the various types of fusion are highly dependent on temperature, within the current Sun elements as heavy as carbon, nitrogen and oxygen can be made, although at much reduced rates. Many of these fusion reactions can produce neutrinos, with the heavier element fusion in general more likely to generate higher energy neutrinos. Thus the trick in measuring neutrinos from the Sun became finding a workable compromise between the much-reduced flux of heavier element fusion and the increased neutrino energy.

Table F.1 shows the richness of the various fusion processes in the Sun. Since it is moderately complicated, I have relegated it to the appendix. Basically, it lists the many ways that one can take two protons from the Sun and fuse them together into helium-4 (^{4}He). Professionals have to deal with all of this, in all of its complexity, and the curious reader is invited to peruse Appendix F. But for us, we can see a few interesting things. To begin with, most (99.6%) of the hydrogen-2 (^{2}H) in the Sun comes from proton-proton fusion. This ^{2}H, which consists of a proton and a neutron and is also called a deuteron (and written d^+), is fused with a proton to make helium-3 (^{3}He). Most (85%) of the ^{3}He is fused with itself, not making neutrinos. However, 15% of the time, helium-3 and helium-4 are fused to make beryllium-7, which then very rarely (0.13% of the time), fuses with a proton to make boron-8 (^{8}B). We see how we can claim that the number of ^{8}B neutrinos is 0.02% compared to the flux that one gets from proton-proton fusion ($15\% \times 0.13\% = 0.02\%$). This will be crucial in a moment.

The study of neutrinos produced from the Sun (i.e. solar neutrinos) would answer many questions. In addition to the simple verification of a neutrino flux from the Sun (as must be present if the Sun is powered by nuclear fusion), one could use neutrinos as a probe of the solar interior. It was already well known that photons produced in the center of the Sun take a very long time to work their way to the Sun's surface (about 10,000 years!). Neutrinos, on the other hand, escape the Sun immediately. On the photons' journey out of the Sun, they change properties through multiple interactions with all of the atoms they pass by in their voyage through the Sun's interior. With their very low interaction probability, the neutrinos that we could observe would basically be the same as they were the moment that they were created.

In 1958, at the New York Meeting of the American Physical Society, Harry Holmgren and Richard Johnston, both then at the US Naval Research Laboratory, announced the result of an experiment that they had performed in which helium-3 and helium-4 fused into beryllium-7 (^{3}He + ^{4}He → ^{7}Be). Much to their surprise, the measured probability for this interaction to occur was about 1000 times higher than previously thought. I actually know Harry from my graduate student days. When I contacted him for his recollection of the event, he told me that they were unaware of the low theoretical prediction and might not have done the experiment had they known. As you will see as you read further, such a failure could have been disastrous for subsequent neutrino physics studies. While there was no clear link to neutrino physics, this measurement was pivotal. This is a clear example of the inter-relationship of science and reminds us that you never know what research will prove crucial. It also shows that theoretical predictions should be trusted only so far.

Obviously, a consequence of this discovery was the fact that the amount of ^{7}Be in the Sun must be much higher than previously believed. This information led William "Willy" Fowler and Al Cameron to suggest that the ^{7}Be might fuse with a proton to form boron-8 (^{7}Be + p$^+$ → ^{8}B). This reaction does not yield a neutrino, but ^{8}B is unstable and it beta-decays into an "excited state" of beryllium-8, which is simply a state of ^{8}Be in which the protons and neutrons are jiggling madly (^{8}B → ^{8}Be* + e$^+$ + ν_e, note "*" means excited state). The neutrino from this interaction is very energetic (with an average energy of 7 MeV) and is sufficient to be detectable. Although this neutrino has enough energy to be interesting, there was the problem that there were very many fewer reactions producing them. In fact, the rate of neutrinos from the ^{8}B decay was 0.02% the rate from the dominant proton-proton fusion.

With the realization that the ^{7}Be and thus the ^{8}B rate was 1000 times original expectations, the search was on for a suitable detection mechanism. In late 1963, John Bahcall realized that the interaction of neutrinos with chlorine was higher than expected (by a factor of 20). Chlorine-37 would combine with an electron neutrino and the resultant products would be an electron and an excited state of argon-38 (^{37}Cl + ν_e → e$^-$ + ^{38}Ar*). With this observation, the stage was set for a measurement of the neutrinos from the Sun.

In 1964, two rather focused scientists, John Bahcall and Raymond Davis, published back-to-back articles in *Physical Review Letters,* "the" American physics journal. These articles proposed that a detector be constructed, containing 100,000 gallons of per-chloroethylene (C_2Cl_4). Perchloroethylene is the fluid that dry clean-ers use that so troubles environmentalists. While these two guys have been rightfully considered as pioneers in the field of solar neutrino research, as is usual in research, they did not work in a vacuum and utilized the insights of earlier researchers. I won't discuss this earlier work for reasons of brevity. Bahcall was a young physicist at the Kellogg Radiation Laboratory at the California Institute of Technology and was to go on to become one of the mainstays in solar neutrino physics, while Ray Davis was "only" a chemist at Brookhaven National Laboratory. I say "only" a chemist because physicists tease chemists at least as much as experimentalists tease theorists. However, in this case, what was being suggested was so impressive as to seem impossible. Together, Bahcall and Davis predicted that about 1.5 neu-trino interactions would occur in their detector each day. They would let the solar neutrinos interact with the perchloroethylene for about two months and then extract the argon. A little math shows that they expected 90 interactions over that time and thus naively, one would expect to extract 90 atoms of argon. In fact, the real number was 48, because ^{37}Ar has a half-life of only 35 days and thus some of the atoms decay during the data-taking period. Further, their efficiency for extracting the argon was only 90%. So let's put this in perspective. They had 100,000 gallons of C_2Cl_4. If you do the arithmetic, this works out to be about 10^{31} atoms of chlorine and they expected to extract and identify 48 argon atoms. That just sounds *really* hard. So maybe chemists are pretty smart after all…Ray Davis' part of the 2002 Nobel Prize in physics shows with what regard the physics com-munity has held this effort.

Davis and Bahcall first pitched the idea to the Brookhaven chemistry department which, after a few weeks of consideration, approved the experiment and booted the idea upstairs to the AEC (the forerunner to

Figure 7.1 Left: Ray Davis (left) and John Bahcall (right). Right: Homestake perchloroethylene tank. (Figure courtesy of John Bahcall and Raymond Davis, Jr.)

today's Department of Energy). The idea was approved and for the relatively modest sum of $600,000 (in 1965 dollars) they built the detector, depicted in Figure 7.1.

Because of the need to protect the experiment against the much more common cosmic rays, the experiment was located in the Homestake gold mine in South Dakota at a depth of 4850 feet. A special cavern was excavated and the experiment performed. The initial results indicated rather than the expected interaction rate of 1.5 interactions per day, Davis and Bahcall measured 0.5 interactions per day, or about 1/3 the expected value. They announced their measurements in another pair of back-to-back articles in *Physical Review Letters*, one detailing the measurement, while the other discussed the standard prediction.

When such a large discrepancy exists between data and theory, there are many possible solutions. The data could be wrong, the theory could be wrong or there could be an unexpected phenomenon manifesting itself (or combinations of the three!) It's a sad fact that the third option is by far the least likely, at least when the experiment and calculation are new. Davis' job was nothing short of heroic. He had to pull a few atoms out of a sea of countless others. A tiny error could explain the discrepancy. However, when they created ^{37}Ar by sticking an intense neutron source in the tank, or when they inserted ^{36}Cl (chlorine-36) into a tank of perchloroethylene, they extracted precisely the amount of atoms that they expected. So it appeared that the experimental technique was solid. Bahcall's task was no less difficult. He had to calculate the flux of neutrinos from an obscure piece of the total fusion budget of the Sun (0.02%). Further, the result depends roughly on the 25th power of the Sun's core temperature. With such sensitivities, one can easily imagine a tiny error in this small rate. Recall that it's difficult to do experiments on the Sun. The air conditioning bill is prohibitive. So in order to verify that their models are correct physicists can only look at the amount of light from the Sun that gets to the Earth, as well as a few other indirect measurements. It doesn't stretch one's imagination that a small error, either theoretical or experimental, could move the flux of ^{8}B neutrinos from 0.02% to 0.007%. And, I admit, this was my explanation for this discrepancy for many years.

My skepticism aside, Davis and Bahcall's results have actually stood the test of time. Since that 1968 article, countless tests of that experiment and theory have been performed. The detected flux of ^{8}B neutrinos from the Sun is about 1/3 the expected value (as measured in chlorine-based detectors). With such a large discrepancy, the possibility remained that there could be an undiscovered error. Clearly, a confirming experiment was needed.

During the period of 1968–1988, Davis and Bahcall continued their work on the solar neutrino problem, but with little additional help. Each had only one co-investigator: Bruce Cleveland (chemistry) and Roger Ulrich (solar model calculations). As the years passed, other experiments were considered. At a minimum, the new experiment should be an independent effort, repeating the original Homestake technique, but with different people. An even better experiment would use completely different techniques, preferably with sensitivity to the dominant fusion reaction in the Sun ($p + p \rightarrow d + e^+ + \nu_e$). For reasons unclear to me, there was no success in trying to reproduce the original experiment, although George Zatsepin tried to have one built in the Soviet Union. This does not mean that the experiment was repeated and showed null results, but rather that the experiment was simply never repeated. Zatsepin's efforts did not result in a repeat of the Homestake experiment, although they helped in the creation of the next generation of solar neutrino detectors.

The only reaction that seemed to be remotely promising in detecting the low energy neutrinos from proton-proton fusion was a conversion of gallium-71 into germanium-71 ($\nu_e + {}^{71}Ga \rightarrow e^- + {}^{71}Ge$). Originally proposed in 1965 by a Russian theorist, V.A Kuzmin, the detector didn't seem to be feasible (or at least was prohibitively expensive) as a useful experiment required three times the world's annual production of gallium. Bahcall and Davis wrote another *Physical Review Letter*, this time proposing that a serious effort be made to construct a gallium experiment, based in the United States. The reaction to the proposal was generally favorable, but turf wars caused considerable grief in getting the necessary funding. Astronomers thought that the proposed experiment was great physics, while physicists thought that it was great astronomy. But in either case, they both thought that the other group should fund the experiment. Within the physics community, even the proponents disagreed, with particle physicists thinking that the experiment was great nuclear physics, while the nuclear physicists really liked the great particle physics proposal. The proposed experiment was a bit of an orphan. Bahcall even tried to get the National Science Foundation (NSF) to fund the experiment. However, as the proposal came from a collaboration that included many scientists from Brookhaven National Laboratory, the NSF rejected the experiment. (BNL is a Department of Energy laboratory and the NSF and DoE typically do not fund each other's experiments.) Like I said … turf wars. In the end, the United States never did fund a big gallium experiment.

A pilot experiment was performed by an international collaboration including BNL, the University of Pennsylvania, the Max Planck Institute in Heidelberg, the Institute for Advanced Study in Princeton and the Weizmann Institute. This experiment used 1.3 tons of gallium. While no conclusive results were forthcoming from this initial effort, the experiment did the engineering work necessary to understand the needed extraction techniques. Like the earlier Homestake experiment, the idea was that they would have a few tens of tons of gallium and extract only a few atoms of germanium.

In contrast to the American experience, the Soviet Union took the idea much more seriously. Moissey Markov, who was then the head of the Division of Nuclear Physics of the Soviet Academy of Sciences, enthusiastically supported the idea and was instrumental in borrowing 60 *tons* of gallium for the duration of the experiment and also in building the Baksan Neutrino Observatory under Mount Andyrchi in Soviet Georgia. Soviet (and now Russian) scientists enjoy a greater degree of prestige in the eyes of their fellow citizens than do American ones.

I once attended a conference at the Budker Institute of Nuclear Physics in Novosibirsk, Russia in February of 1996. It was a very cool conference. (Yes, the pun was intended. ... Siberia in February ... Brrrr.) At these conferences, they usually reserve one evening for a local cultural event. This time we were bused to the local theater for an evening of truly splendid singing. On the way back, we passed the Novosibirsk State Opera House, the largest opera house in Russia and an interesting piece of architecture. It seems to owe its inspiration to a melding of Lawrence Berkeley Laboratory's signature domed building and classical Greek colonnades. Of course, no Soviet-era construction would be complete without the required statues of the workers, soldiers and Lenin standing in solidarity. Because the Russians take their art extremely seriously, we expressed an interest in attending the opera, as it was certain to be good. Unfortunately, we were informed that it was closed for the season. While we were disappointed, we understood and forgot the idea. The next day though, we were informed at the conference that arrangements had been made and that there would be a command performance of the opera for the conference participants. That's clout. I can just imagine what would happen if the director of Fermilab made a similar call to the Chicago Symphony Orchestra or the Chicago Lyric Opera House. The conversation would go something like this: "Hi, my name is Mike Witherell and I'm the director of Fermilab, the highest energy particle physics laboratory in the world. I know you're closed for the season, but I have about one hundred foreign scientists in town and I was wondering if you would open up tomorrow for a performance just for them. Hello? Hello???" While Mike's a very persuasive guy, the United States is just different from Russia. At any rate, in the Russian system, scientists are respected and they have a corresponding degree of clout. So I'm not surprised that a strong personality like Markov was able to procure so much gallium.

The experiment was called SAGE (for the Soviet-American Gallium Experiment) and was headed up by Vladimir Gavrin and George Zatsepin (from the Institute for Nuclear Research in Moscow) and Tom Bowles of the Los Alamos National Laboratory. It initially consisted of 30 tons of gallium metal. Begun in 1988, the experiment was about 1000 meters underground and was upgraded to 57 tons in 1991. A similar experiment began a little later (1991) and involved mostly European scientists. This experiment was called GALLEX and it was located in the Gran Sasso Laboratory in the Italian Alps. GALLEX was led by Til Kirsten of the Max Planck Institute in Germany and consisted of a tank containing 30.3 tons of gallium in the form of a water solution containing gallium chloride ($GaCl_3$). The weight of the entire detector is about 100 tons, when one includes the weight of the water.

In 1990, SAGE announced their measurement of the flux of solar neutrinos and they found about $52 \pm 7\%$ of expectation. Recall that the $\pm 7\%$ means that they are pretty sure that the real number is between 45–59%. So while they were somewhat uncertain, they were pretty sure the real number wasn't the 100% that they would get if the experiment measured exactly what was predicted. GALLEX produced a similar result, observing $60 \pm 7\%$ of the expected neutrinos. Both experiments checked their equipment by injecting known numbers of neutrinos from an intense chromium-59 (^{59}Cr) neutrino source. Both experiments saw in their tests that the measured number of interactions was within 5% of expectations, proving that they understood their detector and techniques. Just to be safe, the GALLEX experiment injected a known amount of arsenic-71 (^{71}As), which decays into ^{71}Ge. This time, the extracted amount of ^{71}Ge was within 1% of expectations. Clearly, failure to understand their detector was not the explanation.

In addition, recall that the gallium experiments were sensitive to the main proton-proton fusion reaction in the Sun. So we couldn't even argue that the effect was a small error in a fringe fusion process. While one might hold out hope that the calculation could still be in error, the probability of this being the explanation is much reduced. Mainstream physicists began thinking seriously about what was happening to the neutrinos from the Sun. Could they be decaying? Interacting? What?

Oh Where, Oh Where, Have My Neutrinos Gone?

Soon after the initial experimental results were available, theorists were proposing answers to the conundrum. In 1969, Vladimir Gribov and Bruno Pontecorvo proposed that the solution to the solar neutrino problem was that the electron neutrinos from the Sun oscillated into other flavors of neutrinos (e.g. a muon or tau neutrino), which were much more difficult to detect. For instance, in the reaction used in the Homestake detector, the neutrinos interacted via ^{37}Cl $+ \nu_e \rightarrow$ e$^-$ $+ ^{38}$Ar*. The Feynman diagram given in Figure 7.2a shows how this occurs.

If somehow the electron neutrino had changed flavor into a muon neutrino, then a reaction like that shown in Figure 7.2b might happen. However, the mass of the electron is 0.5 MeV, while the mass of the muon is about 200 times heavier at 106 MeV. Since the neutrinos from ^{8}B have a maximum energy of 15 MeV, they simply don't have enough energy to make muons. Therefore 15 MeV muon neutrinos

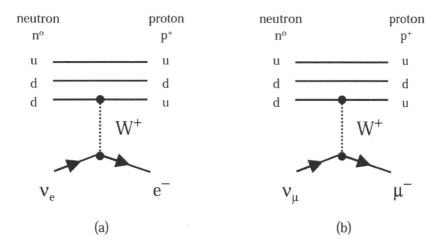

Figure 7.2 A neutrino emits a *W* boson, converting into a charged lepton. The *W* boson converts a neutron into a proton.

do not convert chlorine into argon or gallium into germanium and the Homestake, SAGE and GALLEX detectors cannot detect them. So the neutrino oscillation hypothesis is consistent with the data of both the chlorine and the gallium experiments, in which fewer electron neutrinos were detected than expected.

Initially, this oscillation hypothesis was disbelieved by the majority of physicists (although it now is the consensus view). The biggest problem with the idea is that in particle physics experiments, electron neutrinos and muon neutrinos act quite differently. Recall in Chapter 2 that we discussed the 1962 experiment of Lederman, Schwartz and Steinberger, in which muon neutrinos were shown to only create muons, never electrons. Yet according to Pontecorvo and Gribov, muon neutrinos could change into electron neutrinos (and thus could presumably then create electrons). Something isn't hanging together.

The idea of neutrino oscillations is relatively difficult to explain, so we'll describe it in general terms first and then have an optional Appendix (Appendix F) which talks about the trickier details. Neutrino oscillations occur when a neutrino travels over a considerable distance. Say you have an electron neutrino. If it travels a certain distance it has a certain probability of changing into (say) a muon neutrino. At other distances, the neutrino can have a 100% chance of changing into a muon neutrino. Later the neutrino has a chance of converting back into an electron neutrino. In quantum mechanics, you can't calculate what a particular particle will do, just the probabilities. However, when you have lots of particles, it's easier to see. If the oscillation probability is 50%, then half of the particles will change. This point is shown in Figure 7.3.

In Figure 7.3, the changing point is shown as a "poof," implying that all neutrinos change at a particular time, but the reality is that the probability of converting neutrino flavor changes smoothly. This more realistic behavior is shown in Figure 7.4.

At the beginning, which we call distance #1 (d_1), your beam of neutrinos is 100% ν_e and 0% ν_μ. At the second distance d_2, the neutrinos have oscillated to 0% ν_e and 100% ν_μ. At the distance d_3 the mix is 50/50, while at d_4 we are back to 100% ν_e. This pattern repeats itself over and over, so that a detector placed at different distances from the source will measure different mixtures of muon and electron neutrinos, even if at the beginning, you were certain that you had only ν_e.

So how does one reconcile this oscillation with the observation in 1962 that neutrinos from decaying muons did not create electrons? The answer to this stems from another question. What is the distance between oscillations? If the distance is long compared to the approximately one mile size of accelerator experiments, that experiment will not observe oscillations (as the neutrinos haven't traveled long enough to oscillate). It turns out that the distance needed to oscillate is dependent on the energy of the neutrino (among other things). Taking all the relevant parameters into consideration, we can calculate that the distance needed for oscillations to occur can be hundreds to thousands of miles. The details on how we can calculate these things are given in Appendix F.

Of course, thus far in our reading we haven't proven that neutrinos actually oscillate (although we're pretty sure that they do). We

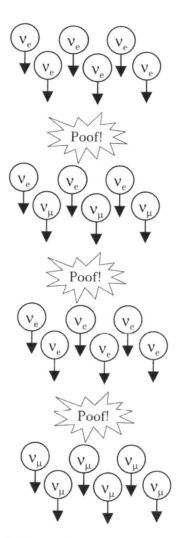

Figure 7.3 Cartoon of a beam of pure electron neutrinos, which convert into a mixture of electron and muon type neutrinos, then back into electron neutrinos and then into muon neutrinos. While the "poof" is non-realistic, it gives the basic idea as to how neutrinos oscillate.

could explain the missing neutrinos from the Sun as solar physicists not doing their calculations correctly, or by something exotic like neutrino decay. Oscillations are just one of a number of competitive explanations. For proof, we must turn to a different set of experiments.

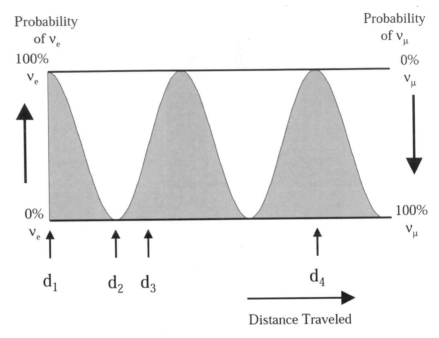

Figure 7.4 A more realistic description on how neutrinos oscillate. Initially (d_1) a pure beam of electron neutrinos smoothly converts into a mix of electron and muon neutrinos (the region between d_1 and d_2). There are spots where the beam is entirely converted into muon neutrinos (d_2).

Neutrinos from Thin Air

At this point in our discussion, we must take a detour and a hop in a time machine and return to the late 1970s. Physicists at the time were keenly interested in how one could unify the various forces into a single coherent theory. Theories of this nature were available and a common theme of the more successful flavors of the theories was that they predicted that protons were ultimately unstable and would decay. The predicted lifetime of the proton was far longer than the lifetime of the universe, so the theory agreed with the observation that we are still here, but perhaps a sufficiently clever experiment might be sensitive enough to observe proton decay.

In the late 1970s, three large detectors were proposed to investigate this pressing question. Two were large tanks filled with water. The third consisted of huge modules made of iron and gas (which was used for the actual detection). The two water detectors were called IMB (for Irvine-Michigan-Brookhaven), located in the Morton salt mine in Cleveland, Ohio and Kamiokande (for Kamioka Nucleon Decay Experiment), located in the Kamioka mine in Japan. The gas and iron based detector was Soudan 2, located in the Soudan mine in Minnesota. To simplify our discussion, let's concentrate on the Kamiokande experiment. Like the earlier solar neutrino detectors, it was imperative that all three detectors be located deep underground to shield them from the much more common types of particle interactions that one sees above ground. The Kamiokande detector was to be 3,000 tons of pure water, surrounded by photomultipliers which, as we recall from Chapter 6, can convert light into electrical signals. The idea was that if a proton did decay, one thing it could turn into would be a positron and a neutral pion ($p^+ \rightarrow e^+ + \pi^0$). When the positron moved through the water, a small blip of light would occur. This light would travel to the photomultiplier tubes and be detected. In addition to providing plenty of protons to decay, the water was an important factor in detecting the light. This is because an electron or positron traveling quickly through water can generate Cerenkov light as we discussed in Chapter 6. The Kamiokande collaborators then sat and waited. After about ten years of near-continual running, they observed no proton decays and in doing so, they set a limit on the lifetime of the proton as greater than 10^{33} years, although it could be much greater than that. In fact, we have no experimental evidence to suggest that a proton can ever decay. Since the current age of the universe is about 10^{10} years, such a huge lifetime for the proton allows us to rest well at night.

We have concentrated on the Kamiokande experiment, mostly because of the extraordinary success of its follow-on experiment, Super Kamiokande. However, the IMB experiment was quite competitive and, in fact, quite a bit larger. It consisted of 8,000 tons of water and was located about 600 meters underground outside Cleveland. The Soudan 2 detector was about 960 tons and located about 690 meters underground on the 27th level of an old iron mine. While Soudan 2 turned on a little later than IMB and Kamiokande, it has nonetheless provided complementary evidence to the discussion. I have a special fondness for the Soudan 2 detector. While I never worked on it, it was the first particle physics experiment that I ever saw "in the flesh." In the spring of 1986, I visited Argonne National Laboratory for a tour of their particle physics facilities and they were building the Soudan 2 modules there. We got to go up to the modules and touch them. Each module was one-meter square and 2.5 meters tall. I could imagine the 224 modules stacked in a vast array and got a sense of the scale of modern particle physics experiments. That visit validated my earlier decision to be a practicing particle physicist. It's been fun ever since.

Because proton decay is so infrequent (which one can infer directly from the long lifetime), unambiguously detecting such decays becomes much more difficult. The reason is that no detector detects just one type of physics process. While one tries very hard to make the detector only sensitive to the measurement in which you are interested, often the detector will also be able to observe other phenomena. Sometimes the "other" phenomena, whatever they may be, look quite a bit like the type of interaction that you're trying to observe. Thus it is imperative to understand these "backgrounds" or copycat false physics events.

One of the backgrounds with which the physicists had to contend is cosmic ray muons. Even though the detector is buried deep underground, still occasionally a few cosmic ray muons pass through the detector. Luckily this type of unwanted event is easy to identify and reject, as they usually consist of a charged muon entering the detector from the top and leaving out the bottom. The way the muon is detected is by Cerenkov radiation (at least in IMB and Kamiokande...Soudan 2 was different), which we discussed in Chapter 6. The signature is a big cone of light which one sees as a circle of light that illuminates the light-detecting phototubes. Because the cosmic ray muons enter at the top and exit at the bottom, the circle of light is observed near the bottom of the detector.

While muons are easy to identify, neutrinos are trickier. After all, the neutrino enters undetectably, interacts with the water atoms and leaves a signal. This looks a lot like the expected decay signature of a proton (nothing entering the detector and an interaction occurring spontaneously within the detector). So understanding neutrino interactions which occur very rarely, but far more frequently than proton decay, is critical to being able to say anything about the lifetime of the proton.

It turns out that solar neutrinos aren't a large background, because their energy is so low (i.e. it is easy to separate solar neutrino interactions from the signal from proton decay). However, there exists another source of neutrinos which causes tremendous problems. This is atmospheric neutrinos. These neutrinos have a moderately confusing name, as the atmosphere itself is not directly the source of the neutrinos. The actual source is cosmic rays, typically protons, but could be other atomic nuclei. As you will recall, cosmic rays are charged particles from outer space, which hit the Earth at great energies. While cosmic rays span a great range of energies (a fact which will be very useful in a little while) there exist cosmic rays that have just the perfect energy to look like proton decay.

Cosmic rays create neutrinos in the following way. A cosmic ray hits the atmosphere and interacts with the nucleus of some material in the air (oxygen, nitrogen, etc.) When a proton hits an atomic nucleus, many particles come out of the collision. Each collision is unique and one cannot predict which particles and energies will exit the collision. However, what is true is that the particles exiting the collision are predominantly pions. While the neutral pion decays very quickly into two photons ($\pi^0 \rightarrow 2\gamma$), the charged pions live much longer. They live so long that they can hit another air nucleus (although many don't and so decay in flight, a crucial fact that will be discussed in a moment) and make even more pions. This pattern repeats itself again and again, with each collision increasing the total number of pions; by this mechanism, each interaction can cause (say)

5–30 pions from the original interaction eventually resulting in thou-sands or even more pions. As an example, let's make the simplifying assumption that exactly 10 charged pions are made in each collision. After the first collision, there are 10 charged pions. If each pion hits an air nucleus, there will then be 100 pions. If each of these repeats the process, there will be 1000 and so on. So each single particle from outer space can make a huge (and highly variable) number of pions. Given that each "shower" as these are called can be so incredibly vari-able, how can we hope to understand them? Further, what does this have to do with neutrinos? Note that while this shower occurs in air and therefore spans a great distance, this is identical to the sorts of showers described in Figure 6.17 and the surrounding text.

Recall that charged pions are unstable and eventually decay. Many of these pions decay before subsequent interactions with an air nucleus. A charged pion decays into a charged muon of the same elec-trical charge and an associated muon neutrino or muon antineutrino ($\pi^+ \rightarrow \mu^+ + \nu_\mu$, $\pi^- \rightarrow \mu^- + \bar{\nu}_\mu$). Muons interact only a little bit with the air and so mostly just decay. Muons decay into the electron with the same electrical charge and a pair of neutrinos, one muon-like and the other electron-like ($\mu^+ \rightarrow e^+ + \bar{\nu}_\mu + \nu_e$, $\mu^- \rightarrow e^- + \nu_\mu + \bar{\nu}_e$). So taking for example, the decay of a negative pion, we write

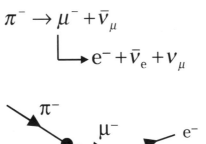

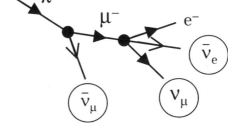

Figure 7.5 A pion from cosmic rays decays into a muon and then an elec-tron. Two muon-type neutrinos are created for each electron-type neutrino.

Not worrying about the fact that it is a two-step process, we can say $\pi^- \to e^- + \bar{\nu}_e + \nu_\mu + \bar{\nu}_\mu$. Thus we see that eventually all charged pions decay into an electron or positron and three neutrinos, always two muon-like and one electron-like. So while we can't know how many neutrinos came from any cosmic ray and further we can't know when and where cosmic rays will hit, we do know one thing without any ambiguity. When you see atmospheric neutrinos, they should occur in a ratio of two ν_μ's to one ν_e, a point illustrated in Figure 7.5. However, when the IMB, Kamiokande and Soudan 2 experiments measured this ratio, they found the rather shocking result that there were just about exactly as many muon neutrinos as there were electron, rather than the twice as much that was expected. At first blush, one could conclude that muon neutrinos are disappearing. With a little more thought, you realize that electron neutrinos could be appearing or both could be changing. So at this point in our reading we don't know what the whole story is, but it certainly appears that something funny is going on with atmospheric neutrinos, just like occurred with solar neutrinos.

Super-K Finds the Truth

A lot of work was done to try to sort out what was going on, but the real answer came in 1998 from the Super-Kamiokande experiment, lead by Professor Yoji Totsuka of the University of Tokyo. Super-Kamiokande (or Super-K, as it is known) is a scaled up version of the original Kamiokande experiment. Super-K is depicted in Figure 7.6 and is shaped like a cylinder 128' wide and 135' (or 11 stories) high. It too is located in the Kamioka mine. It contains 50,000 tons of pure water. In order to reject events in which a particle first interacted in the surrounding rock, only the 22,500 tons of water at the center of the cylinder can actually be used to detect neutrinos. If there is any signal in the outer 27,500 tons, the event is assumed to have originated in the surrounding rock and thus is not considered to be a neutrino event.

Figure 7.6 The Super-Kamiokande detector. Bottom, physicists in a boat inside the Super-Kamiokande detector, polishing phototubes. (Figure courtesy of The Institute for Cosmic-Ray Research of the University of Tokyo.)

Because Super-K was so large, they would get many more neutrino interactions than their predecessors did. Even more importantly, they could distinguish between events in which the initial neutrino was moving upwards versus neutrinos moving downwards. As we see

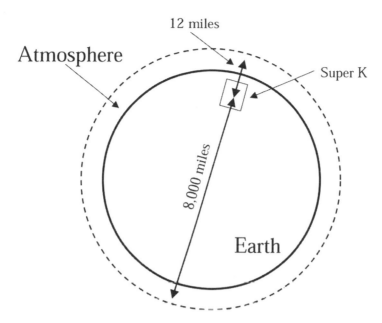

Figure 7.7 An illustration of how neutrinos created in the atmosphere directly above the Super Kamiokande detector travel a much shorter distance than those created on the other side of the world.

in Figure 7.7, if the neutrinos are forming in the atmosphere, the neutrinos coming from above travel only about 12 miles (about 20 kilometers) before they hit the Super-K detector. Upward going neutrinos on the other hand, were created on the other side of the world, about 8,000 miles (about 13,000 kilometers) away. As long as the characteristic oscillation length is significantly greater than 12 miles, but much less than 8,000 miles, then you would expect to see the expected 2:1 ratio for down-going neutrinos and something else for upward-going ones. That is, of course, if neutrino oscillations are real. This technique is especially brilliant, as even if you're wrong about all of your calculations, comparing the number of muon versus electron neutrinos going up versus down, you would expect the ratio to be unchanged, unless some interesting physics phenomenon is going on.

In 1998, the Super-K collaboration announced at the *Neutrino
'98* conference their analysis of up versus down going neutrinos.
They found that the numbers of up-going and down-going electron
neutrinos were exactly as expected if no oscillations existed.
However, the story was much different for muon neutrinos, a result
depicted in Figure 7.8. The Super-K collaboration found that the
number of down-going neutrinos were just as expected. The real
excitement was caused by the fact that they saw only half of the

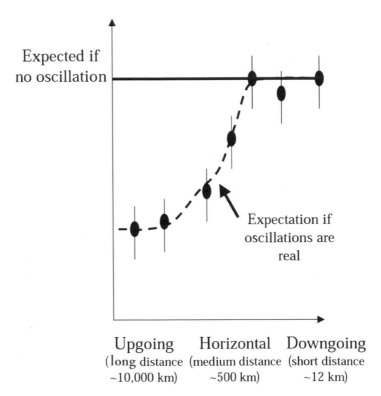

Figure 7.8 A demonstration as to what the typical neutrino oscillation
experiment data looks like. No neutrino oscillations would be a flat line at
the expected value. The fact that the measurement is below expectation for
up-going neutrinos as compared to down-going ones is compelling evidence
for atmospheric neutrino oscillation.

upcoming muon neutrinos as would occur if neutrino oscillations were not present. The basic idea is illustrated in Figure 7.8. So the question was where did these missing muon neutrinos go? We know that they did not oscillate into electron neutrinos. We know this because Super-K saw exactly as many electron neutrinos as they expected. If muon neutrinos had oscillated into electron neutrinos, one would expect to see extra electron neutrinos. So the only options are the (now universally accepted) idea that muon neutrinos oscillated into tau neutrinos or the (now discredited) idea that there might be a fourth kind of neutrino, previously unobserved and with slightly odd properties.

The above discussion is for the higher energy neutrinos from cosmic rays. When Super-K looked at lower energy ones, they saw the same number of up-going and down-going electron neutrinos as they would expect if electron neutrinos don't oscillate. But when looking at the muon neutrinos, the results differed from expectations. Just like in the higher energy measurement, there weren't as many up-going muon neutrinos as expected. However, now they also saw that there were fewer down-going muon neutrinos than expected. According to theory, the distance that neutrinos travel before oscillating depends, among other things, on energy (c.f. Appendix F). By separating the data into high and low energy sets, the Super-K detector was able to provide very strong evidence for neutrino oscillations. For the success of the Super-K effort, Masatoshi Koshiba was awarded part of the 2002 Nobel Prize.

On November 12, 2001, a tragedy befell the Super-K experiment. A technician was walking towards the detector, when he felt a rumble in the ground. Needless to say, feeling a rumble when you are that far underground, especially when you're in earthquake-prone Japan, must be a scary experience. But the rumbling wasn't caused by an earthquake, but rather by the Super-K detector destroying itself. As you may recall, the Super-K detector is basically just a large (50,000 tons) tank of water, being stared at by 11,000 photomultiplier tubes. The Super-K phototubes are just large hollow glass spheres, about

50 centimeters in diameter. Within the course of just a few seconds, fully 2/3 of the phototubes broke. While it will never be possible to know with 100% certainty, the most likely cause was that one of the tubes on the bottom (where the water pressure was greatest) imploded and the resultant shock wave caused its neighbors to implode. The chain reaction of implosion and shock waves spread across the tank, destroying most of the phototubes. In a few seconds, over $20,000,000 worth of damage was done, although luckily no one was hurt. At this time, the Super-K physicists are in the process of understanding what happened and engineering solutions to safeguard against another such disaster. They will rebuild their detector, but in the meantime, they will spread their remaining phototubes over the detector and continue as best as they can.

Why Oscillations?

Another crucial parameter in neutrino oscillations concerns the masses of the neutrinos. If neutrinos are truly massless, they cannot oscillate. We talk about the technical stuff in Appendix F, but we can get the basic idea here. Be warned however that the phenomenon has its cause rooted deeply in quantum mechanics; so all analogies break down if you look at them closely enough.

The bottom line is that if different flavor neutrinos have different masses, then they move at slightly different speeds. Quantum mechanics is all about probability and probability waves. Things are likely or unlikely depending on the height of these waves. If the "electron neutrino wave" is high when the "muon neutrino wave" is high, then they are both equally likely. However, if one wave is moving faster than another, one of them (say the electron neutrino) will pull ahead of the other and then a high point in the "electron neutrino wave" will correspond to a low point in the "muon neutrino wave." This could correspond with a large chance of the neutrino being of the electron type and a low chance of it being of the muon type. Figure 7.9 tries to clarify this point.

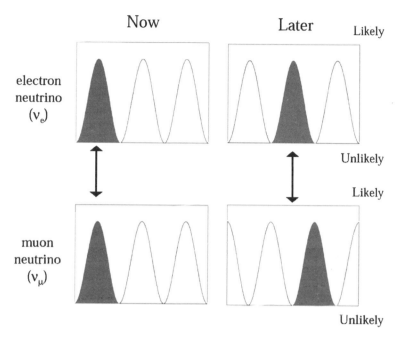

Figure 7.9 A fairly realistic description of how neutrino oscillations work. If at a particular time, muon and electron neutrinos are equally probable, the fact that the two can move at different speeds means that, at a later time, they will not both be equally likely.

I find this a bit difficult to get my head around, so perhaps we can make it a little more concrete by talking about two adjacent lanes of traffic, each filled with cars with the same separation between them. One lane is moving at 60 mph, while the other is moving at 62 mph. In Figure 7.10, you can see how I've arrayed the two lanes of traffic to start out the same. However, if one lane moves slightly faster than the other, it will slowly creep ahead. Some time later, the cars on one lane will line up with the gaps in the other lane. Eventually, they will keep moving until the cars all line up again. This is somewhat analogous to how it goes with neutrino probability oscillation.

We see in our analogy that a bigger speed difference means that the consecutive alignments of the cars occur more frequently. We

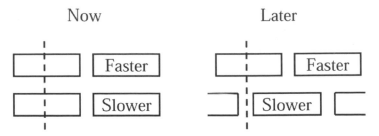

Figure 7.10 An analogy for neutrino oscillations. If two lines of automobiles are lined up at a particular moment, yet they are moving at different speeds, they will eventually no longer be lined up.

could say that a greater speed difference corresponds to a faster oscillation. In neutrino-land, a greater speed difference corresponds to a greater mass difference. So, if different neutrino flavors have large differences in mass, they oscillate more quickly.

A very important aspect of neutrino oscillations is that they prove that the different neutrino flavors have *different* masses. It says *nothing* about the actual masses. If you have two objects, one weighing one pound and the other weighing two pounds, the weight difference is one pound. But this is also true of two objects, one weighing 100 pounds and the other weighing 101 pounds. They have the same weight difference, but very different weights. Neutrino oscillations only show that the masses of the neutrinos are different, it says little about the masses themselves. To get a handle on the mass itself, rather than the mass difference, one must return to the same sorts of experiments that provided the original clues that a neutrino might exist: beta decay. Since this topic is not crucial to the discussion of the phenomenon of oscillating neutrinos, I mention it only in passing. However, the most modern beta decay experiments have established that the mass of the electron neutrino is very small. Exactly how small is currently unknown, but we know that the mass of the electron neutrino is smaller than 2.2 eV. Other experiments have set stringent limits on the mass of the other two types of neutrinos as well. Because there are so many neutrinos in the universe, mostly left over from the Big Bang (about 330 neutrinos in every cubic centimeter of space), the mass of the neutrino could have cosmological implications. We will return to this in Chapter 9.

So while atmospheric neutrino oscillations ($\nu_\mu \to \nu_\tau$) appear to be pretty well established, it took yet another experiment to make the solar neutrino oscillation idea similarly solid. Recall that one of the proposed explanations for the solar neutrino deficit was that electron neutrinos were decaying into some unknown state of matter on their journey from the Sun.

The Sudbury Neutrino Observatory (SNO) is a unique detector. Instead of being a tank filled with water (H_2O, for the two hydrogen and one oxygen atom in water), SNO is filled with D_2O, or two deuterium and one oxygen atom. Recall that deuterium is an isotope of hydrogen, containing one proton and one neutron, rather than the single proton in the nucleus of a hydrogen atom. Because of this difference, SNO can see all types of neutrinos (ν_e, ν_μ and ν_τ). While Homestake and GALLEX could see only electron-type neutrinos and not muon and tau-type neutrinos, SNO sees them all, although only those from boron-8. In the summer of 2001, they announced their first results. Because this is only their initial measurement, their results will improve. By the time that you read this, they will have new data which will lead to a more precise determination of the number of ^8B neutrinos from the Sun (although most likely their conclusions will not change). The bottom line is that they can combine the information from SNO with the more mature (and consequently more precise) measurements of Super-K. In doing so, they find that if they add up all of the three kinds of neutrinos, they measure the same number of ^8B neutrinos as leaves the Sun. Thus the ^8B neutrinos, which leave the Sun as electron neutrinos, hit the Earth as a (detectable) mixture of the three kinds of neutrinos (ν_e, ν_μ and ν_τ). Therefore, they aren't decaying into something undetectable. This strongly supports the neutrino oscillation idea.

Neutrino Detectors, Current Status

As I'm writing this, all of the dozen or so neutrino oscillation experiments all hang together except one. Each experiment had or has its strengths and weaknesses and none of them could measure the whole truth of neutrino oscillations. But they didn't contradict one another, except for the one. We'll get to that experiment in a minute, but first let's talk about what it means for experiments to agree or disagree. Getting agreement between two subsequent measurements is easy for some measurements, like comparing your weight on scales at home and at the health club. But it's more difficult when you're doing frontier research, because even the best experiment has its limitations. Suppose that we have a number of good, but not perfect measurements. How can they agree? Let's answer by way of an analogy. Suppose you have a famous actor in a room. You let several people take a quick glance in the room and use that information to try to identify the actor from the observations. Person 1 might report that they saw a male with dark hair. Person 2 might report that they saw a person with a slim build and blue jeans. A third person reports that they couldn't be certain that they saw a person, but they did see blue pants and a white shirt. A particular observant person (i.e. a good experiment) might report that they saw a guy wearing a white polo shirt and the guy looked familiar, like he might have been in a George Lucas film. Notice that while not all people reported the same things, but assuming that they all made accurate observations, we can determine that the person in the room is a slim male with dark hair, wearing a white polo shirt and blue jeans. Further, the guy may have played in a Lucas film. From that, we might narrow it down and determine who it was. But suppose that another person looks in and reports that they see a blonde woman. This person is reporting something that is clearly inconsistent with all of the other reports. So you could infer that the last observation was in error. Of course, it's also possible that the actor was temporarily dressed in drag and wearing a wig for that observation. In this case, the observation was accurate, but one needs a new theory to explain the data. But in any event, you'd view this observation with some worry and want some kind of confirmation.

Getting back to neutrinos, there is one experiment that reported in 1995 results that were inconsistent with all of the others. This experiment is the Liquid Scintillator Neutrino Detector (or LSND) at Los Alamos National Laboratory. LSND was interesting in that it shot a beam of muon neutrinos at a detector far away. They reported seeing the appearance of electron neutrinos in the detector. Recall that the amount of neutrino oscillation seen is dependent on neutrino energy, the distance between the neutrino source and the detector, the difference in mass between the two neutrino types and finally the "strength" of the mixing (a.k.a. the "mixing angle.") The first two variables were known, so LSND reported that their observation suggested a particular range of mass differences and mixing angles. The only problem is that other experiments looked at those values of the mass difference and mixing angle and saw nothing. So either LSND was simply wrong, or something very strange is going on. The drama is compounded by the fact that the accelerator producing the neutrinos, the Los Alamos Meson Production Facility or LAMPF, was scheduled to be decommissioned. As experiments near the end of their lifetime, there is a tendency to report contentious results in order to get extra beam time needed to either confirm or kill their result (recall the Higgs drama at LEP, discussed in Chapter 5?). So LSND probably reported their result before they might have under less pressured circumstances. Even more confusing, an LSND collaborator, James Hill, defected and published a paper in the same issue of *Physical Review Letters* disputing the official LSND result. Such a choice is not unprecedented, but is extremely rare and cast a pall over the whole situation. LSND shut down for good in 1998, without materially changing their initial result.

Of course, if the LSND result were true, it would be extremely exciting, as our current theory cannot deal with all of the other measurements and LSND's result. This would mean that we would have to rethink our theories. Of course, the theory has shown to be pretty good, so before we pitch the whole thing, we really need a confirming measurement.

Luckily, a new experiment at Fermilab is being built. This is the MiniBoone experiment (for Miniature Booster Neutrino Experiment). MiniBoone is headed by Bill Louis of Los Alamos National Laboratory (and spokesman of LSND) and Janet Conrad, of Columbia University, and was designed specifically to confirm or refute the LSND result. As I am writing this (early spring 2003), the experiment has just started (having received first beam on target in August of 2002). By the time that you read this, the experiment may well have interesting results. I, for one, eagerly wait to hear what they have to say.

In our discussion of neutrino oscillations, there remains a gaping hole. While the evidence for neutrino oscillations is extraordinarily compelling, all of it relies on a calculation of the neutrino source. Now the calculations are pretty robust (both solar and atmospheric), but calculations simply aren't as compelling as good measurements. What you'd like to do is to measure the composition and energy of the neutrino beam near the source (at a "Near Detector") and do it again far away, after the beam has had a chance to oscillate, this time at a "Far Detector." This gives you the ability to compare the near detector measurements with calculations and finally you will have a clear picture of neutrino oscillations. If the neutrino source is a particle physics beam, you can alter the beam energy and repeat the experiment. Since the amount of oscillation changes with beam energy, you can predict how the different neutrino populations will change at the far detector. Finally you will have a good, solid, controllable measurement.

There exist many experiments under construction that will do just this. There is the MINOS (Main Injector Neutrino Oscillation Search) experiment at Fermilab, which should get underway in about 2005. There is the K2K (KEK to Kamioka), in which a beam is sent from the Japanese high energy physics laboratory (KEK) to the Super-K detector. There's also CNGS (CERN to Grand Sasso), scheduled to start in 2005. CNGS really is a beam aimed at a cavern in Italy, which will house experiments, including OPERA and ICARUS. There are a few others.

While K2K is already underway, the damage to the Super-K detector means that they are running below design performance. Nonetheless, they have already released preliminary results and it looks like their data supports the neutrino oscillation hypothesis. Here in America, the MINOS experiment's construction is well underway. A beam of muon neutrinos from Fermilab is aimed at the Soudan mine, 730 kilometers away, in which they've built a whole new detector. The beamline is shown in Figure 7.11. While I've never worked directly on MINOS, I actually played a role in the design of their detector. The MINOS collaboration decided to steal technology from the DØ experiment, of which I am a member. (Of course steal, in this context, is a good thing and a compliment. As the late Victor Weisskopf said, "The real crime is to hear of a good idea and *not* steal it.") This technology consists of long strips of plastic scintillator (see Chapter 6 for a refresher if needed). Along with my immediate collaborators, Alan Bross and Anna Pla-Dalmau, we invented and tested scintillator strips manufactured by this method. (Well... they invented and I made it work...) DØ liked the idea enough to use it and it has performed so well that the idea is being used by several experiments. Scientific research is collaborative on so many levels.

MINOS' near detector is similar in structure to the far detector and currently lags the construction of the far detector, mostly for reasons of civil construction and excavation. They hope to start taking data in 2005, although as I write they just announced that their far detector just observed its first upward-going neutrino. So they're well on their way.

Finally, I've not mentioned a number of experiments that are looking for neutrino oscillations from nuclear reactors. These

Figure 7.11 The MINOS beamline, running from Fermilab to the Soudan mine. (Figure courtesy of Fermilab.)

experiments, like CHOOZ, Palo Verde, Bugey, etc., all have been running and setting limits on other parameters of neutrino oscillation theory. A new experiment, called KAMLAND, intends to use the entire Japanese nuclear power industry as a neutrino source.

There exist dozens of experiments, defunct, current and proposed, to study neutrino oscillations. In fact, there exists a web page, entitled "The Neutrino Oscillation Industry," which is encyclopedic in its listing of the various efforts to understand neutrinos. Many of them have only just begun or will begin in the next few years. The whole field of neutrino oscillation study is extremely active, with many more experiments underway than I can possibly mention (although ICE-CUBE, the one where they want to use a cubic kilometer of ice in Antarctica, is worth plugging, just for being a really cool idea). This is a deeply exciting time for the study of neutrino oscillations and the reader should keep an eye on the popular press for the steady stream of interesting discoveries that will be announced over the next few years.

Mystery #2: Where's the Antimatter?

There is another pressing mystery, which is currently consuming the efforts of over one thousand physicists worldwide. The laboratory nearest and dearest to my heart (Fermilab) and its most direct competitor (CERN) comprise only a small fraction of the people involved, while the Stanford Linear Accelerator Center (SLAC) and its own competitor lab, KEK in Japan, are the current research hotspots. The name of this topic is CP violation, a name that really doesn't clarify why it's interesting. So what is it about CP violation that engenders such enthusiasm among current physics researchers? This topic fascinates because it may provide the answer to the perplexing question of why we're here at all.

The question of existence is not one explored exclusively by scientists. For millennia, men and women, mystics and seers, advocates of religions and philosophies both outlandish and established, as well as the occasional shepherd watching his flock under the clear midnight skies, have pondered this weighty question. Yet while many proposed solutions have been tendered, it is only the knowledge that has been gathered using the scientific method that has provided technical explanations and predictions on the topic. And because science has provided a wealth of insights on a multitude of things, it is natural that one would apply this method to the question of existence. There's only one problem. Given the information you've read so far in this book, our scientific knowledge fails badly on this question. In fact, it predicts that the universe we observe shouldn't exist at all. So is science wrong, or is it time to reveal yet another subtle phenomenon which provides at least the hope of resolving this alarming conundrum?

To understand this seeming failure of science, we must return to Chapters 2–4 and recall what we have learned on the topic of anti-matter. The universe, as best we know, is made essentially entirely of matter. Antimatter is an antagonistic substance that can annihilate with matter and release an enormous amount of energy. In fact, com-bining a single gram of matter with a corresponding amount of anti-matter releases an amount of energy comparable to the atomic explosions that devastated Hiroshima and Nagasaki.

The first antimatter, the positron, which is the antimatter analog of the mundane electron, was discovered in 1932 by Carl David Anderson in the Guggenheim Aeronautical Laboratory at the California Institute of Technology. However, while sub-microscopic amounts of antimatter have been isolated in modern particle physics experiments, no large quantities of antimatter have been observed. So the question one must answer is "Is the overwhelming presence of matter and the corresponding dearth of antimatter anyway surprising? Or is this as one should expect?"

To address this question, one should return to Chapters 3 and 4. In these chapters, we discussed how antimatter is created. The only known way to create antimatter is by the conversion of pure energy into two particles, always one matter and one antimatter.

Representative diagrams are given in Figure 7.12. Examples include when a photon (γ) converts into a quark (q) and an antiquark ($\bar{\text{q}}$), or when the photon converts into a charged lepton (ℓ^-) (an electron, muon or tau) and an antilepton (ℓ^+). Similarly a gluon (g) can convert into a quark-antiquark pair. A Z boson can split in the same ways as a photon. Finally, a W boson can convert into a quark and antiquark pair or into a charged lepton (ℓ^-) and a neutrino (ν). In the case of a W boson, if the charged lepton is a matter particle, the neutrino is an anti-matter particle ($\bar{\nu}$), while if the charged lepton is antimatter (ℓ^+), the neutrino is a matter neutrino. W bosons can also decay into unlike pairs of quarks and antiquarks, e.g. $W^+ \rightarrow \text{u} + \bar{\text{d}}$.

The most important point in the preceding discussion and figures is that matter and antimatter are created in equal amounts. A similarly

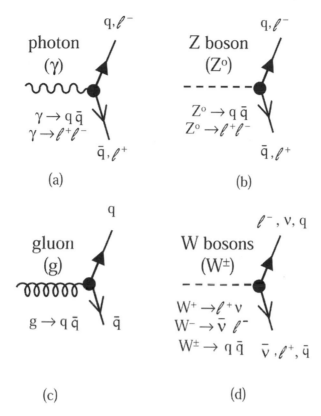

Figure 7.12 All matter is created in pairs; one matter particle for each anti-matter particle.

critical point is that in order to convert matter into energy, one needs an identical antimatter particle (e.g. $e^+ + e^- \rightarrow \gamma$). Thus matter and antimatter are created and destroyed in pairs and so it sure seems like there should be equal amounts of matter and antimatter in the universe. And, with equal amounts of matter and antimatter, the particles would eventually find one another and annihilate into pure energy. The universe would thus ultimately be filled only by a diffuse energy glow, without any matter at all.

So why does this not appear to be the case? There appear to be a couple of possible explanations. The first one is rather radical and might be described as the nagging suspicion that the laws of physics as we understand them are simply wrong. Any reputable scientist should consider this to be a possibility. But the remarkable achievements in technology, improved standard of living and the general advancement of mankind, all initially made possible through the careful application of the scientific method, suggests that our understanding of the world is actually pretty accurate. Besides, the equal footing of matter and antimatter has been reliably established in countless experiments by thousands of experimenters across the continents. So this explanation is unlikely to be the right one.

A second explanation for the observed apparent absence of antimatter is that it is an illusion and there really are equal amounts of matter and antimatter in the universe. A proponent of this explanation would simply state that while the region of the universe in which we reside consists exclusively of matter, there exists in other places in the universe large concentrations of antimatter, with a simultaneous lack of matter. Such an explanation, while reasonable, requires experimental verification. If there exists a concentrated accumulation of antimatter in the universe, where is it? We know that it can't be close. Upon landing on the Moon, Neil Armstrong was able to say "The Eagle has landed," rather than annihilating into a massive fireball that would have been visible from the Earth. Further, we have landed probes on many of the planets without problems. Thus the solar system consists exclusively of matter. Even though we've not visited other stars in our galaxy, we can rule out the possibility that they consist of antimatter using a different set of observations. While interstellar space is empty indeed, it does consist of a thin cloud of gas. We know this from astronomical measurements. This gas permeates the galaxy and within it all the stars are embedded. If this thin gas of matter hit an antimatter star or planet, we would easily be able to see the resultant energy release. So large concentrations of antimatter in our galaxy can be excluded. A similar argument rules out the possibility of a nearby antimatter galaxy. In this instance "nearby" means within a few tens of millions of light-years.

On much larger distance scales the observational evidence for local concentrations of antimatter is quite a bit weaker. However, any theory that has had any success explaining the beginnings of the universe has difficulty explaining local clumping of matter and antimatter on such a large distance scale. So while this explanation is not excluded, it seems unlikely.

So, having considered and rejected two explanations for the mystery of the apparent matter domination of the universe, where does this leave us? Clearly here is a tantalizing puzzle of the highest importance. Surely scientists must have worked on this question and solved it. Or is it possible that this mystery remains unsolved, waiting only for a clever idea from some young physicist to clarify everything? Well, as you might have surmised from the placement of this topic in a chapter called "Short Term Mysteries," modern research physicists believe that they understand the main ideas necessary to coherently explain the phenomenon. However, there are a few measurements still required to nail down our understanding. These experiments, as will be described presently, are underway.

But before we continue our discussion of the matter domination of the universe, we need to first pause and work on our vocabulary. We began our discussion with the observation that energy could be converted only into matter-antimatter pairs of particles. But the two particles must also be the same *kind* of particles, for instance, one never sees a photon decay into a quark and an anti-lepton or vice versa. In fact, we have stumbled onto another important set of conservation laws. Conservation laws embody crucial understanding and observations of the universe. Recall our earlier contact with the law of conservation of energy. This states that the amount of energy in a system under consideration never changes, although it can change forms. In this, it's like a perfectly sealed vessel containing a certain amount of water. One can heat or cool the vessel and the water can convert into steam or ice, but the total amount of water never changes.

There are many conservation laws known to modern science, each one revealing deep insights into the structure of the universe. In the context of our current discussion, we need to understand a seemingly obscure conservation law, called lepton number conservation. Lepton number is really pretty simple. If a particle is a matter lepton, we say that it has a lepton number of +1. An antimatter lepton has a lepton number of −1 and a particle that isn't a lepton at all is said to have a lepton number of zero. Lepton number is somewhat like electric charge, in that a particle can have positive or negative charge, or be neutral. As an example, let's consider the simple case of a photon (lepton number of zero) splitting into an electron (lepton number of +1) and a positron (lepton number of −1) ($\gamma \rightarrow e^+ + e^-$). We can see that in this interaction, lepton number is indeed conserved (before the split the lepton number is zero and after the split the net lepton number of zero is maintained, since $(+1) + (-1) = 0$).

One can make a similar argument for quarks, which would give each matter quark a positive number and an antimatter quark a negative number. Since quarks aren't leptons, their lepton number must be zero. Thus you might expect that there is a corresponding "quark number," and one could easily develop a theory using this idea. However, for historical reasons, an equivalent terminology has been developed. Since baryons, the heavy particles discovered primarily in the 1950s, of which the proton and neutron are the most familiar, were discovered long before the quarks were postulated, what is actually used is a quantity called baryon number. Each matter baryon is assigned a baryon number of +1, while for antimatter baryons, their baryon number is −1. For particles that are not baryons, the baryon number is zero. (Note that since each baryon contains three quarks, the matter quarks are assigned a baryon number of +1/3.)

Since the stable baryons (say protons and neutrons) are much more massive than the lighter leptons (the electrons and neutrinos) (e.g. the mass of the proton is about 2000 times greater than the mass of the electron), historically the debate and indeed even the very language of the discussion of the matter domination of the universe, has centered on the baryon. Since each antimatter baryon can cancel out the baryon number of a matter baryon, in a universe with equal amounts of matter and antimatter baryons, the net baryon number of the universe should be zero. If, as seems to be the case, there is more matter than antimatter, the baryon number of the universe should be positive, while an antimatter dominated universe would have a net negative baryon number. Thus the whole debate on the matter/antimatter asymmetry can be reduced to a single number...the baryon number of the universe.

Sakharov's Three Conditions

In 1967, Soviet physicist and political dissident, Andrei Sakharov, published a paper in which he set out what properties would be required of any theory that could explain how the observed universe could have a positive net baryon number. These criteria can be separated into three categories, the first of which is the condition that there be some sort of physical process that violates baryon number, i.e. some type of interaction that can have a different baryon number before and after the interaction. The second condition is that the laws of nature must somehow be asymmetric in a way we have not yet observed: matter must be preferred over antimatter. The final condition is that whatever physical process turns out to be the one that violates baryon number, it must be out of "thermal equilibrium," an obscure physics term with a very specific and technical meaning. Thermal equilibrium originally described a property of temperature, so we use that much easier topic to describe what we mean here. Suppose that you have a bar of metal a couple of feet long. Simultaneously, you put one end on a block of ice and heat the other end with a flame. After a few minutes, you remove both the flame and ice and touch the two ends of the bar. Rather unsurprisingly, one finds that one end is hot, while the other end is cold. What then happens

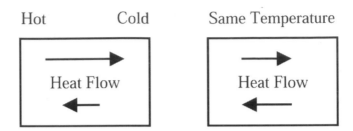

Figure 7.13 To be in thermal equilibrium, there must be no concentration of energy. When there is no concentration of energy, energy flow in all directions is equal (i.e. in equilibrium).

is that heat flows from the hot end to the cold end. The bar is not in equilibrium, as heat moves from one part of the bar to another. Later, the bar is everywhere the same temperature. Because there are no temperature differences in the bar, any heat flow from one end of the bar is exactly cancelled by an equal flow in the other direction. Because these flows of heat energy are the same, the bar is said to be in thermal equilibrium, a fancy way to say that there are no temperature differences (or equivalently differences in concentration of energy) in the bar. This point is illustrated in Figure 7.13.

Within the context of the matter-antimatter asymmetry, if the baryon number violating processes were in thermal equilibrium, this would have the tendency of evening out any momentary excess of baryons as compared to antibaryons. Thus we see that the baryon violating process must not be in equilibrium.

With the clarity of thought offered from Sakharov's conditions, scientists began to construct new theories that satisfied his criteria. In fact, the proton decay experiments described earlier in this chapter were an attempt to measure the baryon violating decay of a proton into lighter particles (none of which can be baryons, as the proton is the lightest baryon). With the null results reported, physicists looked for alternate explanations. It was soon true that they realized that a perfectly viable solution was sitting under their collective noses the entire time. The electroweak force could provide all of the relevant needs, at least in principle.

To demonstrate this not-so-obvious idea, let us consider the third of Sakharov's three criteria, the requirement of thermal equilibrium. As we discussed in detail in Chapter 5, above a certain energy, particles have no mass, while below that energy, the symmetry between electromagnetism and the weak force is broken, thereby giving elementary particles their mass with which we are familiar. Consider how this must have occurred in the first seconds following the Big Bang. As the universe cooled, random fluctuations required that certain portions of the universe cooled first. Where this cooling occurred, "bubbles" formed, inside of which all particles could have mass, surrounded by hotter areas where the electroweak symmetry had not yet been broken. Energy would flow from hotter to cooler areas and thus the boundary between the "bubbles" and the surrounding volume would not be in thermal equilibrium. Thus, Sakharov's third condition is satisfied.

So what about his other two conditions? Initially, the electroweak force doesn't seem to help. For instance, how does the electroweak force violate baryon number? It seems that it can't as any Feynman diagrams we have written involving W or Z bosons do not allow for baryon number violation. Here the vagaries of quantum mechanics help us. Just like in the case of neutrino oscillations, in which an electron neutrino can oscillate into a muon neutrino, baryon number violating oscillations are possible. The problem is that they are suppressed because of the energy needed to allow the oscillation. The non-zero masses of the observed quarks and leptons make the oscillations very unlikely. However, when the universe was much hotter, all the quarks, leptons and electroweak bosons (W and Z) had identically zero mass and so this restriction was not present. This is like a roomful of greedy bankers from different countries, each with an identical amount of money, although in different currencies (dollars, pounds, lire, rubles, etc.) Temporarily we impose the exchange rate to be equal among the currencies (i.e. 1 dollar = 1 lire = 1 pound, etc.) As long as the exchange rate doesn't change, the bankers will swap money freely. But, if one currency all of a sudden becomes much more valuable, or the symmetry is broken, that currency becomes "frozen out" of trading. In physics jargon, we say that trading in this currency is "suppressed," as each banker will hold onto it, because its value (or mass, to stretch the analogy) is greater than the others. Thus, the electroweak force allows for baryon number violating effects, as long as the temperature of the universe is hot enough for the Higgs mechanism to not matter.

The second of Sakharov's conditions deals with the existence of a mechanism that favors matter over antimatter. It is this topic that engrosses so many of my colleagues and we will thus spend some time trying to understand what we know thus far and how we hope current experiments will further increase our knowledge. To understand these points, we need to understand some new conservation laws, called rather mysteriously charge conjugation (C), parity (P) and the combination of the two (CP).

Charge, Parity and All That

Before we attempt to understand these new laws, we should review some ideas we've discussed previously. We are now familiar with the conservation laws of energy, momentum, spin and charge. In general, we have found an enormously important fact regarding conservation laws and the mathematical equations that describe the world. For each conservation law, which we recall is physics jargon for "that quantity doesn't ever change," there is a corresponding requirement on the equations that is what we call a symmetry. In this context, symmetry means that you can change a variable in the equation and have no effect on the outcome. To see an example of this, let's consider one of the simplest experiments one can imagine. We go into a large room with a horizontal floor. We then drop a ball and time how long it takes to hit the floor. While things like how high we release the ball from the floor or whether we throw the ball downwards will affect the outcome, there are variables that we can change that will have no bearing on our measurement. For instance, one thing that will not affect the measurement is whether we do our experiment in the exact center of the room or at a horizontal distance from the center. Walk five feet east, west, north or south and you'll get the same result. Similarly, it doesn't matter what time we do the experiment; today or tomorrow doesn't matter. A third variable that doesn't matter is the direction in which you're facing. Face in any of the infinite directions of the compass and the measured time will always be the same. The fact that these three variables do not affect the measurement means that the equations must have translational, temporal and rotational symmetry. These three symmetries lead inexorably to the laws of conservation of momentum, energy and spin respectively.

In order to understand the symmetries that are relevant to further explore the reasons for the matter dominance of the universe, we must revisit our old friend, quantum mechanical spin. Within the context of particle physics, it is much easier to think of spin if we temporarily (and incorrectly!) think of an elementary particle (say an electron) as an olive with a toothpick stuck through it. (Clearly, I've spent too much time in bars...) Imagine that one end of the toothpick is painted black. If we look along the toothpick with the black side facing us and spin the olive around the toothpick axis, we see that the surface of the olive can move either in the clockwise or counterclockwise direction. We could call these two cases plus and minus, or clockwise and counterclockwise, but instead we call the two cases right-handed and left-handed. This is done because our hands are mirror-images of one another. To understand what "right-handed" means, we put our olive in our right hand, with the black side of the toothpick in the direction our thumb is pointing. As is shown in Figure 7.14, if you let your fingers curl naturally and point the black end of our toothpick at our eyes, we see that our fingers are curling in a counterclockwise direction. Thus a counterclockwise rotation is said to be "right-handed." Repeating the experiment with your other hand shows that clockwise rotation is "left-handed."

Another insight one needs to understand what we need to know about spin is the fact that any clock can be thought of turning clockwise or counterclockwise, depending on how you hold it. Looking at a clock in the normal way, the hands sweep clockwise. However, turn the clock around as we have done in Figure 7.15 and the hands sweep counterclockwise.

Finally, we must recall that quantum mechanical spin is different from our normal understanding of the word, as the spin axis can point

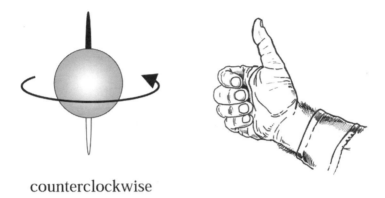

counterclockwise

Figure 7.14 A rotating ball can have a handedness. A counter-clockwise rotating ball has a "right" handedness. (Drawing courtesy of Dan Claes.)

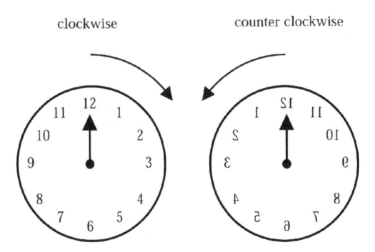

Figure 7.15 A normal clock runs clockwise. Yet when viewed from behind the same clock appears to run counterclockwise. Thus it is important to define how you are looking at your rotating object.

in only two directions... along or against the direction that the particle is moving. If the particle is not moving, then even a quantum
mechanical particle spinning in a void can have its "toothpick" or axis
of rotation pointing in any direction. That's because in a void, there's
nothing to define a direction. However, if a particle is moving, then
one direction, the direction in which the particle is moving, is different. Figure 7.16 shows us how we can use the direction of the particle's motion to define a coordinate system. To take our hand example,
if we use our thumb to point in the direction of motion, we see that
we must use either our right or left hand to simultaneously wrap
our fingers in the way that the particle is rotating. Depending on
which hand is needed, we say that the particle is right- or left-handed.
Figure 7.16 illustrates this point, while Figure 7.17 shows a shorthand way to draw the same situation.

Each and every subatomic particle can be represented in this way,
except for particles that have exactly zero spin. These particles are

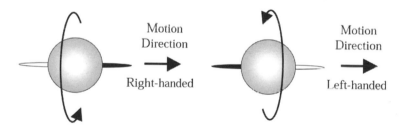

Figure 7.16 We define handedness from the direction of motion. An object
that rotates in a counterclockwise sense with respect to the direction of
motion is called right-handed. A clockwise rotation is left-handed.

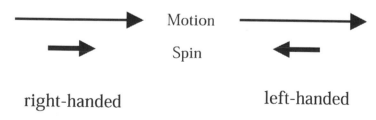

Figure 7.17 A simple way to denote the handedness of particles.

actually much easier to deal with mathematically, but aren't relevant for the present discussion. Nonetheless, to appreciate the following discussion, which we're finally ready to undertake, one must keep Figure 7.17 in mind.

The first of the ideas we must consider is parity or simply P. As is often the case with physics, the term parity has a generally accepted meaning and an extremely specific meaning for physicists. The general meaning of parity implies that things are equivalent or of similar magnitude. However, for physicists, it instead deals with the direction in which a particle is moving. Recall when we spoke of energy, momentum and spin conservation and we saw that this related to the results of a measurement (and hence the laws of physics) being independent of where you were, the direction you were facing and when you performed the experiment. However, another variable one could change is the possibility of whether or not you are moving forward or backward. If your measurement is independent of whether you are moving forward or backward, then this implies parity. For physicists, this means that if you replace backward and forward in your equations (technically, this means you replace every x in your equations with $-x$), your measurement (and predictions) should not be changed. In fact, parity is really a three-dimensional operation, so you should replace left with right, up with down and forward with backward. However, for clarity of explanation, we will simply work in terms of left and right. A second aspect of parity shows up in quantum mechanics. As is usual in quantum mechanics, things are a bit trickier here and so we'll only touch on a particular idea. In quantum mechanics, each particle can be described by a mathematical expression, called the wave function (WF). This wave function is pretty complicated, but one can ask what happens to the wave function if you do the parity operation on it (i.e. replace all directions with their opposites). For some particles, if you do this, you find that the wave function isn't changed. For this particle, it's said to have positive parity ($P = +1$). Alternatively, for some particles, if one does the parity operation on the wave function, one gets back the negative of the original wave function. Such a particle is said to have a negative parity ($P = -1$). One can see this pretty easily by looking at a simple wave function equal simply to x, the position variable (i.e. WF(no swap) $= x$). Swapping left and right simply means that you replace every x with a $-x$. We see that after the replacement $x \to -x$, the new wave function is negative as compared to the original one, WF(swap) $= -x = -$WF(no swap). This would be a negative parity wave function. If, on the other hand, the wave function was WF(no swap) $= x^2$, then replacing $x \to -x$ would have no effect (after all $(-x)^2 = x^2$). Such a wave function would have a positive parity, because WF(swap) $=$ WF(no swap).

The property of parity became interesting in the 1950s, when strange particles began to be made in large quantities. Naturally, the first thing people tried to do was to measure how these strange (i.e. long lived) particles could decay. Of special note were two particles carrying strangeness, called the tau (τ) and the theta (θ). (Note that both of these names are no longer in use and specifically, this tau particle is utterly unrelated to the tau lepton discovered in 1974.) The tau and the theta particle had identical mass, lifetime and charge, however one of them (the theta) decayed into two pions, whereas the tau always decayed into 3 pions. Parity is one of those quantities that is conserved, meaning that the parity should be the same before and after the decay. If we knew the parity of the particles after the decay, we could infer the parity of the parent particle. Further, we need to combine the individual particles' parities in order to get a total parity. One does this simply by multiplying the parities of the particles (i.e. Parity(all) $=$ Parity(particle 1) $\times$ Parity(particle 2)).

From earlier work, physicists knew that the parity of a pion was -1. Thus for the theta, with its two decay pions, the parity was $(-1) \times (-1) = +1$, or positive parity. In contrast, the tau decayed into three pions, so the tau's parity was $(-1) \times (-1) \times (-1) = -1$, or negative. If parity were conserved in the decay, then the tau and theta had to be different particles, as they decayed into different final states, with different parities. If, on the other hand, parity were not conserved in the decay, then possibly the tau and theta were the same particle. Because parity was not conserved (in this hypothesis), a particle could decay into either parity configuration. It was simply our ignorance about the role of parity in strange particle decay that was tripping us up.

The Death of Weak Parity Conservation

The exploration of the tau-theta mystery led two young and bright scientists to an insight that would shake the physics world. In April 1956, Chen Ning Yang attended the so-called Rochester Conference, a biennial conference which in recent years has been held many places, none of which are in fact Rochester, New York. The Rochester conferences were designed to bring theorists and experimenters together to try to better understand the pressing physics mysteries of the day. At conferences of this sort, typically each morning or afternoon is devoted to a particular subject, with several subjects being covered over the course of the week-long conference. In the session entitled "Theoretical Interpretations of New Particles," Yang gave the first talk, which summarized the tau-theta situation. Several talks followed, each giving different proposals to solve the problem. One of the many proposed solutions was that the tau and theta were, in fact, two ways a particular and unnamed particle could decay. Since this particle could decay into combinations of particles with parity of -1 and $+1$, this would imply that the original particle didn't have a unique parity or, if it did, then clearly parity was not conserved. If this explanation turned out to be true, it would be very odd. The discussion at the conference did not solve the question, but it did feed Yang's curiosity.

While Yang's permanent position was at Princeton University's Institute for Advanced Study, he was spending the summer at Brookhaven National Laboratory on Long Island. Back in New York City, Yang's friend Tsung-Dao (T.D.) Lee was a professor at Columbia University. Because of their close proximity and longtime friendship, Lee and Yang would meet twice a week to work on theoretical questions of great interest, including the tau-theta problem. Together, they were inexorably drawn back to the question of the particle's parity. The tau and theta both decayed via the weak force. If one particle could decay via the weak force in two different ways, with each way having a different parity, this would imply that parity wasn't conserved in the weak interaction. Recall that "conserved" in physicist's lingo means that parity would be the same before and after something occurred (like particle decay). At the time, the general physics community believed that parity was always conserved, as many experiments had proven that parity was conserved in the strong and electromagnetic interactions. Lee and Yang asked the correct question... Had anyone ever verified that the weak force conserved parity?

To answer this question the two theorists turned to another Columbia professor, Chien-Shiung Wu. Wu was an impressively competent experimenter who was the local expert in beta decay and the weak force. She gave Yang and Lee a single book, comprising some thousand pages, which compiled the results of all reputable beta decay experiments in the previous 40 years. After Lee and Yang had perused the existing data, they found the rather surprising fact that the question of parity symmetry for the weak force simply had not been experimentally investigated. They prepared a paper "Question of Parity Conservation in Weak Interactions," which arrived at *The Physical Review* on June 22, 1956. While their paper does not argue that parity conservation does not occur, it rather notifies their experimental colleagues that the question was entirely open. They said:

> ...It will become clear that existing experiments do indicate parity conservation in strong and electromagnetic interactions to a high degree of accuracy, but that for the weak interactions (i.e., decay interactions for the mesons and hyperons, and various Fermi interactions) parity conservation is so far only an extrapolated hypothesis unsupported by experimental evidence.

Even prior to Yang and Lee's submission of their paper, their colleague Wu had decided to make the measurement. While the most likely result of the experiment was that the weak force conserved parity, the possibility remained that it didn't. And if it didn't, this result would change everything. Wu had intended to take a cruise on the Queen Elizabeth II back to the Orient to celebrate the 10th anniversary of her coming to America. With such a cool and rewarding experiment, she cancelled her tickets, sending her husband alone. Her choice may seem a bit unusual, but the idea of discovering parity violation was just too appealing. Wu got to work.

The essential things she needed to perform the experiment were the following. She needed to use an atomic nucleus that decayed due to the weak force (i.e. beta decay). The second requirement was that the nucleus had to have an intrinsic quantum mechanical spin. Luckily, this is true of very many nuclei. So far, so good. But the really tricky thing was that she had to cause the spins of the nuclei (the little black end of the toothpick in the olive) to all point in the same direction. So why is this? This is because the way you test parity violation is that you define a direction and look at the decay particles that come out of the nucleus, for instance do the decay products come out aligned with the nuclei's spin, in the opposite direction or at 90°. If parity symmetry isn't violated, you shouldn't be able to tell if you flip all directions to their opposite. Following this train of thought, we need to remember that a parity flip changes the direction, but not the spins. But Figure 7.18 shows that keeping the spin direction the same and flipping the coordinate system is equivalent to keeping the coordinate system unchanged and flipping the spin. In both cases, the direction and spin go from pointing in the same direction into opposite ones. It doesn't really matter what direction the "direction" arrow points, what matters is the relative orientation of the direction and spin.

So Wu's experiment was in principle simple. She would take a sample of highly radioactive Cobalt-60 (Co^{60}) and align the spins of the atom's nuclei. Co^{60} beta decays spontaneously into Nickel-60 (Ni^{60}), a positive electron (or positron, e^+), and an electron neutrino (ν_e). Basically, a proton (p^+) converts into a neutron (n^0), ($p^+ \rightarrow n^0 + e^+ + \nu_e$). The cobalt nucleus has a quantum mechanical spin of 5,

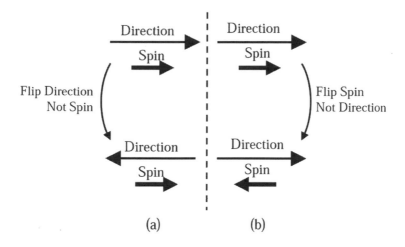

Figure 7.18 Flipping spin is equivalent to flipping direction of motion. In both cases, if the spin and motion are pointing in the same direction, flipping only one will result in the two pointing in opposite directions.

while nickel has a spin of 4. Since quantum mechanical spin is strictly conserved (recall that this means the spin after an event, say a decay, is the same as before that event), this means that the sum of the spins of the positron and neutrino must sum to $+1$, as illustrated in Figure 7.19.

So far, this is fairly mundane physics. The cool stuff is soon to come. One final thing that you need to know is that due to a subtle physics fact, the particles tend to come out along or against the direction of nickel's spin and rarely at 90°, a point illustrated in Figure 7.20.

Finally, one sees that basically there are two situations possible, given in Figure 7.21. In the decay, either the positron or the neutrino can travel in the direction of the spin of the nickel nucleus, while the other travels in the opposite direction.

Concentrating on just the positron for a moment, you see how the right-hand cartoon of Figure 7.21 can be obtained by doing the parity operation on the left-hand cartoon. This point is made more clearly by taking another look at Figure 7.18. The big difference in the left and right sides of Figure 7.21 is that on the left side, the

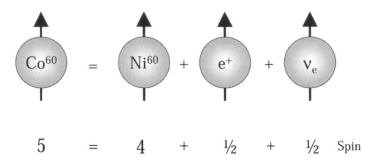

$$5 \quad = \quad 4 \quad + \quad \tfrac{1}{2} \quad + \quad \tfrac{1}{2} \quad \text{Spin}$$

Figure 7.19 Spin conservation in Cobalt-60. Both the positron and the electron must spin in the same direction (because $5 = 4 + 1/2 + 1/2$).

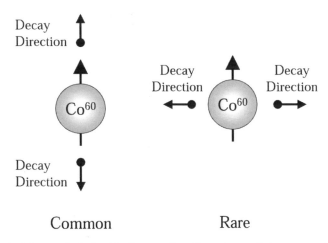

Common   Rare

Figure 7.20 In Cobalt-60 decay, the direction of the outgoing particles tends to be parallel to the original nuclei's spin. Particle emission at 90° with respect to the spin direction is rare.

positron has right-handed spin and the neutrino is left-handed, while on the right-hand side, the neutrino is right-handed and the positron is left-handed. If the weak force that governs beta decay doesn't care about parity (i.e. the weak force exhibits symmetry under the parity operation), both of the different possible decays in Figure 7.21

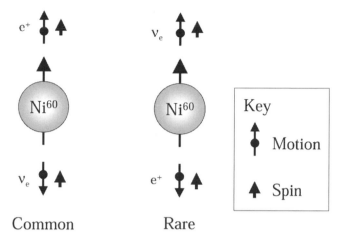

Figure 7.21 Compelling evidence for parity violation. The fact that a left handed neutrino is common, while a right handed neutrino is rare killed the idea of parity symmetry in the weak interaction.

should occur equally often. If they occur with different frequency, parity is not a symmetry of the weak force.

Professor Wu and her collaborators had to do a very difficult experiment. We'll skip the technical details here, except to say that even 45 years later, it's impressive that they got it together so quickly. They aligned the spins of the cobalt nuclei; they arranged their detectors in the direction that the particles would come out and they simply started counting. If one defines the direction in which the cobalt or nickel nuclei pointed as the positive direction, the core measurement of the experiment is to compare the number of times the positron goes in the positive direction as compared to the negative one. If the numbers are the same, parity is a symmetry of the weak force. If they aren't, it isn't.

So what's the answer? The answer is...maestro, a drum roll please...all through the latter half of 1956, they had some indication that the rates were *not* equal, but all of the difficulties they had with their experiment made it unwise to make any definitive statement (although even before they were sure, Wu and company informed

Lee and Yang of their progress). Finally, in early January of 1957, they were certain. Definitely the positrons flew in the positive direction much more often than in the negative direction. The effect was huge. In one of the very finest physics traditions, at 2 AM on January 7, 1957, R.P. Hudson, one of Wu's collaborators from the National Bureau of Standards, opened a bottle of 1949 Chateau Lafite-Rothschild and toasted with his colleagues their achievement. As seems to be required, they used paper cups. In fact, in my experience, one can claim to be a physicist when one has celebrated a discovery or technical achievement by drinking good wine out of bad glasses between midnight and six in the morning. It's an important rite of passage.

In 1956, Columbia University had more professors than Lee and Wu and the other faculty were of similar caliber and drive. Further, physicists live in a small world. Lee and Yang's paper had been published and Wu's preliminary results were being discussed over the lunch table. Other physicists started to think how they might contribute to the effort. One thing that was important to understand was whether parity symmetry violation was a feature of the weak force, or just in beta decay. By that time, people believed that Fermi's Universal Theory of weak interactions explained all weak interactions, from beta decay, pion decay, muon decay, strangeness, etc. So testing the universality of parity symmetry violation was a crucial test.

On Friday, January 4, 1957, a group of physicists from Columbia University went for lunch at the Shanghai Café. Physics gossip and egg rolls were consumed in equal quantities. The hottest topic of conversation was the rumors surrounding Wu's work. A young professor by the name of Leon Lederman was mulling over how he and his colleagues could contribute. Over the course of the day he thought about pion decay, which was a weak force decay. The accelerator at Nevis Lab in the physics department at Columbia University could copiously produce pions. These pions would decay into a muon and a neutrino. If parity symmetry were violated in pion decay, like in beta decay, the muons would have a particular handedness. These muons then decayed into an electron and neutrinos (another weak decay, although at the time there were subtle points of this decay not yet understood). The idea was that they would make muons (which had aligned spins) and look to see if the electron preferred to decay in one direction. After a couple of days work, they were able to show that indeed the electrons decayed from muons in a preferred direction. By the following Tuesday, weak parity symmetry was dead.

Lederman and company's signal was actually much more significant (i.e. compelling) than Wu's, although she came first. Both groups published seminal papers and on February 6, 1957, they announced their results to the meeting of the American Physical Society. People hung onto every word during the talks, as history was being made.

Not everyone believed the initial results. Wolfgang Pauli, in a letter to Victor Weisskopf, wrote "I can't believe that the Lord is a weak left-hander." But quickly the experimental results were confirmed by groups around the world. As usual, some people realized that data they had taken years earlier had shown that parity was not respected in the weak interaction. Yet another group of physicists joined the "If only ..." club. But it was Wu and Lederman and associates that got there first. Lee and Yang, for their prescient insight and pivotal paper, shared the 1957 Nobel Prize in Physics. The Nobel Prize committee doesn't like to make mistakes, as it would be very embarrassing to award the prize to someone, only to realize that their discovery was wrong. The speed with which they awarded the prize for parity violation shows the discovery's importance and the strength of the effect.

Parity Symmetry is Dead! Long Live CP!

With the death of weak parity conservation, theorists scrambled to find out what symmetry actually was preserved. The experimental results had shown one unequivocal result; neutrinos always had left-handed quantum mechanical spin and antineutrinos only were right-handed. A right-handed neutrino and a left-handed antineutrino were never observed. Because parity, or P, required that spin could be flipped, this meant that the parity operation would turn a left-handed neutrino into a right-handed neutrino. Since such a thing was never observed, parity symmetry was not a property of the weak. However, there was another operation that people thought that nature should respect and that was charge conjugation, or C.

Charge conjugation converted a particle into its antiparticle. So it would convert a left-handed neutrino into a left-handed antineutrino, again something never observed. So no go.

However, theorists realized that when both operations were performed, first parity and then charge conjugation, or CP, then this worked well. For instance, starting with a left-handed neutrino, parity converted it into a right-handed neutrino, which we now know is impossible. But subsequent charge conjugation converted that into a right-handed antineutrino, which was OK. So, while the weak force wasn't invariant under C or P separately, it was invariant under the combined operation of CP. This basic idea is shown in Figure 7.22. Such an observation settled the nerves of most physicists, who were rather upset at the prospect that the universe would not respect parity symmetry. They were content with the idea that the "correct" way to make the universe right-left symmetric was to include the swap of matter and antimatter. Order had been restored.

As we might recall, we started this digression in an attempt to find a process that preferred matter to antimatter. In this, we have not yet succeeded. Our next story begins the final chapter in the saga.

The year was 1954 and Murray Gell-Mann was on leave from the University of Chicago and spending some time at Columbia University. He and Abraham Pais, of the Institute for Advanced Study at Princeton University, wrote an interesting paper. Pais was the older physicist, having spent his early 20s on the run from the Gestapo in Amsterdam. Gell-Mann was young and brash…a star on the rise. Their paper discussed the neutral theta meson (θ^0). In it, they realized an amazing thing. This theta particle was one of the strange particles that were so fascinating at the time. We no longer use the terminology of θ^0, instead calling the same particle the K^0. We will continue the discussion using the more modern terminology. Because their idea predated Gell-Mann

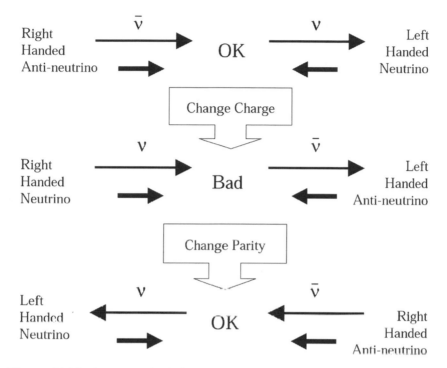

Figure 7.22 Cartoon which highlights the fact that while changing only charge or parity will bring one to a forbidden configuration, changing both results in allowed configurations.

and Zweig's 1964 quark hypothesis, they couched the discussion in terms of the K^0 mesons decay properties.

Gell-Mann and Pais knew that the K^0 meson (which we now know to contain a down and antistrange quark ($d\bar{s}$)) could be copiously produced in particle accelerators using the strong force. However, because the strong force conserved "strangeness," this means that if you originally had no strange quarks, then each time you created a strange quark, you had to simultaneously produce an antistrange quark. Because one can't see bare quarks, this means that when a K^0 ($d\bar{s}$) was created, simultaneously an anti K^0 meson, the $\bar{K}^0$ ($d\bar{s}$), was created. (Note, in reality, a K^0 could be created with many other hadrons, as long as the other hadrons had a strange quark in them, but we'll only talk about the $K^0 - \bar{K}^0$ case.) Since a strange and anti-strange quark have opposite strangeness, added together their strangeness cancels out and the result was zero strangeness. The bottom line is that particles carrying the strange quark can be created via the strong force (i.e. in great quantities) as long as they are created simultaneously with a particle carrying an antistrange quark.

On the other hand, after the K^0 and $\bar{K}^0$ are created, they move away from one another. Since afterwards each meson (say the K^0) carries a single strange quark, without a partner with which to decay, the K^0 cannot decay via the strong or electromagnetic force, as both of these forces require matter-antimatter pairs to annihilate. Thus, the K^0 meson can decay only via the weak force, as this force does not conserve strangeness. The consequences of these facts, creation by the strong force and subsequent decay via the weak force are peculiar. Coupled with the additional fact that the K^0 and $\bar{K}^0$ mesons are neutral and they have different strange quark content, leads to profound consequences.

The first of these consequences is called $K^0 - \bar{K}^0$ mixing. Basically, this means that a K^0 meson can convert into its own antiparticle and back again. This behavior stems from quantum mechanics. Because the K^0 & $\bar{K}^0$ can each decay into two pions ($K^0 \rightarrow \pi\pi$, $\bar{K}^0 \rightarrow \pi\pi$), quantum mechanics says that before the K^0 meson can "really" decay, it can "temporarily" decay into a pair of pions, before re-emerging as the original K^0 meson. This pattern can repeat many times before the K^0 meson irrevocably decays into the two pions that fly off to be detected. So far, this is a peculiar quantum mechanical behavior, but one that isn't really so weird, once you get used to the idea. The really weird behavior comes when you realize that the $\bar{K}^0$ meson can also temporarily decay into two pions before re-coalescing back into a $\bar{K}^0$ meson. However, you might ask yourself "How do the pions know which kind of K^0 or $\bar{K}^0$ meson to re-emerge as?" The answer is, they don't. A K^0 meson can temporarily decay into two pions, which can then recombine into a $\bar{K}^0$ meson. Through this chain, a K^0 meson can change into a $\bar{K}^0$ meson (and back again). This is really odd and can be seen in Figure 7.23.

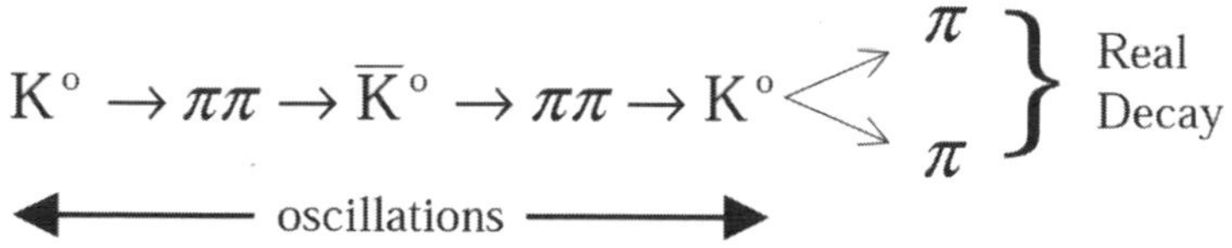

$$K^\circ \rightarrow \pi\pi \rightarrow \overline{K}^\circ \rightarrow \pi\pi \rightarrow K^\circ \begin{smallmatrix} \nearrow \pi \\ \searrow \pi \end{smallmatrix} \Big\} \begin{smallmatrix} \text{Real} \\ \text{Decay} \end{smallmatrix}$$

$\longleftarrow$ oscillations $\longrightarrow$

Figure 7.23 Kaon oscillation. Kaons oscillate into anti-kaons and back again before they finally decay into a 2 or 3 pion final state.

While $K^0 - \overline{K}^0$ oscillation is pretty weird, there's another, deeper, insight suggested by Gell-Mann and Pais. This is the realization of the fact that the K^0 (or $\overline{K}^0$) meson can decay into two or three pions. This is a revisit of the tau-theta puzzle. Since a state containing two pions has a parity of $+1$ and a state with three pions has a parity of -1, then a single particle cannot decay in both ways unless the particle doesn't have a unique parity.

Quantum mechanics purists will cringe slightly at the following discussion, but for most of us, the explanation will do. Basically, since the strong force does not distinguish between right and left (i.e. it is parity symmetric), each K^0 meson has a 50% chance of having a parity of $+1$ or a parity of -1, so you would expect half of the K^0 mesons to decay each way. Thus we can see that we could write the population of K^0 mesons as (50% of parity $+1$ K^0 mesons) and (50% of parity -1 K^0 mesons). Now for a crucial insight. For the decay $K^0 \rightarrow \pi\pi$, there are only two pions after the decay. This means that less energy is tied up in pion mass and so it can decay quickly. In contrast, the decay $K^0 \rightarrow \pi\pi\pi$ ties up more energy in pion mass, so it decays more slowly. In fact, the slow decaying (parity -1) component lives 100 times longer than the quick decaying (parity $+1$) component.

So here's the crucial idea. If you make a beam of K^0 mesons and let them decay, the fraction of them with parity of $+1$ will decay very quickly into two pions. Because the fraction of K^0 mesons with parity -1 lives so much longer, eventually all you will have is three pion decay ($K^0 \rightarrow \pi\pi\pi$) left. We have called the long-lived component of the K^0 meson with the name K^0_L (for K^0 long) and the short-lived component K^0_S (for K^0 short). Leon Lederman and collaborators (they were a busy bunch) discovered the K^0_L, with its characteristic long life and three pion decay, in 1956, two years after its prediction.

So far, so good. K^0 mesons seemed to decay, with the quick decaying component decaying into two pions and the slow decaying component into three pions. Everything was as expected if the CP symmetry was strictly respected by the weak force. Of course, if a long-lived (i.e. parity -1) kaon (K meson) ever decayed into only two pions (parity $+1$), then the CP symmetry would be violated. In 1964, Jim Cronin and Val Fitch led an experiment, accompanied by Jim Christenson and René Turlay, to address this question. They found that after they let the beam travel a long distance (long enough for all K^0_S's to have decayed away), a small fraction of the K^0_L mesons decayed in the wrong way into two pions. While the effect was small, occurring only 0.2% of the time, the result was earth-shaking. This was because CP had been violated and thus, finally, this provided us with a mechanism that prefers matter to antimatter. In 1980, Cronin and Fitch shared the Nobel Prize for this discovery.

While I've spoken only of the three pion decay of the long-lived K meson K^0_L, they can also decay into

$$K^0_L \rightarrow \pi^+ + e^- + \bar{\nu}_e \qquad \text{(a)}$$

$$K^0_L \rightarrow \pi^- + e^+ + \nu_e \qquad \text{(b)}$$

Since equation (b) can be derived from (a) by doing the CP operation (i.e. flip spins and replace matter with antimatter), they should occur with equal probability, if K^0_L mesons were only of the parity -1 variety. But measurement shows the K^0_L decays into a positron 0.33% of the time more than an electron. Finally, we have a reaction that distinguishes between matter and antimatter. In fact, we can now make the definitive statement as to what is defined to be antimatter. It is the charged lepton with the electrical charge that is preferentially produced in this particular decay of the long-lived neutral kaon, K^0_L.

With the realization that there exist processes that distinguish matter from antimatter, we have accomplished that for which we set out.

We have satisfied Sakharov's second criterion and possibly have solved the question of matter dominance in the universe. As you'd imagine, with such an important question, there has been an enormous amount of work on the topic. Recently, two experiments have been competing to make the best measurements, KTeV at Fermilab and NA48 at CERN. Both experiments have made very careful measurements that explore the topic of CP violation with impressive precision.

Of late, however, interest has shifted from the behavior of K (i.e. strange quark carrying) mesons to B (i.e. bottom quark carrying) mesons. The reason for this is that while the observation of CP violation in the behavior of K^0's proves that the Standard Model can explain some matter-antimatter asymmetry, the theoretical framework (discussed soon), in conjunction with the K^0 measurements thus far discussed, does not predict *enough* asymmetry to explain our world. Either the phenomena discussed in the last few pages are interesting, but unrelated, facts, or there's more to the story. So which is it? The goal of current research is to answer just this question.

In 1963, a theorist of some note, Nicola Cabibbo, had the first idea on how to explain how the strong force and the weak force can treat the K^0 mesons so differently. Recall that the strong force conserves strangeness and since a K^0 contains an antistrange quark and a $\bar{K}^0$ contains a strange quark, the strong force can never look at a particle and see that it's simultaneously a K^0 and a $\bar{K}^0$. The weak force however, doesn't worry about strange quark content; in fact it's through the weak force that the strange quark decays. Thus the weak force can see a particle as simultaneously having a probability of being either a K^0 or a $\bar{K}^0$. This behavior is pretty tricky and comes from quantum mechanics, but we can see the most relevant behavior by looking at Figure 7.24. This figure takes a little explaining. Suppose that you take two pencils and tape them together with a 90° angle between them. Call one a K^0 and the other one a $\bar{K}^0$. If you were only allowed to look along the axis of one of the pencils, you would only see the K^0 or the $\bar{K}^0$ pencil. The other pencil will be viewed end-on and will be essentially invisible. This is in analogy with the strong

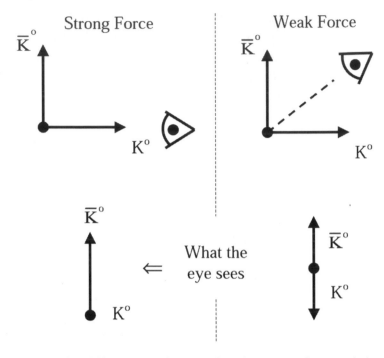

Figure 7.24 The difference in the way that the strong force and the weak force will view any particular kaon. The strong force is required to explicitly observe that the particle carries strangeness or antistrangeness. The weak force will see the same particle as being a mix of both.

force, which can only see the K^0 or the $\bar{K}^0$ nature of the particle, but never both. Now consider what happens if you rotate your point of view so that you are no longer looking along a pencil's axis (see Figure 7.24b). Now you can see both pencils. In a similar way, the weak force can simultaneously see the K^0 and the $\bar{K}^0$ nature of a particle. The strong force views the two types of kaons as kind of like the blind men feeling the elephant. The one touching the trunk describes the elephant as "snake-like," while the one touching the elephant's legs describes the elephant as being "tree-like." Both are right, but since each one only has a limited perspective, neither has the full picture.

Similarly, Nicola Cabibbo realized that the strong and weak force simply "saw" different aspects of the "real" particle. This crucial insight was expanded in 1972–1973 by Makoto Kobayashi and Toshihide Maskawa, when they realized that Cabibbo's idea could be extended to explain CP violation. There was a problem though. In 1972, Gell-Mann and Zweig's quark theory included only the up, down and strange quarks. Kobayashi and Maskawa's extension required that there be three more quarks. Since even the up, down and strange quarks had not yet been firmly established by that point, their idea was greeted as only a mild curiosity. The picture changed dramatically in 1974, with the discovery of the charm quark and in 1977 with the bottom quark. Discovery of the top quark waited until 1995, but even before its observation, Kobayashi and Maskawa's idea was taken most seriously. The new so-called CKM theory (for Cabibbo-Kobayashi-Maskawa) was able, in principle, to predict CP violation and ultimately (perhaps) the matter domination of the universe.

At its core, the CKM theory contains nine crucial parameters. These were the probabilities that an "up-like" quark (up, charm, top) would emit a W particle and become a "down-like" quark (down, strange, bottom). Since there are nine possible combinations (u → d, u → s, etc.), these nine numbers told the whole story. The current state of the art is that not all of these numbers are determined with equal precision and measuring these numbers is currently a program of intense study at several accelerators in the world. If CKM theory is right, then not all of these nine parameters are independent and interlinks are predicted. In the end there are only four independent parameters in CKM theory.

So why are there so many scientists exploring CP violation using bottom quarks? This is because of the guidance provided by the CKM theory. Recall that the whole beginning of the field came from the prediction of oscillation of a K^0 into a $\overline{K}^0$ and back again. Like the neutrino oscillation discussed earlier, for the $K^0 - \overline{K}^0$ oscillation to occur, this required a small $K^0 - \overline{K}^0$ mass difference. Further exploration of this idea shows that the whole CKM idea requires that the six quarks not have the same mass, or the whole approach falls apart. Because we've been talking about using electroweak symmetry breaking to explain the matter excess of the universe, the relevant mass scale is that of the electroweak bosons, the W and Z, with their characteristic mass of about 100 GeV. Looking at Tables 3.3 and 4.2, we see that all the quarks, save the top quark, have a mass very much smaller than that of the W and Z. Nonetheless, the bottom quark is very much larger than the other, lighter, quarks and thus the amount of CP violation in B^0 mesons is expected to be larger than that measured so far. So rather than looking at $K^0 - \overline{K}^0$ mesons, with the paltry 0.2% CP violation, scientists use $B^0 - \overline{B}^0$ mesons, which are expected to exhibit this much larger effect. Thus the $B^0 - \overline{B}^0$ mesons may well reveal a lot about CP violation.

As we consider the import of the knowledge we've been discussing, we see that what we have are the ingredients and even a recipe for producing an excess of baryons (i.e. matter) in the universe. However, the writing on the recipe is a bit smudged and we therefore need to understand it a bit better. The current experimental program studying mesons containing bottom quarks is a crucial effort to help us understand what's going on.

As I write, the PEP-II accelerator in the SLAC laboratory in Palo Alto, California is delivering electrons and positrons to the BaBar (a takeoff on $b\overline{b}$ or b-b-bar and serendipitously a famous elephant of children's books) detector. Simultaneously, the KEK-B accelerator is providing electrons and positrons to the Belle detector at the KEK laboratory in Tsukuba, Japan. In both of these detectors, the beam energies are tuned to produce an enormous number of B^0 and $\overline{B}^0$ mesons and, with this data, they hope to observe CP violation with bottom quarks. Meanwhile, the DØ and CDF detectors at Fermilab are also trying to make similar measurements.

In the autumn of 2002, the state of knowledge in the topic is the following. $B^0 - \overline{B}^0$ mixing (where a B^0 oscillates into a $\overline{B}^0$ and back) was first observed in 1987 by the Argus and UA1 experiments. Observation of CP violation is on much shakier grounds. CDF, Belle and BaBar have made measurements that are consistent with CP violation, but are insufficiently precise to claim discovery. This is not to be construed as a criticism on any of the participants, after all, the measurement is extremely difficult and only excruciatingly careful scientists could have accomplished what's been done so far. Unfortunately, to claim discovery they need more data and it is hoped that by the end of 2004 or 2005, they will have made a much more precise measurement. Stay tuned … this is an exciting time in CP violation studies.

As we close out this topic, it's worthwhile to take another look on the subject that started this journey, the fact that antimatter does not seem to exist in quantities anywhere in the observed universe. What I'm about to tell you starts with informed speculation and gradually becomes more concrete. In the beginning, as they say, the entire universe was packed into a tiny space like an egg waiting to hatch. For some reason (quantum fluctuations?, the hand of God?, who's to say?), the universe exploded. Initially the universe was unimaginably hot, some 10^{30}°C. The universe was a measly 10^{-41} seconds old and the number of particles and antiparticles were equal to each other and about equal to the number of photons. As the universe expanded and cooled to the relatively cooler temperature of 10^{26}°C, the mechanism that made matter slightly more likely than antimatter kicked in. For every billion antimatter particles, there were one billion and one matter particles. Still, the number of photons was similar. The universe was only 10^{-34} seconds old and the tiny matter excess that would become the universe was now in place. As the universe aged even further, getting to the ripe old age (in the particle physics world anyway) of one second, the process was complete and the universe looked grossly similar to the one which we inhabit. The universe had a temperature of 10^{10}°C and most of the excitement was over. Each billion antimatter particles had annihilated with a billion matter particles, leaving one matter particle left over. The photons were not affected, so for each matter particle, there were about one billion photons. Hence we reveal the real truth. While we observe a world that seems dominated by matter, really it is the photons (and neutrinos, discussed more in Chapter 9) that dominate the universe, at least in terms of gross numbers.

It is that crucial transition at 10^{-34} seconds after the Big Bang that allows us to exist. Without it... that is, without that one extra matter particle for every billion matter-antimatter pairs... we wouldn't be here at all. Thus the current studies of B^0 mesons hope to better reveal a pivotal period in the history of the universe. While we think that perhaps we understand what happened in the crucible that formed the universe, it's only with more data and better measurements that we can be sure.

In this chapter, we have chosen two very interesting questions that are currently under intense scrutiny by physicists. In both cases, much is understood and, even better, reasonably sophisticated mathematical theories have been devised to explain the phenomena thus far observed. Thus while it is proper to characterize these phenomena as existing at the fringe of the spectacularly successful theory of the universe that we call the Standard Model, significant data and complex theories to understand that data have been devised. One hopes that within the next decade, physicists will crack these problems once and for all. However there exist even greater mysteries to explore with no experimental evidence and only mostly-unconstrained theoretical speculation to guide us. For the Standard Model of particle physics, with all of its extraordinary success, is an incomplete theory. There are questions that the theory raises and can't answer. For these much deeper mysteries, we must turn to the next chapter.

chapter 8

Exotic Physics (The Next Frontier)

We shall not cease from exploration
And the end of all our exploring
Will be to arrive where we started
And know the place for the first time.

— T.S. Eliot

In the preceding chapters, we've discussed much of what is known about fundamental physics. The Standard Model of particle physics does a brilliant job of explaining all measurements thus far made over a vast range of temperatures and energies. Taken with its sister field of chemistry (which is really nothing more than very complicated atomic physics) and its cousin field of biology (which is essentially complex chemistry), one can explain all of the phenomena with which you are familiar. If one folds in Einstein's moderately mind-bending theory of general relativity, one can describe with good accuracy all phenomena ever observed.

Normally such a success would make someone rather cocky, but physicists learned their lesson at the end of the 19th century. While explaining experimental data is a necessary requirement for an ultimate theory, it is not sufficient. To be an ultimate theory, the theory must leave no questions unanswered. And for all the Standard Model's success, this is one criterion that it does not satisfy. By now, you must have some questions of your own. It may surprise you that you are in good company. Physicists have many such questions. While each of us may be more curious about one conundrum or another, the sorts of questions that are asked include the following:

- Why are there quarks and leptons and what is it that makes them different?
- Why are there three generations, each containing a pair of quarks and a pair of leptons?
- Could there be more generations?
- Why are there four forces and why do they have the relative strengths that they do?
- If the electromagnetic and weak forces can be shown to be two manifestations of a single and more fundamental electroweak force, is it possible that further effort and thought might show that there is only one real force, with the four apparent forces being simply different facets?
- Why are matter particles fermions, while force-carrying particles are bosons?
- The Higgs mechanism can explain *how* the top quark can be so much more massive than the other quarks, but it is silent about *why*.
- For that matter, will the Higgs hypothesis prove to be correct?
- Will it ever be possible to reconcile general relativity and quantum mechanics?
- Why is it we seem to live in three spatial dimensions and one time dimension?
- What is it that makes time different from space?
- If quantum mechanics and general relativity can be reconciled, does it stand to reason that there can be a smallest bit of space or even time?

The answers to each and every one of these questions, as well as many, many more, are a complete mystery. But the bottom line is that there are questions that are easy to ask, but for which the Standard Model is unable to provide the answer. In this chapter, we will discuss attempts to address some of these questions. As we do so, you should be aware that we've departed from the realm of comfortable knowledge and entered the realm of the unknown. Such a transition is often accompanied with peril, this time of an intellectual sort, rather than the mortal variety experienced by the intrepid explorers of a bygone era. But it is useful to recall that we are entering *terra incognita* and have little guidance to shepherd us along the path of truth. However, this ignorance does not provide us with total freedom to speculate wildly. Such a path is followed by modern day mystics, who invoke outlandish theories to fill in our ignorance. Science, unlike its competitor worldview of mysticism, is constrained by the vast body of knowledge described in earlier chapters. At a minimum, any new theory will have to explain all experimental data at least as well as current theories and preferably better. In addition, in order to supercede the existing theories, the new theory will have to correctly predict a new phenomenon or explain a fundamental link between two aspects of the data that the old theory did not illuminate.

A good theory reduces the number of variables (or bits of information) that must be put in "by hand." As an example, current theory predicts that matter and antimatter particles have the same mass. An antimatter electron, the positron, must have the same mass as its more common sibling, the electron. So there's no reason to measure both. Of course, we do, as any mass difference between the two would signal a breakdown of the theory.

While existing theory predicts a symmetry between matter-antimatter pairs, it is silent on the mass of the electron itself. We know of no way to calculate this without experimental input. In fact, the Standard Model of particle physics now requires 20 parameters to be inserted, with no understanding of the interlink of these parameters. These experimentally measured (and not derived) parameters are: the masses of the quarks (6), the masses of the charged leptons (3), the Higgs boson mass and the "vacuum expectation value" which is the amount of energy the Higgs field adds to empty space (2), four independent numbers that make up the Cabibbo-Kobayashi-Maskawa (CKM) matrix (which is described in Chapter 7) (4), the "coupling constants" for the electromagnetic, strong and weak forces (essentially how strong each force is) (3) and two more obscure parameters, the "phase for the quantum chromodynamics vacuum," which we won't discuss and the cosmological constant, a variable that describes gravity, which is currently outside the scope of the Standard Model (2). If the data discussed in Chapter 7 indicating that neutrinos have mass turns out to be correct, the number of input parameters is increased by 7. These additional parameters are the masses of the neutrinos (3) and a matrix similar to the CKM matrix that describes neutrino mixing (4).

Fiddling with the Parameters

None of these parameters seem like they have much to do with our world of common experience, yet extraordinarily, they do. Consider how the world would be different if the theory remained unchanged and we left the parameters untouched, save one. Let's simply think about how the world would be different if the mass of the electron were as large as the mass of the muon. Such a small change would seem to be unimportant, but we will see that this is not so. The first change would be to the size of atoms. The size of an atom is inversely related to the mass of the electron. Double the mass of the electron and you halve the size of atoms. Since the muon is about 200 times larger than the electron, this means the size of the atom would be 200 times smaller. Such a decrease sounds drastic, but it's really not. When Alice (in Lewis Carroll's "Adventures in Wonderland") shrank to a tiny size, she nearly drowned in the puddle of tears she had shed when she was large. In contrast, in our heavy electron world, *all* atoms are shrunk by the same factor, so a mini-you would sit in a mini chair, eat mini-food, drive a mini-car (some mini-Brits would drive a mini-Mini). A mini-Dr. Evil would even adopt a mini-Mini-Me. A mini-person living in a mini-universe would not have a very different worldview than the one that you experience each day.

As we see in the next paragraph, the following point is moot, but one difference would be your weight. A mini-you, sitting on a mini-Earth, each having the same number of atoms as the real you and Earth, would experience a weight 40,000 times greater than you do now. This is because while your and the Earth's mass would only change by less than 10%, the Earth would be 200 times smaller. Because the separation between you and the center of the Earth would be smaller, the net effect would be that you would be much heavier.

However, a world containing very heavy electrons would look nothing like the world we inhabit. While muons would not decay in our hypothetical world (and therefore be stable), atoms would be very unstable. As is currently the case with muons, heavy electrons swirling around the nucleus of the atom would occasionally penetrate the nucleus and occasionally an electron would combine with a proton to make a neutron and a neutrino ($e^- + p^+ \rightarrow n^0 + \nu_e$). Thus all atoms would quickly decay into only neutral particles (neutrons and neutrinos). Hydrogen would not exist, stars wouldn't burn, and life would not form. The universe would be stable, but boring.

So why didn't this occur in our world? This is because the neutron has a slightly higher mass than the proton. A real (i.e. our universe) electron does not provide enough energy to allow this complexity-destroying reaction to proceed. So we're safe.

We can make this point a little more clearly by looking at Figure 8.1. In order to make a particular subatomic particle, one needs to start with at least as much energy as the particle carries. Since energy and mass are equivalent, we will couch the discussion in terms of mass. In order for a proton and electron to be able to combine to form a neutron, their combined mass must exceed the neutron's mass. We see from the figure that in the hypothetical universe, this condition is satisfied, allowing

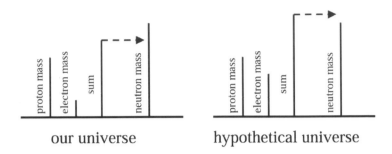

Figure 8.1 Diagram showing how a more massive electron would change the universe. In our universe, a light electron, combined with a proton, does not have enough energy to create a neutron. In contrast, for a hypothetical heavy electron, they will combine to make a neutron. This will substantively change the universe.

neutron formation. In our universe, neutron formation from an electron and a proton is forbidden.

If a little thing like changing the mass of the electron could vastly alter the nature of the universe, so could a little thing like altering the mass of the quarks. As you recall, all ordinary matter is made of up and down quarks, with a proton consisting of two up quarks and one down, and a neutron containing the opposite. Suppose the mass of the up and down quarks were reversed. Currently the down quark carries slightly higher mass and since the neutron contains two down quarks and only one up quark, the neutron is ever so slightly more massive than the proton. While neutrons can exist quite happily in the nucleus of an atom, a lone neutron is unstable, decaying in about 15 minutes into a proton, an electron and an electron anti-neutrino $(n^0 \rightarrow p^+ + e^- + \bar{\nu}_e)$. At the quark level, what is really happening is a down quark is changing into an up quark $(d^{-1/3} \rightarrow u^{+2/3} + e^- + \bar{\nu}_e)$.

In our converse world, the opposite would be true. Protons would be unstable while neutrons would exist forever. Normally hydrogen consists of a lone proton, surrounded by a lone electron. However, in our hypothetical universe these protons would decay into neutrons. So the hydrogen gas that permeates the universe would not be present. Since the proton would decay into a neutron, a positron and an electron neutrino ($p^+ \rightarrow n^0 + e^+ + \nu_e$), the positron would wander around and eventually meet and annihilate with an electron that would have been surrounding the hydrogen nucleus, had the proton not decayed (i.e. like in our universe), thus yielding two photons ($e^+ + e^- \rightarrow 2\gamma$). We're a little luckier with helium. As we will discuss in Chapter 9, the helium in the universe was largely already created a scant three minutes after the Big Bang. Since the lifetime of one of these hypothetical protons would be 15 minutes, it would exist long enough (three minutes) for helium to form. As is the case for neutrons in our universe, in atomic nuclei, the proton would be stable, so the helium would not decay. In fact, the swapped-quark-mass universe would consist of about as much helium as we see in ours, but without any hydrogen. So our swapped-quark-mass universe would consist of helium, neutrons, neutrinos and photons. Life as we understand it, indeed probably any life, would not exist.

So far, we've monkeyed with the electron and the up and down quarks, but these particles are all obviously common in the universe in which we exist. What if we change the top quark's mass? The lifetime of the top quark is about 10^{-24} seconds. Surely changing this would have no effect? The bottom quark is about three times more massive than the charm quark which, in turn, is about 2–3 times more massive than the strange quark (although the mass of the strange quark has not been directly measured, so this assertion is a bit uncertain). In contrast, the top quark is about 40 times more massive than the bottom quark. How would the universe be different if the top quark's mass were lower, say only three times the mass of the bottom quark, continuing the previous pattern (i.e. instead of being 175 GeV, the top quark would have a mass of about 17 GeV)? Since the top quark is so ephemeral and existed in quantities only tiny fractions of a second after the Big Bang, surely this change cannot matter... correct?

The details of how one calculates the effect of such a change are a bit complicated, but when the effort is complete, one obtains a surprising result. Rather than just being an ephemeral particle, with no effect on the everyday world, the mass of the top quark has observable consequences. If the top quark had a mass of about 17 GeV, rather than the measured 175 GeV, the masses of the proton and neutron would be about 80% of their current values. While not as dramatic a difference as some of the other changes, this result suggests that the effects of the top quark are not as insignificant to the world that we experience as one might think. I rather like this consequence of a lower top quark mass, as its net effect would be to reduce the weight of everything in the universe to 64% of its current value. A 200-pound guy would suddenly weigh 128 pounds. I'd be rich! Move over Atkins, grapefruit and whatnot… Don's top quark diet is here!

One could spend a long time considering what would happen if one twiddled each of the 20 unexplained parameters. Some combinations would change the universe only slightly, while others would entirely change the complexion of the universe. If one were able to change the parameters two at a time, the situation would be even more complicated.

This short exercise underscores the real questions. Why are there 20 independent parameters? Is it possible that a better (i.e. more complete) physics theory would reveal links between the various parameters and reduce the number of parameters that are truly independent? Why do the parameters take the values that they do? Is it possible to follow historical precedent and somehow comprehend this complexity and replace it with a simpler and more fundamental understanding?

While I don't know the answer to any of the above questions, the goal of modern fundamental physics is to simplify. We hope that all forces will be shown to be simply different facets of a single primordial force. It is also hoped that the 12 quarks and leptons will eventually be revealed as different aspects of a single particle or, even more tantalizing, different ways in which space itself can vibrate. In the following pages, I will discuss several theories that have been suggested as offering a more complete understanding of the world than that given by the Standard Model. The first few theories discussed are not complete theories, in that they do not attempt to explain everything, rather they try to add a new truth to our understanding of the universe. They have the nice feature that each of these two theories has potentially observable consequences for current experiments. In other words, these theories predict a phenomenon or phenomena, not yet observed, that modern experiments could in principle find. Thus as we proceed, we will include some discussion of the relevant state of the art experimental results. As of this writing, none of these new ideas have been proven to be true, but my colleagues at DØ and CDF are working extremely hard to support or refute the theories.

While these theories are manifestly not "ultimate" theories, in that they don't even attempt to explain everything, we end the chapter with a theory that does aspire to "ultimate" status. This theory makes predictions, but unfortunately, modern experiments are not sensitive enough to prove the theory or rule it out. Nonetheless, it's a cool idea and we talk about it to get an idea of what our ultimate theory might look like.

I talk about "ultimate" theories and "effective" theories (i.e. ones that are simply "better"). How are these two different? Essentially, the two are different in both the degree to which a theory requires input that is impossible to derive from within the theory itself and the degree to which the relationships between the various components of the theory are explained. We call theories that require input from experiment to be "effective theories." Effective theories don't sound as good as "ultimate" theories, which require no input from anywhere; all quantities are derivable from within the theory. Nonetheless, effective theories are very useful. An example might be Newton's Law of Gravity, as applied to the solar system. Newton's law requires as input the masses of the Sun, the planets and the sundry bits of rock and ice floating nearby. Once these are given, Newton's laws can completely describe the motion of the solar system, but the theory is silent on the reason for the various input masses. There exist better (i.e. more complete) theories of planetary formation, which can predict, for example, why the inner planets are smaller and rocky, while the outer ones are more massive and gaseous. These more complete theories are still effective theories in that they require as input the original distribution of gas that made up the solar system. For each effective theory, there exists, at least in principle, a more complete theory that explains what, for the first theory, were arbitrary inputs. Physicists would like to invent a theory that requires no external inputs.

As we continue our discussion, you will hear about lots of "Planck" things; e.g. Planck length, Planck time and Planck mass. Max Planck was one of the very early pioneers of quantum mechanics. He was addressing an age-old question ... "Are there 'right' units of measurements?" The foot was defined to be the length of the foot of the king, while the meter was originally defined to be 1/10,000,000 of the distance between the North Pole and the equator, along a line through Paris, France. Thus these definitions were a historical accident. Change the king or the size of the Earth and the length of the foot or meter changes. What is needed is an absolute method for determining length, time and mass. Planck realized that one could do this using universal, fundamental constants. He started with three universal constants: the speed of light (c), Newton's universal gravitational constant (G) and Planck's constant (h), which plays an important role in quantum theory. The units of each of the three constants are given in Table 8.1.

Table 8.1 Units of fundamental constants.

Constant	Symbol	Units
speed of light	c	$\dfrac{\text{length}}{\text{time}}$
gravitational constant	G	$\dfrac{(\text{length})^3}{(\text{time})^2(\text{mass})}$
Planck's constant	h	$\dfrac{(\text{mass})(\text{length})^2}{\text{time}}$

Since these three parameters are thought to be universal across the vast expanse of the universe and presumably across time, by taking ratios of these three parameters, one could determine universal values for length, time and mass. The Planck length is $(hG/c^3)^{1/2}$ and has a value of about 10^{-35} meters. The Planck time is $(hG/c^5)^{1/2}$ or about 10^{-44} seconds and the Planck mass is $(hc/G)^{1/2}$ or about 10^{-8} kilograms. On the particle physics scale, the Planck length and time are unimaginably short, while the Planck mass is extremely high. To give a perspective, recall that the Planck length is just a new unit, like a foot or mile, a meter or a kilometer. A proton is about 10^{-15} meters wide, which sounds rather small, but that's about 10^{20} (that's 100,000,000,000,000,000,000, or one hundred quintillion) Planck lengths long.

The Planck parameters, so far removed from human experience, have a real significance in addition to their desired universality. For instance, if one has a Planck mass in the space the size of the Planck length, one has satisfied the conditions necessary to create a black hole. Perhaps most importantly, physicists suspect that at lengths and times comparable to the Planck length and time, and energies comparable to the Planck mass (remember $E = mc^2$?), one will be able to write an "ultimate" theory. Before we discuss attempts to write such a theory, we first turn to some effective theories that are somewhat more complete than the current Standard Model.

If You Knew SUSY ...

Intrinsic to all theories currently under study is the idea of symmetry. Symmetry, as we recall from earlier discussions, has a particular meaning when theoretical physicists use the term. Symmetry in a physics theory means that you can change something and something else doesn't change. For instance your height, defined to be the distance from your head to your feet, is always measured to be the same, no matter where it's measured. No matter where I go, I'll always be 5' 11" (and hence my membership in the APS and not the NBA).

We say that a person's height is invariant (i.e. unchanging) and therefore the equation of the theory describing height must be "symmetric"
under the operation of picking new places to make the measurement.

As we recall, all force-carrying particles (photon, W, Z, gluon and
graviton) are of the boson variety, which means that they have a quantum mechanical spin of 0, ± 1, ± 2, ± 3, and so on. Bosons are gregarious creatures and it is possible to have identical bosons in the
same place and they all get along just fine. In contrast, matter particles (quarks, charged leptons and neutrinos) are fermions, with their
characteristic quantum mechanical spin of $\pm 1/2$, $\pm 3/2$, $\pm 5/2$ and so
on (in fact, the known fundamental particles only have a quantum
mechanical spin of $\pm 1/2$). Fermions are the loners of the particle
world, in that it is impossible to have two identical fermions in the
same place. This difference between fermions and bosons is very
important, as if one were to swap them (i.e. make force-carrying particles fermions and matter-carrying particles bosons), the universe
would be unrecognizably different. With this observation, we return
to one of our "mystery questions." Why do force-carrying and matter particles have the opposite kind of quantum mechanical spin?

One can phrase the question in a different, but equivalent, way.
What sort of symmetry is exhibited by the theory if you everywhere
swap fermions and bosons? Clearly now the theory has no symmetry
under this operation. If you have in your equation say an up quark
(a fermion), you'd simply make that a bosonic up quark. Since these
aren't observed, the equation (and universe!) cannot have this
symmetry.

However, in 1982 Howard Georgi of Harvard University and
Savas Dimopoulos of Stanford University had an idea. Suppose the
more accurate theory did exhibit a symmetry under the operation of
swapping fermions and bosons. Then one would have to add extra
mathematical terms to account for the unobserved bosonic matter
particles and fermionic force carriers. While we have to deal with the
non-observation of these particles eventually, let's just think about
what would be the implications of this idea. This new symmetry
needed a name and since it was devised in the 1970s (predating

Georgi and Dimopoulos' paper), with its superstars, etc., the new theory was labeled Supersymmetry, or SUSY for short. Technically, Georgi and Dimopoulos' work was to integrate supersymmetry with the Standard Model. Their particular version was called the Minimal Supersymmetric Model, or MSSM.

SUSY is a very simple idea. Hundreds, if not thousands of new theories (i.e. models) have been proposed, which contain SUSY as a crucial component. Some theorists have spent vast fractions of their professional lives trying to elucidate the inner workings of these models and experimentalists have sought to observe the predicted physical consequences. In the following pages, we will discuss just why SUSY is such a popular idea.

Supersymmetry was initially devised as a sort of addition to certain physical theories known to not correspond to reality, but which were being studied for their interesting mathematical properties. Essentially, physicists were playing a mathematical game ... an intellectual exercise that SUSY made even more interesting. Later, after the mathematical ideas and techniques were well developed, theorists began to see their applicability to particle physics. While as of this writing (December 2003), there is zero direct evidence that SUSY is a property of the universe, there is a strong suspicion that it might be, as evidenced by the approximately 10,000 scientific papers written on the subject. A reasonable person might inquire as to why the physics community exhibits such enthusiasm for the idea. This is because theories incorporating SUSY have been able to provide explanations to questions on which the Standard Model remains silent. I would like to remind you at this point that *an* explanation is not necessarily *the* explanation. SUSY may yet prove to be an interesting, yet ultimately wrong, idea. We'll discuss ways in which we could determine if SUSY is true, i.e. what are the expected experimental signatures for which physicists are looking? However, before we turn to that, we should discuss some of SUSY's theoretical successes.

One of the most perplexing questions dealing with the unification of the three forces, the strong, weak and electromagnetism, is why it appears to occur at so high an energy. The energy at which this occurs is the so-called grand unification theory, or GUT, scale. Later unification at the Planck energy of the three forces with the thus-far intractable gravity is even more perplexing. The fundamental curiosity is the following. If the electroweak unification scale is about 100, or 10^2, GeV, why are the GUT scale more like 10^{15}–10^{16} GeV and the Planck scale at 10^{19} GeV? What physical processes cause such a vast separation in the respective unification scales? The Standard Model is silent on this question, which has been termed the hierarchy problem. Because there exist 20 parameters for which we cannot calculate values, but rather require experimental input, we are forced to tune our theory to agree with all observations from current experiments. This tuning requires ridiculously perfect precision … in fact, one must tune the theory to one part in 10^{32}. That's kind of like measuring something from here to the next star (4 light years away) and needing to do it to an accuracy of about the size of a proton. Such a need for that kind of precision is suspicious. Typically the need for such precise tuning indicates that there is a physical phenomenon operating which is not included in your theory.

If SUSY is true, then it is possible to conjecture the existence of new particles, the existence of which would solve this problem. We will soon learn about the properties of these hypothetical particles. If the mass of the new (and thus far unobserved) particles predicted by supersymmetric theory were identical to their already-observed counterparts, the cancellation is perfect. If the supersymmetric particles' masses are too large, the cancellation doesn't work. In fact, it is this property of SUSY theory that makes theorists confident that experiments that will operate within the decade will observe the newly predicted particles. If they are not observed, some serious head scratching will result.

In Chapter 5, we discussed the Higgs mechanism, which provided an explanation whereby the elementary particles get their mass. We must recall that Peter Higgs' idea did not follow naturally from the Standard Model, rather it was imposed artificially after the fact. In developing a new theory that incorporates all of the Standard Model's successes with the new principle of SUSY, theorists were able to derive the Higgs mechanism in a simple and natural way. One consequence of this success was that the new theory predicted not a single Higgs boson, but an entire menagerie of Higgs bosons, including some with electric charge. While no Higgs bosons have thus far been observed, observation of several kinds of Higgs bosons would strengthen the case for SUSY.

There is one theoretical success of the SUSY idea that is especially impressive. There is a disconcerting mystery involving the Higgs boson that we've not mentioned thus far. If you recall from Chapter 5, the Higgs boson idea works by assuming that there exists everywhere in the universe a Higgs field. All fields contain energy. One can calculate the "energy density," i.e. the amount of energy per unit volume, which the Higgs field should have and compare it to the amount measured from cosmological experiments and one observes a little discrepancy. Well, maybe not so small; in fact, the Higgs field predicts a 10^{54} times greater energy density than what is observed. To give you a sense of just how huge a difference this is, if the measurement was equivalent to the unimaginably small Planck length, the Higgs theory predicts something much closer to the size of the universe. Oops. ...

This disparity was first realized in the mid 1970s and regarded to be a mystery, but one that could be put aside for the moment. The Higgs idea worked so well, that it was explored in great detail. "Mystery be damned" was the attitude of the day. This attitude isn't as foolishly cavalier as it appears at first thought. The history of science is replete with theories that worked very well, but yet made a single silly prediction. The silly prediction was acknowledged and put aside, to await a fresh insight that would resolve the conundrum. Enrico Fermi's theory of weak interactions in the 1930s is one such example. As we have seen, that mystery was eventually resolved.

Supersymmetry has provided a possible answer to this lingering worry. Because of subtle mathematical facts, supersymmetry can neatly cancel out the large predicted energy density obtained from the

Higgs theory. However, this is true only if the newly predicted particles have exactly the same mass as their normal matter analogs. We know that this supposition isn't true, because we've not observed any of these hypothetical particles. Given that we know that, even if they exist, these new particles must have a large mass, it follows that SUSY cannot simply solve the Higgs energy field problem. However, this property is interesting enough that some theorists continue to make theories for which this disparity in masses isn't a showstopper.

Another success of SUSY deals with the unification of forces. As you may recall, one goal of particle physics research is to unify the forces, i.e. show that the strong, weak and electromagnetic forces are all aspects of a single, more fundamental, force. The problem obviously is that the three forces have hugely different strengths. Thus the three forces can only be united if they somehow change in strength. Experiments have established that the strength of the forces vary as the energy of the collision increases; the strong and weak forces grow weaker with increasing energy, while electromagnetism increases in strength. The strengths of the forces have been measured only for a small range of energies, but we can extrapolate the trends to greater energies and see if the forces' strengths ever become the same. Following this approach, we find that the forces do eventually end up with the same strengths, but the three forces do not attain the same strength at the same energy. This observation is illustrated in Figure 8.2a.

One question that is interesting to ask is "At what energy do the three forces gain equal strength?" We see that this occurs at an energy of about 10^{15} GeV (recall that the highest energy accelerator in the world has an energy of about 10^3 GeV). Even more interesting is the realization that the Planck energy (the energy equivalent to the Planck mass) is about 10^{19} GeV, thus the projected "force unification" energy is tantalizing close to the "ultimate theory" Planck energy.

While the three forces eventually attain the same strength, we see that they do not merge at the same point. When the same extrapolation is done using the principle of SUSY, naturally the details of the

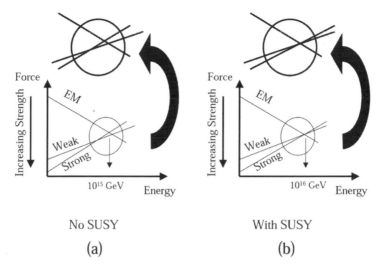

No SUSY   With SUSY

(a)   (b)

Figure 8.2 In the absence of supersymmetry, the weak, strong and electro-magnetic force become more similar in strength, yet the energy at which they unify is not the same. If supersymmetry is valid, the three forces unify at the same point, something that seems, mostly from an elegance argument, to support supersymmetry. This is theoretically appealing, but experimentally not compelling.

trends of the three forces change, but remarkably, the three forces become equal in strength at the same energy! The "equal force" energy is changed a little, this time about 10^{16} GeV, somewhat closer to the Planck energy, as shown in Figure 8.2b. The fact that the three forces unify at the same energy is not a priori required of a theory, but it is certainly intriguing that SUSY accomplishes just that.

SUSY has many other theoretical successes; the interested reader is invited to peruse the bibliography for suggested reading to extend the discussion of this interesting theoretical idea. However, we should recall that thus far we have had no direct experimental evidence that SUSY is, in fact, true. Let's now turn our attention to how my exper-imental colleagues are trying to establish the truth of the theory or kill the idea once and for all.

The centerpiece of the SUSY idea is that fermions and bosons should enter the theory with equal weight. This means that if the theory predicts mass-carrying fermions, there must also exist mass-carrying bosons. Similarly, there must exist fermionic analogs to the force-carrying bosons. Essentially, for each particle that we know to exist, SUSY predicts another particle, thus far not observed. There is a simple recipe for naming the new particles. For each type of matter fermions (quarks, leptons and neutrinos), there exists a boson with a similar name, each preceded by the letter "s" (squarks, sleptons and sneutrinos, generally called sparticles). For each force-carrying boson (photon, W, Z, gluon and graviton), there exists an analog fermion with a similar name, this time followed by the phrase "ino" (photino, wino, zino, gluino and gravitino, generally called bosinos). The fermion analog to the Higgs boson is called the higgsino. We denote these supersymmetric particles with a "~" over them, thus a squark is a $\tilde{q}$, a sneutrino is a $\tilde{\nu}$, a wino is a $\tilde{W}$, and so on. Verification of the idea of SUSY would require the observation of some of these supersymmetric partner particles.

We know something about these supersymmetric particles. They have not been observed in any experiments thus far undertaken. This clearly shows that they must be very massive, otherwise they would have already been observed. In addition, we know either that supersymmetric particles are unstable or that they do not interact very strongly with ordinary matter. This is because although the energy necessary to create such supersymmetric particles is large, energy of this magnitude was available at the time of the Big Bang and in cosmic ray collisions in the atmosphere every day. Since supersymmetric particles should be produced, then their observed absence suggests that they must decay rapidly or if they exist, our failure to observe them suggests they don't interact much. While it's certainly true that there may exist massive particles that don't interact with ordinary matter (except through gravity), these are not the supersymmetric particles. We know this because the supersymmetric particles are predicted to be created in fairly ordinary ways, about which we will learn soon. Thus we are left with the option that supersymmetric particles must decay in some fashion into ordinary particles. With the knowledge described in this paragraph, we begin to see how we might experimentally verify SUSY.

Desperately Seeking SUSY

Before we give a concrete example, we need to know that the rules whereby supersymmetric particles are made are similar to those in the Standard Model. Supersymmetric particles (i.e. sparticles) are made in pairs, much like a strange quark. Recall in Chapter 4 how we showed that one could annihilate an electron and a positron to form a Z particle, which in turn decayed into a muon/antimuon pair ($e^+e^- \to Z \to \mu^+\mu^-$)? Well, supersymmetric particle production looks pretty similar. If supersymmetry were true, the same Z boson could have decayed into a supersymmetric electron and positron, the selectron ($\tilde{e}^-$) and the spositron ($\tilde{e}^+$). Figure 8.3b shows the basic idea.

You'd think that all you'd have to do would be to look for the selectron/spositron pair, but you'd be forgetting that the supersymmetric particles are unstable. Now one must ask into what particles might they decay? Since a selectron is a supersymmetric particle, one cannot simply let it decay into an electron ($\tilde{e}^- \to e^-$). Somehow the supersymmetric nature of the initial particle must be reflected after the decay. At this time, we must introduce an important new concept, the lightest supersymmetric particle, the LSP.

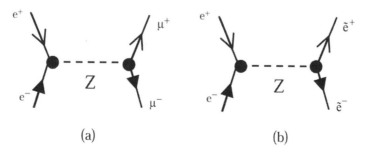

(a)   (b)

Figure 8.3 The creation of selectron/spositron pairs occurs via a mechanism similar to simple muon/antimuon creation.

The LSP is one of the neutral supersymmetric particles. It is unknown precisely which particle makes up the LSP, it could be the gravitino, the photino, etc. Because this is, by definition, the *lightest* supersymmetric particle, there are no other supersymmetric particles that are lighter. Since there is a supersymmetric quantity that is conserved across the sparticle decay (analogous to charge, strangeness, etc.), each supersymmetric particle decay must have a lighter supersymmetric particle as one of its decay products. Since there are no supersymmetric particles lighter than the LSP, we are left with the conclusion that, alone among the supersymmetric particles, the LSP is stable. Because it is stable and given that it hasn't been observed, one is forced to realize that the LSP must have no electrical charge and have no "color" or strong force charge. This is because if it did feel these forces, we would have seen it interact with ordinary matter by now. Since we don't know if SUSY is true or, if it is, the masses of the various sparticles, the identity of the LSP is a mystery and we therefore refer to it in the generic sense.

We see then that Figure 8.3b is incomplete. The selectron and spositron must decay, each into their respective normal matter electron or positron and an LSP. A more correct diagram of the production process is given in Figure 8.4. Since the LSP is stable, electrically neutral and doesn't experience the strong force, it acts somewhat like a neutrino and escapes the detector. Thus the LSP is detected only by its absence.

The astute reader will be alarmed at this point. The "experimental signature" (i.e. what you see after the interaction) of the interaction shown in Figure 8.4 is an electron, a positron and two missing LSP's. So really what one sees is an electron, a positron and missing energy. With a little thought, one can see how one might get just such an event using the ordinary physics with which we are now so familiar. Such a possibility is given in Figure 8.5.

The neutrinos escape the detector like the LSPs and thus this event looks a lot like the more exotic supersymmetric event in question. I'm assuming at this point that you are asking yourself "What clever thing do they do to distinguish between the two possibilities?"

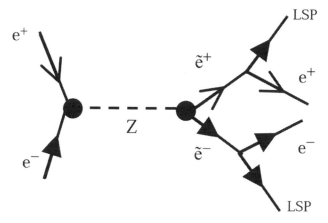

Figure 8.4 Since selectrons and spositrons must be unstable, they decay into a lightest supersymmetric particle (LSP) and a normal matter electron or positron. All supersymmetric particles are currently undiscovered theoretical constructs.

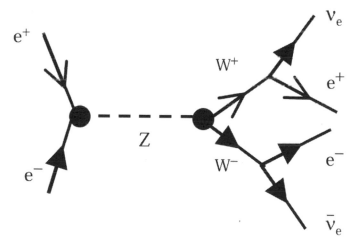

Figure 8.5 Background for supersymmetric particle creation. The neutrinos are not observable, just as the LSPs are predicted to be. Disentangling the two can be challenging.

The short answer is we don't. Instead we use our excellent knowledge of the Standard Model to predict how many events of the variety shown in Figure 8.5 we expect. We then do the experiment and count how many we see. If we see too many, we begin to believe that

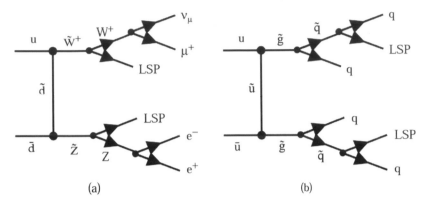

Figure 8.6 Two supersymmetric particle interactions that are predicted to dominate at the Fermilab Tevatron. The tri-lepton signature in (a) is especially appealing, due to its small backgrounds.

perhaps it's because we're starting to see a few events of the type shown in Figure 8.4.

While the sorts of events we draw in Figures 8.3–8.5 are useful because of their simplicity, these events are unlikely to be observed at Fermilab, currently the only accelerator running that might be able to detect SUSY. Because Fermilab collides protons and antiprotons, the initial particles must be quarks or gluons. While it is impossible to know what sorts of interactions are more probable, theorists do have a few favorites, shown in Figure 8.6. Figure 8.6a is an event in which an up quark from the proton exchanges a sdown squark with an anti-down quark from the antiproton, making a wino and a zino. The wino decays into a LSP and a W boson, which decays in the usual way. The zino to Z boson decay proceeds similarly. Thus the final state particles are a muon, a muon neutrino, an electron and a positron, accompanied by two invisible LSP's. This so-called "tri-lepton" signature, named for its three charged leptons and missing energy, is a favorite one for Fermilab experimenters, as it is a very striking event, which DØ and CDF can easily see. Even more important is the fact that it is very difficult for known physics processes to make an event with similar characteristics. Observation of even a single such event, while not conclusive, would be regarded as one of great interest.

Figure 8.6b is another type of interaction that some theorists predict will occur in great (well relatively) quantities at Fermilab. Here an up quark exchanges a sup squark with an up antiquark, creating two gluinos. Each gluino decays into a quark and a squark, which decays in turn to a quark and an LSP. Each quark eventually fragments into a jet (see Chapter 4) and the LSPs escape undetected. In this sort of event, what a detector sees is four jets plus missing energy. Unfortunately, such an event is relatively easily created by known

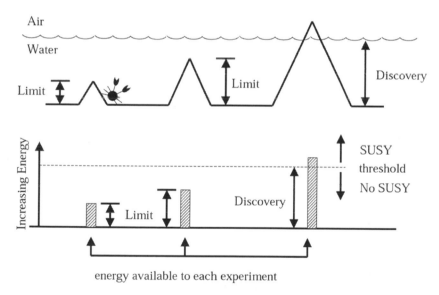

energy available to each experiment

Figure 8.7 An analogy of how particle searches occur. Until one crosses an energy threshold, it is difficult to state whether or not a particle exists. A crab, climbing an underwater mountain, will not observe air and thus cannot state what is the precise depth of the ocean. By climbing higher mountains, he can only set a lower limit on the depth of the ocean. Particle experiments are similar. More energy will allow an experiment to cross a production threshold and observe a rare particle. Below that threshold, you can only say that the mass of a particle is higher than your "energy mountain."

physics processes, especially when one includes realistic detector performance. Thus establishing the existence of SUSY using this sort of interaction will be more challenging, yet my colleagues on DØ and CDF are hot on the trail.

You might wonder, with all the effort going into understanding and attempting to detect SUSY, what's happened so far? All four of the LEP experiments (see Chapter 6) looked and found nothing, as did DØ and CDF in our "Run I" (i.e. 1992–1996) data taking period. No supersymmetric particles were observed (or at least recognized). Since March of 2001, both DØ and CDF have resumed operations, with greatly improved detectors. Neither group has seen anything yet, but the effort is intense. Both experiments will search for events that look similar to those shown in Figure 8.6, as well as many others not mentioned here. Either we will find enough examples of the desired kinds of events or we won't. If we do not see enough of the desired events to claim discovery, we will "set a limit" on the various particles. This means that we will be able to say that while we can't definitively rule out SUSY, we know that the as-yet-unobserved supersymmetric particles have a mass greater than some value, which is typically on the order of a couple of hundred GeV. It's kind of like a crab climbing underwater mountains to see when the water stops and air begins. Each time he climbs a higher mountain, he can set a higher limit on the sea level, but he can't say for sure if the water ever ends. Finally, he may climb a mountain that crosses the water's surface (i.e. he exits onto the beach of an island) and pass through to the atmosphere. Then he can finally measure the sea level. To make our analogy link with particle physics, each group (crab) tries to do an experiment at higher energy (climb a higher mountain), with the express goal of crossing the energy threshold above which supersymmetric particles are formed (cross from water to air). So far, no experiment has crossed that threshold; although since they know the energy they have attained, they each have determined limits on the masses of the supersymmetric particles… they must be higher than each experiment's "energy mountain." Both DØ and CDF have improved their detectors and are set to scale even higher "mountains." Time will tell if the crab passes the threshold or remains all wet. The results of the first experiments by the Fermilab detectors became available in 2003 and improved results (i.e. higher limits or possibly a discovery) will be available over the next few years. Towards the end of this decade, the Large Hadron Collider (LHC) will commence operations and probe energies with which the Fermilab Tevatron simply can't hope to compete. If neither the Tevatron nor the LHC discovers SUSY, many theorists think that SUSY may prove to have been a fascinating theory, but one that simply doesn't describe the universe. Time will tell and, at any rate, the search will be fun, perhaps even revealing something entirely unexpected.

Before we leave the interesting possibility of SUSY, we must raise a final issue. Recall that the fundamental point of SUSY is that one treats the fermions and bosons symmetrically (i.e. identically). If you think about it, we've just proven in our above discussion that this is impossible. This is because we know that, for example, the selectron is much more massive than the more prosaic electron. Similarly, the photino and photon have different masses. As best as we can tell, none of the supersymmetric particles have the same mass as their normal-matter analogs. This difference in mass means that there clearly isn't a symmetry between the matter and their hypothetical super partners. So what gives?

With these musings, we have begun to appreciate a more sophisticated thought. SUSY, if it exists, must be a broken symmetry. Just like we discussed in Chapter 5, where the Higgs mechanism is thought to break the electroweak symmetry, giving the photon and the W and Z bosons different mass, there must exist another mechanism that breaks the symmetry between fermions and bosons. The situation is different in that the symmetry between electromagnetism and the weak force was not appreciated until long after the two forces were understood independently. With SUSY, the symmetry (and symmetry breaking!) was postulated before the discovery of any particles unique to the theory.

If SUSY is a good symmetry at higher energy, but broken at the energies available to modern particle physics experiments, what is the mechanism that breaks the symmetry? This is unknown, although certainly ideas have been discussed. Before the question can be resolved, physicists must detect at least some of the particles predicted by the imperfect symmetry that is SUSY.

Given that SUSY seems to be only partially correct, how is this an improvement over the broken electroweak symmetry discussed earlier? Well, supersymmetry, if true, explains some mysteries not addressed by the Standard Model. Thus while SUSY is still an effective theory, if true, it is more complete and therefore better. So although SUSY is unproven, it will continue to garner considerable attention for the next decade (and beyond, if it is proven to be true).

Before we leave the idea of SUSY for good, I'd like to take some time to underscore a point that needs greater emphasis. Even though I've described some of the types of events that might indicate the discovery of supersymmetry, these are just a particular set of events that could be evident for a particular model. In fact, SUSY is a much greater idea than any model. SUSY is a *principle* that models incorporate. SUSY is the idea that bosons and fermions must come into the theory with equal weight. Any theory that incorporates this principle is supersymmetric. However, it is possible for two models incorporating the idea of supersymmetry to make quite different predictions.

One might offer an analogy to this idea, taking from the field of evolution theory. A principle might be "survival of the fittest." Different species have evolved via this principle to many different solutions to the question of how they might best propagate their genes. Some species evolve mighty armor to keep from being eaten, while others forgo armor in favor of speed. Still others beget many offspring in the hope that some may survive. All of these evolutionary solutions are analogous to the many disparate supersymmetric theories.

We thus can see that proving that SUSY doesn't exist is quite a difficult thing to do. While any particular theory may be disproved, disproving the underlying principle is much more difficult. Just as in the case whereby one observes an armored species becoming extinct, this doesn't disprove the principle of "survival of the fittest," as it might be that the correct manifestation of that principle was actually speed.

While SUSY is certainly an extremely popular idea that is consuming the attention of many physicists, both theoretical and experimental, it is by no means the only idea on the market. We now turn to another very interesting question: why is it that we seem to inhabit three spatial and one time dimensions? Are other dimensions possible?

Large Extra Dimensions: Fact or Fiction?

We recall from our discussion of SUSY that one great theoretical mystery is the so-called hierarchy problem, which is a fancy way to say that we don't understand why the electromagnetic and weak forces unify at a much lower energy than when the strong force and gravity join the unification. One answer to the hierarchy problem is the previously-described precise tuning, but other answers have also been proposed. Suppose that rather than merging at the GUT and Planck scales of 10^{16} and 10^{19} GeV respectively, one found that the real GUT and Planck scales occurred at about 1,000 (or 10^3) GeV, only about ten times larger than the electroweak scale. Because the electroweak unification, GUT and Planck scales would be so similar, there would no longer be a hierarchy problem.

In order for the GUT and Planck scales to be so much lower than those one would predict by extrapolating from their behavior at low (i.e. already measured) energy, this means that some new physical process must manifest itself to change the trend that we observe. What sort of physics might suit this purpose? Recalling that everything that follows on this topic is pure (if informed) speculation; let's explore the following fascinating idea.

In February 1998, Nima Arkani-Hamed, Savas Dimopoulos and Gia (George) Dvali (ADD), all then at Stanford University (Dvali being a visitor, normally resident at ICTP, a theoretical center in Trieste, Italy), proposed a rather counterintuitive idea. Suppose that the familiar three spatial and one time dimension are not the only dimensions. What if there are more dimensions than our familiar four? The idea of additional dimensions wasn't new, dating back to the 1920s, with the work of Theodor Kaluza and Oskar Klein, with a cameo appearance by Albert Einstein. ADD's notable contribution was the idea that gravity might span more dimensions than the other forces.

Extra dimensions and parallel universes have been extensively used in science fiction literature. A separate universe, perhaps with quite different physical laws, might exist right where you are now. Because of some sort of barrier, we are unable to perceive the denizens of this other world and we are fated to pass ghostlike through one another, unaware of each other's existence. Unaware, that is, until some mad scientist, brilliant and feeling persecuted (does the phrase "Fools ... I'll show them all ..." ring a bell?), penetrates the veil and sets loose some sort of havoc in our universe. Along comes a studly young man, accompanied by the scientist's beautiful and equally brilliant daughter, who sets things right. Well, while you've all read a book with this basic story line, that's not what is really meant here by extra dimensions.

To get a better idea of what is meant by extra dimensions, one might turn to a marvelous book, written over 100 years ago. In 1884, Edwin A. Abbott, a noted Shakespearean scholar with a hobby of the study of higher mathematics, published a book "Flatland: A Romance of Many Dimensions." In this marvelous tale, the main character, A. Square, is a geometric figure consisting of only two dimensions, length and width. He travels his "Flatland" two-dimensional world, encountering other figures ... pentagons, hexagons and so on. As the book closes, A. Square's two-dimensional world is visited by a spherical creature from the three dimensional "Spaceland." This creature takes A. Square into the third dimension and A. Square is able to view his world from a perspective not available to his fellow Flatland dwellers. As A. Square begins to grasp the import of this extra dimension, he speculates that perhaps Spaceland is a small sub-space of a larger four-dimensional space. What Arkani-Hamed, Dimopoulos and

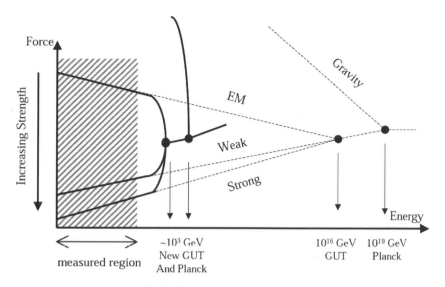

Figure 8.8 The hypothesis of large extra dimensions allows for the various forces to unify at a much lower energy than one would predict from extrapolating the low energy behavior. The dashed lines denote extrapolation from the measured region. The solid lines outside the measured region shows the behavior of ADD's theory.

Dvali (ADD) have proposed is that perhaps the universe we inhabit has additional dimensions, which we are unable to perceive. In doing so, they can solve the hierarchy problem, as illustrated in Figure 8.8.

Talking about extra dimensions is always a mind-bending experience, after all, how do you get your mind to think about such a counterintuitive thought? Accordingly, we will come at this idea from a variety of directions, so as to better resolve the confusion. Since we can perceive three spatial dimensions, we will start our study by thinking of one and two-dimensional worlds, as we might gain by analogy an insight of what would happen as one adds dimensions. As we pursue this discussion, we should also consider an important thought. We perceive only three spatial dimensions. If there are additional dimensions, our inability to experience them implies that there is some sort of barrier to our ability to observe their existence. Alternatively, it is possible that these additional dimensions are somehow different from

Figure 8.9 Each of the four lands, Pointland, Lineland, Flatland and Spaceland are simply spaces with a larger number of dimensions.

our familiar ones. We will consider these ideas first in the abstract and later come back and make contact with the theory proposed in 1998 by ADD.

Let's talk about four different worlds, depicted in Figure 8.9. Stealing from Abbott's language, a place containing zero dimensions is Pointland, a one-dimensional world is Lineland. Flatland contains two dimensions, while our own three-dimensional universe is Spaceland. As each dimension is added, a new direction of motion is allowed. Further, each dimension can be thought of as an infinite number of the next lower dimension. Consider our zero-dimensional Pointland. An inhabitant of this universe cannot move, as there is only one place in the universe. If one takes an infinite number of zero-dimension points and puts them near one another, it becomes a line, as illustrated in Figure 8.10. In the one-dimensional Lineland, one is now allowed to move left and right only. Similarly, an infinite number of parallel lines can make up a two-dimensional space. While an inhabitant of Lineland can only move left-right, a Flatland denizen can also move up and down. In effect, as one moves up and down, one jumps between adjacent one-dimensional universes. In fact, a hypothetical inhabitant of a particular left-right Lineland would perceive a two-dimensional creature moving in an up-down direction as appearing and disappearing as it crossed the one-dimensional Lineland. Because

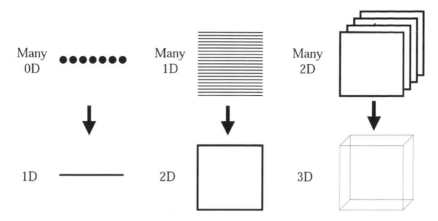

Figure 8.10 Each dimensional space can be made up by an infinite number of the next-lowest dimensionality space.

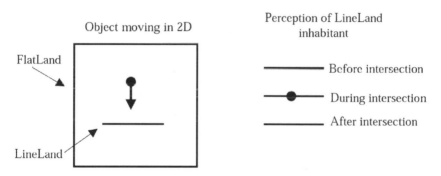

Figure 8.11 A Lineland being will not be able to see a being moving in two dimensions. Only when the two-dimensional being crosses the one-dimensional being's world will he be perceived.

the Lineland inhabitant cannot perceive anything outside that one-dimensional world, he can't see the particle growing closer. He is restricted to only see the object when it is in his dimension, as illustrated in Figure 8.11.

As you'd expect, going from two dimensions to three adds yet another allowed direction of motion, this time into and out of the page. Similarly, one can build a three-dimensional space by stacking an infinite number of two-dimensional planes. This aspect of building a space of a certain dimensionality from an infinite number of spaces of a lower dimensionality is shown in Figure 8.10.

ADD proposed that perhaps there are more dimensions than the three spatial dimensions with which we are so familiar. By extension of the earlier discussion, this new space can be made by an infinite number of our three-dimensional worlds. Similarly, rather than only being able to move in the three familiar directions (left and right, up and down, in and out,) we should be able to move in the fourth dimension in two more directions, say blith and blath, two terms which I just made up. The problem is that we know from common experience that there is no blith and blath. So either the idea of higher dimensionality is stillborn, or ADD have some "splainin" to do (with apologies to Desi and Lucy).

One explanation that is popular with science fiction buffs everywhere is the idea that there is somehow a barrier between the various three-dimensional universes. Each universe is similar in basic physical laws, but somehow one cannot easily move in the blith-blath direction. Examples of this sort of barrier are the Star Trek episode "Mirror, Mirror" (a.k.a. the one where Spock has a beard) or in R.A. Heinlein's last few books. Taking "Mirror, Mirror" as an example, in the story, there exists in our universe a political entity called the United Federation of Planets which embodies the ideals of Western Civilization, in which each planet is allowed to live as they choose. In contrast, the normally inaccessible parallel dimension is physically identical, but has a very different political climate, one containing the "Terran Empire" in which one advances through military conquest and assassination. Some main characters of the "right" Star Trek universe (Kirk, Uhura, Scotty and McCoy of the Starship Enterprise, for those of you so uncivilized as to not be an avid Star Trek fan), penetrate the barrier between the two universes and the story unfolds.

However, a barrier between the three-dimensional universes, existing in the four-dimensional space, is not the only explanation as to why we cannot move in the blith-blath directions. Tacit in our explanation of how subsequent higher-dimension spaces can be built from universes of lower level dimension was the assumption that each subsequent dimension had identical properties to those that came before. This doesn't have to be so. Imagine that the fourth dimension were much smaller than our familiar three dimensions. Then movement in the blith-blath direction would be so small as to be imperceptible.

Such an idea seems abstract, but it can be made more concrete by considering an example of lower dimension. Let's consider how one might create a two-dimensional space from a bunch of one-dimensional spaces. In Figure 8.10, we showed how one might take lines to make a plane. However, taking many lines and arranging them into a circular pattern could make a cylinder, a point illustrated in Figure 8.12.

A cylinder is a two-dimensional world, if one can only walk on its outside surface. Think of an ant on the outside of a garden hose. The ant can walk in two directions, along the length and around the perimeter of the hose. The two dimensions are different in their properties, but one cannot dispute the assertion that there are two.

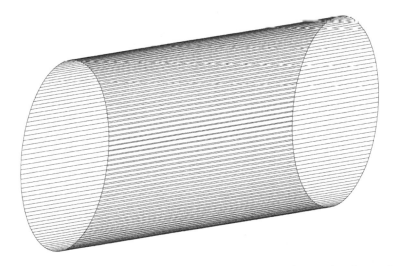

Figure 8.12 A two-dimensional cylindrical space can be made of an infinite number of parallel lines. This two-dimensional space is quite different in behavior than our familiar universe, as the circular dimension is finite, while the long dimension is infinite.

Now let's consider what happens if we shrink the circular dimension. Take a ten-foot long piece of garden hose and replace it with a ten-foot piece of hard spaghetti, then a piano wire and finally a string of individual atoms. The cylinder is still a two-dimensional universe, but with one dimension shrunk so much as to be imperceptible. Thus a two-dimensional space can look like the one-dimensional one shown in Figure 8.13.

Is it possible to have more than one small dimension? Sure. Think now about an inner tube like kids float on in the summer. An ant, walking on the surface, still moves in two dimensions only. Now shrink the inner tube to the size of a donut, then to a Cheerio, and then down to atomic sizes. The two-dimensional inner tube now looks much like the zero-dimensional point of Figure 8.13.

The 1998 paper of Arkani-Hamed, Dimopoulos and Dvali raised the interesting possibility that there were extra dimensions that were "large." Large, in this context, requires some defining. Others had considered theories that required extra dimensions, but these were

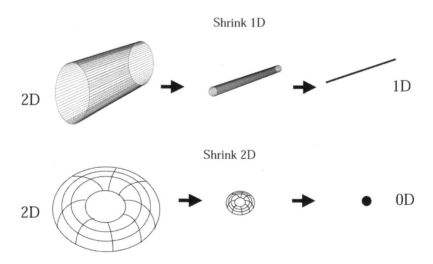

Figure 8.13 A demonstration of how a two-dimensional space, with one dimension sufficiently compacted, will look like a one-dimensional space. Similarly, two compact dimensions will make a two-dimensional space look like a zero-dimensional one.

exceedingly small, perhaps as small as the Planck length. In contrast, large extra dimensions could extend to the respectable size of a millimeter or so, although they may be much smaller.

Recall that the point of ADD's idea was to lower the energy at which the four known forces unify. While gravity is thus far the most mysterious and least known at the quantum level, it is gravity that provides us with the best method for validating or refuting their idea. Let's discuss why this is. There are at least two experimental approaches that we will discuss.

First, let's think of gravity in the conventional sense, first proposed by Isaac Newton in the 17th century. Perhaps the most striking feature is the fact that he found that the force of gravity would get weaker as two objects were separated. Specifically, the force drops off as the square of the distance, or mathematically $1/r^2$. As we recall from our discussion in Chapters 1 and 4, if two objects are separated by a certain distance and then that distance is doubled, the force is reduced by a factor of four ($1/2^2 = 1/4$). Similarly, if the distance is trebled, the force is reduced by a factor of 9 ($1/3^2 = 1/9$). Most students do not appreciate just where that factor of $1/r^2$ comes from, but it turns out to be a fundamental property of space, or the number and character of the relevant dimensions. In this instance, what is relevant is the surface area of a sphere, centered on the mass. The area of the sphere increases as the square of the radius. While our space is three-dimensional, this paper is only two-dimensional, so for clarity let's switch the remaining discussion to two dimensions.

Basically, each object can generate a certain amount of gravitational "flux," which is a measure of how much mass it has (and therefore how much gravitational force it can generate). This flux is radiated equally in all directions. Since the amount of flux is constant, the concentration of flux is reduced by distance, as the same amount of flux must pass through circles of increasing circumference. As we see in Figure 8.14, as an object moves to a larger radial distance from the gravitation-generating mass, it subtends a smaller angle (by being made to move away) and thus feels a smaller force (because it sees a

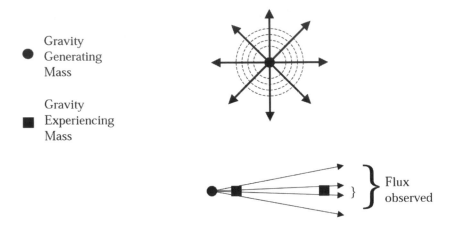

Figure 8.14 The concept of gravitational flux shows the behavior of the gravitational force as a function of the distance from the source. Since a particular mass has a corresponding amount of flux, the amount of flux passes through an increasing surface area as the radius increases. Thus the surface of an n-dimensional sphere increases as a function of radius. The gravitational force decreases as the inverse of the sphere's surface area. An object sees a decreasing amount of the original particle's flux.

smaller fraction of the total flux). In two dimensions, the relevant area is the circumference of a circle, which is proportional to the radius. Since gravitational force is equal to flux divided by area, the increasing area reduces the force. In the two-dimensional universe, Newton's law of gravity would say that gravitational force falls off as $1/r$.

In a hypothetical 4-dimensional world, the area of a sphere would grow as the radius cubed. Since the basic idea of gravitational force is unchanged (force = flux/area), in four dimensions, gravitational force would fall as the cube of the distance ($1/r^3$). Since the force due to gravity has been measured with great precision and found to fall off as the square of the distance ($1/r^2$), we have proven that at least gravity exists in only three dimensions. Or have we?

The above discussion is valid only if the fourth dimension expands in the same manner as our more familiar three. In the event that the fourth and higher dimensions are smaller, the situation changes. Since the force due to gravity has only been measured for distances greater than a millimeter, it is possible that at smaller distances than that, Newton's $1/r^2$ behavior might change into something else (say $1/r^3$). This is a question that can only be answered through experiment.

In order to see how it is that the behavior of the force of gravity can have one behavior at small scales, with a different behavior at large ones, let's consider two-dimensional gravity, not on a plane, as seen in Figure 8.14, but rather on the surface of a cylinder, as shown in Figure 8.15. In this case, this two-dimensional surface consists of one

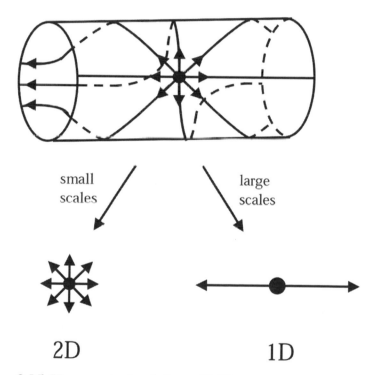

Figure 8.15 How gravitational flux will fill up a space with one infinite dimension and one finite one. The finite space is quickly filled up. As long as one looks at the gravitational behavior for distances small compared to the size of the finite dimension, it behaves in a familiar way. After the smaller dimension is filled up with gravitational flux, the remaining behavior appears to act as a one-lower dimensional space.

one-dimensional direction that is infinite in extent, parallel to the cylinder's axis, with another dimension that is of limited size (along the circumference).

We see from Figure 8.15 that when one is near the mass, the pattern of gravitational flux is indistinguishable from that one finds in a plane, with flux lines pointing radially from the mass. However, the flux lines rapidly "fill up" the small dimension, quickly becoming parallel in the infinite direction. Later expansion can only occur in the infinite dimension. The size at which the transition from one behavior (two-dimensional) to the other (one-dimensional) occurs is about the radius of the small dimension. Thus we see that measuring the force due to gravity at ever decreasing sizes could, when the experimental precision reaches the size of the extra dimensions, reveal variation from Newton's venerable law of universal gravitation and support ADD's hypothesis.

Before we proceed, we should remind ourselves what problem ADD's theory was designed to solve. This was the so-called hierarchy problem, whereby the GUT and Planck energies are very much higher than the energy at which electroweak symmetry breaking becomes relevant. Two important parameters of ADD's theory can play a role in solving the hierarchy problem. These parameters are the number of extra dimensions and their size. The fewer the number of extra dimensions, the larger they must be.

We are able to rule out the possibility that there exists only one additional dimension, because in order for this sole extra dimension to solve the hierarchy problem, it would have to be approximately the size of the solar system. Since Newton's $1/r^2$ dependence of gravity governs very well the motion of the planets, this hypothesis is excluded. So what about two extra dimensions? In this case, they each would need to be about one millimeter in size, just beyond modern experimental limits. If there exist more additional dimensions, they must be smaller still, and for access to dimensions of this diminutive (although still large) size, we must turn to particle physics experimental techniques. If there are three extra dimensions, their size is as much as 3×10^{-9} meters, while for four extra dimensions, they would have a size more like 6×10^{-12} meters.

What are needed are experiments that can measure the behavior of the force of gravity on the distance scale of one millimeter. Experiments that can do this are relatively small...so-called "table top" experiments...in contrast with the large detectors of particle physics. The most sensitive of these small experiments are called "torsion balances," which have their antecedents in experiments performed by Loránd Eötvös in the period surrounding 1890. While there are many groups measuring the behavior of gravity for small distances, the group with the best result (as of December 2002) is at the University of Washington, in Seattle. Through careful analysis, they have measured the behavior of the gravitational force down to a distance of 0.2 millimeters and find no deviation from Newton's $1/r^2$ behavior. As we recall, experiments of this type are only sensitive to the case where there are only two extra dimensions. However, for this particular case, they are able to set a lower limit on the energy at which the four forces unify and this limit is about 3.5 TeV, about 3–4 times better than the best accelerator-based limits (discussed soon). Luckily for the particle physics community, accelerator-based searches can probe situations in which there are more than two extra dimensions. To give a sense of scale, 3.5 TeV is about 35 times greater than the electroweak unification scale (and about 3500 times more massive than the proton), so this is an impressive result. It does not rule out the large extra dimension hypothesis (both due to the limited number of extra dimensions probed and the fact that the limit is within those possibilities allowed by the theory), but it is an important bit of information, as it directly measures gravity at small distances. The University of Washington group is trying to improve their apparatus, with an ultimate goal of about 0.05 millimeters. As always with research at the frontier, we await their next results with great anticipation. Even more, we await the next brilliant idea that allows us to extend the direct limit even further.

While the direct methods described thus far are the easiest to understand, when the number of extra dimensions gets high, their size shrinks and these methods no longer work. To proceed, one must use different techniques, for instance those of particle physics. The dominant feature of this sort of approach is the fact that the three forces, strong, weak and electromagnetism, have been studied at small distance scales (about 10^{-18} meters). Gravity, on the other hand, remains tested only at the millimeter size and thus it remains possible that gravity is allowed to extend into the additional dimensions, while the other forces are constrained to move only in the familiar three. Such a behavior has odd, but observable consequences.

Before we continue our discussion of particle physics and large extra dimensions, let's try to understand how it is possible to have some phenomena constrained to a certain number of dimensions, while other phenomena are allowed to extend into a greater number of dimensions. Let's think of a pool table. The balls are constrained to move only on the two-dimensional surface. However, sound is not. Sound radiates into all three dimensions. Suppose you're a mouse physicist, trying to understand the physics of billiards. Dr. Mouse is only allowed to work on the surface of the table. Further, all of her instruments can only sit on the table as well. Dr. Mouse's hypothesis is that energy is conserved. To test her hypothesis, she takes two balls and lets them hit one another. She carefully measures the energy of the two balls before and after the collision. They should be the same. However, she finds that the energy of the balls after the collision is lower than that before. A lesser physicist would give up, but luckily Dr. Mouse is brilliant (having attended the prestigious Swiss Academy for Cheese and Physics Research). She hears the noise of the impact and recalls that noise is energy. When she adds in the energy of the noise, she finds that the energy balance is better, although the final energy is still less than the energy before the collision. Finally Dr. Mouse is required to conclude that either energy isn't conserved or that some energy is going somewhere that she can't measure.

At this point, we intervene. Dr. Mouse can only put her noise-measuring microphones on the surface of the table and thus she only measures some of the noise. We are not constrained to work in two dimensions and can put microphones in a spherical array surrounding the collision, taking advantage of our access to the third dimension. When we use our measurements, along with those of Dr. Mouse, we find that the energy before and after the collision is identical. The reason that we get the right answer, while Dr. Mouse is in error, is because we can measure the energy flow (i.e. sound) in a dimension inaccessible by Dr. Mouse.

In particle physics experiments, we use the same approach. The strong, weak and electromagnetic forces are constrained to work only in three dimensions, as are all of our instruments. Since gravity can in principle move in four or more dimensions, we will not see the gravitational energy that moves into the "extra" dimensions. Measuring less energy after the collision as compared to before is thus consistent with the idea of the existence of extra dimensions. Note that measuring collisions with missing energy doesn't clinch the extra dimensions idea (after all, neutrinos also manifest themselves as missing energy), so you have to be careful. One must look for particular types of collisions with the right characteristics. It's all fairly tricky.

At the core of many particle physics searches for large extra dimensions is the idea that gravitational energy can enter the higher dimensions. Gravitational energy is carried by a hypothetical particle, the graviton, which is analogous to the more familiar photon, gluon and W and Z bosons. The graviton has not been observed and may or may not exist. In principle, it is a bosonic particle that carries the gravitational force. Since it is the only purely gravitational particle, it alone is free to enter the higher dimensions. Like Dr. Mouse's three-dimensional sound energy, the gravitons can leave our familiar three dimensions entirely, carrying away energy. One of the signatures of events in which gravitons are produced is missing energy. Another signature would be to see the decay of a graviton.

ADD's paper came out in February of 1998, although insiders had knowledge of it for about six months in advance (this is common practice), and was received with considerable interest. The four LEP experiments at CERN (Aleph, Delphi, L3 and Opal), the two Fermilab

Tevatron experiments (DØ and CDF) and at the two major HERA experiments at DESY (H1 and Zeus) immediately started efforts to investigate the idea. By 1999, they were talking about preliminary results at conferences. The usual order for announcing scientific research is to talk to yourself about it in front of a mirror, then with colleagues working closely on similar projects (say 10–20 people). When you gain confidence, you present it to your collaboration (about 500 people). If your results are robust enough to survive any criticism that group can offer, you finally start talking about things publicly at conferences. Finally, when you've made the result as bulletproof as you can, you submit the result for publication. Then, if warranted, you might notify the media. Thus the fact that the experiments were talking publicly about their efforts by 1999 shows how much interest the idea engendered, how quickly people started working on their analyses and just how hard they worked. The first published results came out in 1999 (L3 and Opal), 2000 (H1, Delphi, DØ) and 2001 (CDF).

Basically the experiments looked for two classes of experimental signatures. In the first, a graviton (G) was produced and it then decayed into two objects, with the different experiments looking for different decay chains. Typical graviton searches looked for decays into pairs of muons, taus, electrons, photons or Z bosons. In addition, "associated production" was considered, in which a graviton was created "in association" (i.e. at the same time) with a particle, for instance at the Tevatron a quark/antiquark pair interacts and the result is either a graviton and a photon or a graviton and a parton (e.g. $q\bar{q} \rightarrow G\gamma$ or $q\bar{q} \rightarrow Gg$). This process is depicted in Figure 8.16. Since the graviton can escape undetected into the higher dimensions, what one sees is an event with a photon and missing energy, or an event with a single jet (from a gluon, say) and missing energy.

The searches have been performed and events with the above-described characteristics have been observed. Unfortunately, when one asks and answers the question "How many events of this nature do we expect from mundane Standard Model predictions?", one finds exactly what one predicts. Thus all experiments are forced to conclude that they see no evidence for large extra dimensions.

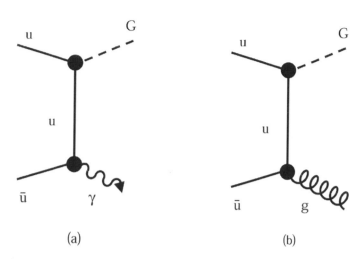

Figure 8.16 Two cases of graviton emission. In (a) the graviton is created in association with a photon, while in (b) the graviton is created in association with a gluon.

Does this mean the idea of large extra dimensions has been killed? Not at all, as it is difficult to prove a negative. As we discussed in the section on supersymmetry, what we can do is to set a limit, which at least rules out some possible answers. The experiments have chosen to cite their results in what energy the new unification scale would be. Recall that conventional theory suggests the GUT scale is 10^{15}–10^{16} GeV, while the Planck scale is about 10^{19} GeV and the electroweak unification scale is about 100 (10^2) GeV. ADD's idea was intended to solve the hierarchy problem and would reduce the GUT and Planck scales to about 1 TeV (or 1000 GeV or 10^3 GeV, all ways to say the same thing). Each experiment can set different limits according to the experiment's strengths and weaknesses, and also based on the experimental signature that they chose to pursue. Further, the answer each experiment quotes depends somewhat on the number of extra dimensions they were considering, but when everything is taken into account, they find that a low energy unification scale may not exist, but if it does exist, it is higher than about 1,000 GeV. Thus current limits remain interesting. Consequently, both DØ and CDF (the only two experiments currently taking data that can weigh in on the topic and improve the limit or make a discovery) continue to collect data. It is expected that we will be able to discover large extra dimensions if the low energy unification scale is less than 2,000 GeV. Further detectors at the LHC can push the limit up to 8–10 TeV (8,000–10,000 GeV). Failure to find large extra dimensions below this limit would cast doubt on the whole idea. Thus the possible payoff of finding large extra dimensions and the relative ease by which we can either discover or refute the idea guarantees that theoretical and experimental effort will continue.

One final idea that could be a consequence of large extra dimensions needs to be mentioned. Imagine, for the moment, that the large extra dimension idea is true and the unification scale is within the reach of the Fermilab Tevatron. A natural question one might ask would be "How do the laws of physics change as we pass this threshold?" Essentially, we're curious as to whether or not crossing that energy threshold would presage new physical phenomena. The answer, probably, is a most emphatic "yes." As we cross this threshold, we would have concentrated an amount of energy equivalent to the (new) Planck mass into a volume characterized by the (new) Planck length, and this, we may recall, is the criterion necessary for the formation of a black hole.

Black holes are typically thought of as dead stellar remnants… concentrations of so much mass in so little space that the strength of gravity is so large as to allow nothing to escape its deadly grip, not even light. The black hole devours all matter and energy which is so unlucky as to come near its irresistible grasp, growing larger and more dangerous, until no matter remains near it.

Letting your imagination run a little wild, you could imagine that as a large accelerator like Fermilab's Tevatron crosses the energy boundary above which large extra dimensions become relevant, the force of gravity would increase rapidly to approximately the strength of the strong nuclear force (recall that the whole point of invoking large extra dimensions is to unify the four forces at low energy). With the force of gravity so much enhanced, a sub-atomic black hole would be formed. In analogy with stellar black holes, it would gobble up surrounding matter, starting first with the detector (and me, as I would probably be on shift!), followed by the laboratory, all of Illinois, the United States and finally the entire Earth, creating a medium sized black hole, all in the blink of an eye. And that, as my teenage children would say, would suck (although they probably wouldn't be aware of the funny pun).

If such an event were possible, it would be imperative that we stop all such experiments immediately. The director of the laboratory would be really grumpy…can you imagine the paperwork such an incident would engender? Luckily, we can prove that this won't happen. And contrary to the rejoinder of some uninformed critics, we are 100% certain in this matter. The reason has nothing to do with arguments from nebulous particle physics theories…it has to do with the fact that we're here at all.

As we discussed first in Chapter 2 and again in Chapter 7, the Earth and indeed all of the planets in the solar system are constantly bombarded by cosmic rays. Cosmic rays are subatomic particles which speed through space at a range of energies, with some of the particles carrying energy so much larger than those created in modern accelerators as to make a particle physicist green with envy. These cosmic rays have been bombarding all of the objects of the solar system for literally billions of years, undergoing countless interactions at energies far above anything we could possibly create in accelerators. We're still here; ergo there is no danger.

Nonetheless, if subatomic black holes could be made, they would provide a great laboratory in which to study the behavior of strong gravity. Being able to study the physics surrounding some of the most awesome forces in the universe, like the super-massive black hole at the center of our galaxy, or possibly even the primordial black hole that spawned the universe itself, would be of enormous excitement to everyone. But, of course, this only works if large extra dimensions prove to be true…something so far not established.

While the last two ideas we have discussed (supersymmetry and large extra dimensions) are intended to solve mysteries not addressed by the Standard Model, neither of these ideas have pretensions of having all of the answers. For instance, the questions of what breaks supersymmetry or why there are the number of extra dimensions that there turns out to be are not addressed by the respective theories. To try to answer all questions, one needs to try different approaches. In the following, we discuss one of a few possible ideas; one in which the idea of subatomic particles becomes obsolete.

We've discussed in this book the various forces and the mechanisms that have been developed to better understand them. While the strong, weak and electromagnetic forces have been understood at the quantum level as an exchange of a force-carrying boson, we've avoided gravity, as it is very weak and thus far has resisted any sort of quantum mechanical treatment. Gravity has been described at the astronomical level by Einstein's Theory of General Relativity. However, as we increase our collision energy in our experiments, we incrementally approach the point where the strength of the gravitational force becomes comparable to the other three. While this unification occurs at the Planck scale, probably at a much higher energy than any we can hope to achieve in any realistic accelerator, intellectual aesthetics requires that we try to provide a theoretical framework in which we view gravity on an equal footing with the other three forces. We need to understand gravity on very small distance and very high energy scales. Essentially, we need to somehow merge general relativity and quantum mechanics. Most attempts to do this, including a multi-decade effort by Einstein, have failed. However, there is a relatively new idea that is showing some promise. We now turn our discussion to the proposal that all particles are, in fact, subatomic strings all vibrating in a melodious cosmic symphony.

Does Superman's Cat Play with Superstrings?

In 1916, Albert Einstein astounded the world (actually only the few physicists who could understand his work) with an audacious new idea. He proposed a new theory for gravity. While Newton's gravitational theory had performed superlatively since its own formulation in 1687, Einstein didn't like it. Einstein had earlier (1905) proposed his theory of special relativity, in which he deduced how one perceived space and indeed time itself depended on an observer's motion. He realized that his new theory and Newton's were fundamentally incompatible and set out to bridge the gap. Beginning with essentially a philosophical premise (but one that required experimental confirmation) he reasoned that the acceleration that one experiences due to gravitational forces was indistinguishable from any other sorts of acceleration. From this humble thought, he formulated his general theory of relativity (so-called because his "special" theory only applied to the special situation where an observer experienced no acceleration, the "general" theory allowed for non-zero acceleration).

At its core, Einstein's general theory of relativity cast the force due to gravity as a bending of space itself. People had studied the properties of space without much consideration for the alternatives. If you've taken high school geometry, the equations that you learned (e.g. the relationship between the circumference and the radius of a circle, $C = 2\pi r$, the Pythagorean theorem, $c^2 = a^2 + b^2$, and the fact that if you add the three angles of a triangle together, you always get $180°$) have embedded within them the assumption that space is "flat." Flat space is called "Euclidean" after the Greek mathematician Euclid who is generally acknowledged as one of the architects of the kind of geometry that one learns in high school.

Examples of flat space might be the surface of your kitchen table, or the ground on which you walk (especially near Fermilab's northern Illinois). However, while the ground seems flat enough, we know that the surface of the Earth is a sphere. So a big triangle painted on the surface of the Earth can be said to exist not in flat or planar space, but rather in a spherical one. On a spherical surface, the sum of the three angles of a triangle no longer add to $180°$. Figure 8.17 illustrates this point.

The net result of the observation that gravity can be viewed as a distortion of the space in which we live means that one needs to learn

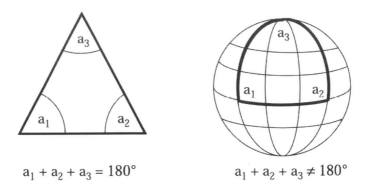

$$a_1 + a_2 + a_3 = 180°$$ $$a_1 + a_2 + a_3 \neq 180°$$

Figure 8.17 The angles of a triangle sum to 180° in a flat space. In a spherical space, the sum of the angles is greater than 180°.

a lot of complicated geometry, far beyond the scope of this book. But geometry is only a tool to understand the underlying physics, so let's return to science rather than pure mathematics. Basically, one can think of space as a trampoline, which is essentially flat. Put a mass on the trampoline, say a bowling ball, and the surface of the trampoline is distorted. Just as a bowling ball distorts the trampoline's surface, Einstein's theory of general relativity shows how a mass can distort space.

As we see in Figure 8.18, the distortion of space is a gradual one, one describable by mathematics called differential equations and difficult, but straightforward, geometry. The smoothness of the spatial distortion that general relativity is designed to describe is important and we will return to this point presently.

While general relativity is designed to describe the behavior of gravity on large size scales, say the intense gravitational field surrounding a black hole, particle physics is concerned with the behavior of matter at the smallest size scales. Describing how matter behaves at the size of an atom and smaller requires quantum mechanics. Quantum mechanics, like relativity, is a rich field which would require its own book (or books!) to adequately describe, so instead we will discuss its most relevant elements. Appendix D parallels the discussion below, but with more detail.

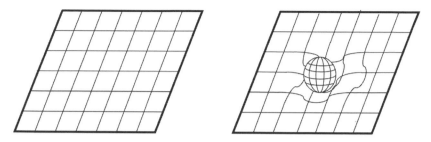

Figure 8.18 A flat space and the same flat space distorted by the presence of mass. Often people draw an analogy with a trampoline and a bowling ball on it.

One of the many counterintuitive elements of quantum mechanics is the Heisenberg Uncertainty Principle, which states that the principle of conservation of energy, of which we have spoken often, doesn't strictly apply at the quantum level. The energy balance can actually not equal out, as long as the imbalance only occurs for a short time. It's kind of like a compulsive borrower and lender...not someone who is trying to make money...but rather someone who likes to lend and borrow money for the sheer sport.

Imagine, as in Figure 8.19, such a guy, named Mr. Compulsive or MC for short. He is surrounded by four neighbors, all of whom are aware of his compulsion, but play along. MC lends neighbor 1 a dollar and consequently MC is one dollar down. This isn't very much money, so this loan goes along for quite a while. Meanwhile, MC borrows $500 from neighbor 2. MC is now up $499, but since $500 is a relatively large amount of money, MC must pay this loan back quickly. He pays it off and lends $100 to neighbor 3 and quickly $5 to neighbor 4. MC is now down $106. He quickly gets the $100 back from neighbor 3 and lets the other two loans float for a while. MC goes round and round, borrowing and lending, never gaining a long-term excess or deficit of money. In fact, in the end, he has the same amount of money as with which he started, but there has been a flurry of lending and borrowing, so that at any particular time he probably has more or less money than with which he began.

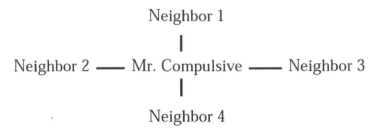

Figure 8.19 A compulsive borrower surrounded by his four helpful neighbors.

Another important point is that when he borrows or lends money, both he and the neighbors are willing to let small loans ride for quite a while, but they expect bigger loans to be paid back quickly. In fact, there is an inverse relationship between the size of the loan and the amount of time it is allowed to stand … the bigger the amount, the shorter the time.

Subatomic particles don't traffic in money, after all, nobody is willing to drive them to the bank. Since we recall that energy is equivalent to matter ($E = mc^2$), a particle that temporarily gains energy can create particles within it. If this idea seems a bit fuzzy, you should review the discussion surrounding Figures 4.23 and 4.24. The most counterintuitive addition to that discussion is the fact that the creation and annihilation of particles occurs in empty space. At a particular place, there may be, on average, no energy. But through the wonders of the Heisenberg Uncertainty Principle (i.e. the idea that energy doesn't have to be conserved, as long as the "borrowing" is short enough) at an empty spot in space pairs of particles (say electron-positron or quark-antiquark pairs) are constantly being created for a brief instant, before being annihilated to balance the energy books. The upshot is that the quantum realm is an active place with particles popping into and out of existence in a mad frenzy. John Wheeler coined the great term "quantum foam" to describe the situation. The term is apt, as anyone who has stared into his or her beer can attest. The bubbles form and pop with marvelous confusion. (Heck, there have even been physics Ph.D. dissertations written on the physics of beer bubbles … some guys have all the luck.)

Now that we have an idea of what phenomena the ideas of quantum mechanics predict, we can return to our original question. How can one merge the ideas of general relativity and quantum mechanics, thus producing a quantum theory of gravity? If we recall, general relativity predicts that the shape of space is smooth and varies relatively slowly. In contrast, quantum mechanics predicts that while general relativity's picture is OK at big sizes, in the quantum realm, space is chaotic, varying unpredictably … the opposite of smooth and slowly varying.

To illustrate the idea that something that looks smooth at large size scales can reveal a much rougher nature, one should look at the surface of carefully polished gold. Gold polishes wonderfully smoothly, but under huge magnification, the surface exhibits the mountainous mien illustrated in Figure 8.20.

The idea of merging general relativity and quantum mechanics was an early one, with a scientist as illustrious as Einstein trying his hand at it. Einstein was not alone in his attempts, yet every effort failed. The tumultuous quantum realm defeated the equations of general relativity. All predictions, rather than yielding a friendly number like 6 or 3.9, ended up being infinity. Infinity, in quantum calculations, is nature's friendly way of saying "try again." Mangling a line from Rudyard Kipling, it appeared that quantum mechanics is quantum mechanics and general relativity is general relativity and never the twain shall meet. This unpleasant state of affairs largely persists to this day, although a contender idea exists that hopes to span this vast chasm.

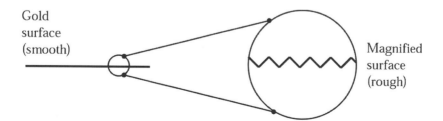

Figure 8.20 A gold surface seems smooth at one distance scale but can seem much rougher at a greater magnification.

In 1968, Gabriele Veneziano made a striking observation. He realized that an obscure mathematical formula, called the Euler beta function, correctly described several aspects of the strong force. Veneziano didn't know why the agreement was so good, but he thought that in this agreement perhaps lay a great truth. Naturally, others followed this train of thought and explored cousins of the Euler beta function with varying degrees of success. This success was encouraging, yet there was a huge problem…no one had the faintest idea as to why these obscure functions described the strong force so well.

Such was the unsatisfactory state of things until 1970, when Yoichiro Nambu, Holger Nielsen and Leonard Susskind made a separate observation that was thought to possibly have been related. They found that if they replaced in their theories a point-like particle with a short one-dimensional "string," then the solution to this new equation turned out to be the Euler beta function. Perhaps the link between equations and the world was revealed? We will discuss the nature of these strings in more detail in a bit, but let's now continue with the history for a moment.

While the string idea was great in that it at least provided an intuitive picture of what was going on at the quantum level, it was quickly shown to make predictions in direct conflict with experiments. If you recall from Chapters 3 and 4, the period of time following 1970 was when quarks and gluons were being shown to be real, as opposed to strictly mathematical, objects. With the success of the quark model and quantum chromodynamics and the failure of the string model, string research was tabled for a while.

One of the problems with the string model was that each string could be thought of as something like a short piano wire that could vibrate in many ways, in fact in more ways than there were particles. The idea was that a string could replace each of the force-carrying bosons. A gluon would be the string vibrating in one way, while the photon was a different vibration pattern. While it was relatively easy to show that one vibration pattern had properties identifiable to those expected for the gluon, other vibration patterns did not correspond to any known or anticipated particle.

In 1974, John Schwarz of the California Institute of Technology and Joël Scherk of the École Normale Supérieure made a crucial connection. The graviton, which is a hypothetical force-carrying boson that carries the gravitational force, might be found in string calculations. While the graviton has never been observed, from the properties of gravity ($1/r^2$ dependence, infinite range, only attractive nature), it is possible to infer some of the properties of the graviton. What Schwarz and Scherk realized was that one of the mysterious vibrational patterns had properties identical to those the graviton must possess. A quantum mechanical theory, which at least had the possibility of successfully including gravity, had been concocted.

Schwarz' and Scherk's paper was essentially ignored by the community. Subsequent attempts to further explore the string idea started coming up with inconsistencies. While the road to hell may be littered with good intentions, one might just as well say that the road to quantum gravity is littered by the carcasses of many failed theories and string theory was beginning to develop a very bad cough.

Not until 1984, when Michael Green and John Schwarz were able to argue very carefully that the apparent problems afflicting string theory could be resolved, did the theory start to show signs of improved health. Even more exciting was the fact that the resulting theory had sufficient diversity as to encompass all of these four forces. Things were beginning to shape up in the search for the "ultimate" theory.

The name for this new idea is "superstring" theory. We might understand from our discussion to this point where the "string" part comes from, but what is the root of the "super?" It turns out that the name comes from our old friend supersymmetry. Superstrings is just a short hand way to say supersymmetric string theory.

Veneziano's original theory actually was only intended to describe bosons (i.e. force-carrying particles). This was a problem, as a theory of everything needed to include the mass-particle fermions. In 1971,

Pierre Ramond, then one of Fermilab's first theoretical post docs, took up the challenge of including fermions in string theory. Through his work and that of many others, this was accomplished and, to everyone's surprise, they found that fermion and boson oscillations occurred in pairs (sound familiar?). By 1977, this pairing was understood to be supersymmetry and superstrings were born. Ramond's intelligence was demonstrated by his theoretical successes, but even more by his subsequent move to the University of Florida. His brilliance is always most apparent to me every January as I sit through yet another northern Illinois winter.

It's not an uncommon feature of mathematics that a theory could be so complicated that one uses approximations to find a solution that is "close enough." For instance, if an Illinois farmer knew the perimeter of his square plot of land and wanted to know the plot's area, he would use the geometry of the plane rather than the more complex geometry that would accurately describe the Earth's spherical surface. However, in superstring theory, the situation is even more difficult. Not only are the solutions approximate, the equations themselves are not known. Superstring theorists can only calculate approximate solutions to approximate equations. To understand this field requires extensive effort in very difficult mathematics. We will forgo these highly technical details and try to understand superstrings at a more qualitative level.

Before we launch into superstrings and particle physics, let's think about strings with a more familiar feel. As illustrated in Figure 8.21, if one takes a string and stretches it, it can vibrate. Think about an electric guitar. The guitar strings are stretched tightly and held fixed at both ends. If one plucks the string gently, it will vibrate gently so that the ends don't move, while the center moves a great deal. We could say that there is only one vibrating section of string in this case. It's also possible for the same string to be made to vibrate so the ends and the center don't move while the "quarter points" move. We now have two vibrating sections. Other vibrational patterns are possible, each with an ever-increasing number of vibrating sections. As the number

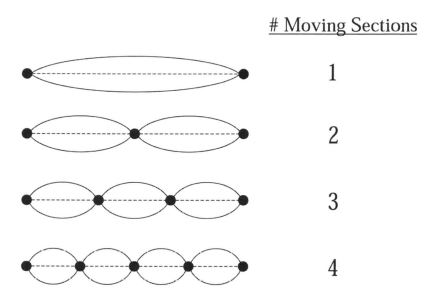

Figure 8.21 The number of moving sections for a guitar string can only be an integer. This is due to the restrictions that the two ends not move.

of vibrating sections increase, the pitch also increases. Figure 8.21 refreshes our memory of the physics of guitar strings.

In particle physics, the strings are thought of as little vibrating loops. A familiar example might be the "triangle" that one sees in cowboy movies, which the cook hits to signal that it's food time. The vibrations manifest themselves in the sound that the cowboys hear. The particle physics strings are loops that can vibrate with ever-increasing frequencies. The lowest frequency is simply the case where the radius of the string varies ... first smaller than average, then larger. The second highest frequency is simply the rhythmic distortion of the circle into an ellipse, first oriented horizontally, then vertically. Higher vibration modes take on a more star-like appearance, as illustrated in Figure 8.22.

To excite the more "wiggly" vibrations takes more energy. Using Einstein's venerable $E = mc^2$, we realize that with more energy, we have created a greater mass. Thus we have provided a possible explanation of how to generate a series of particles with ever-increasing

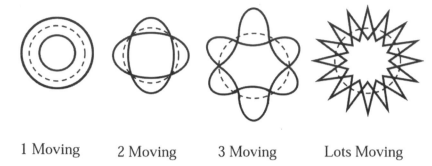

1 Moving   2 Moving   3 Moving   Lots Moving

Figure 8.22 Vibrations of a circular string have many possible options. The simplest vibration is the "breathing" mode, in which the radius of the string simply varies. Each mode becomes more complex. In the figure, the dashed line indicates "neutral," while the solid lines depicts the range of motion of particular pieces of the string.

mass (like the quarks, for example). A single kind of string (one of the goals of an "ultimate" theory), with ever-increasing vibrations, corresponds to particles that have more mass. The idea is simple really.

Upon deeper thought, one realizes that there are some unanswered questions. How big are the strings? What is their tension? Given that we know of only 6 flavors of quarks, how many string vibration patterns are possible? Does the mass hierarchy of the quarks and leptons make sense?

It is possible to calculate the tension (which is a force) holding together one of these strings. One finds the ridiculously large number of 10^{39} tons ... that's one thousand billion billion billion billion tons. With such an incredible tension, the little loop of string is compressed to the tiny dimensions of the Planck length. This is a good feature, as all attempts to measure the size of a quark or lepton fail to show any spatial extent down to a size of about 10^{-18}–10^{-19} meters. While string theory predicts that the string loops do have a finite size (thus deviating from the idea of point particles), the 10^{-35} meter Planck length is comfortably below current experimental limits.

However, this large tension has some troubling consequences. Because of the fact that the energy tied up in string vibrations is related to the string tension, with such a large tension, one would expect that the string would have an enormous mass ... on the order of the Planck mass. This mass is far removed from the mundane world of the familiar electron and massless photon. So either this kills the idea of superstrings, or there's some other explanation.

We are saved by the magic of quantum mechanics. While it's true that the obvious mass stemming from the intuitive vibrations described above is enormous, in addition there are the easy-to-forget fluctuations of the quantum foam. Serendipitously, these vibrations yield a large negative energy. The two energies cancel perfectly, resulting in particles with zero mass. However, such particles are also a problem, as we know that observed particles do not have zero mass. So are we better off? The answer is probably yes. We recall from our discussion of the Higgs mechanism, that above a "magic" energy, the particles are massless. Thus we see that the string idea is the following. At very high energy, the "stringy" aspect of nature is manifested. From about the Planck scale to the electroweak symmetry breaking scale (i.e. when the Higgs kicks in), strings are essentially indistinguishable from the massless particles of more conventional theories. Below the "magic energy," at which the Higgs mechanism turns on, we have our familiar world. The beauty of string theory is not in how it is so different from our current understanding of the universe. Its beauty lies in the fact that it qualitatively represents our observed world, while providing an underlying unifying principle.

If we recall where we started our journey, it was in the attempt to create a successful theory of gravity that included all of the quantum mechanical properties exhibited by the other parts of the Standard Model. Previous attempts always resulted in infinities. At the core of this problem was the fact that as the size probed became ever smaller, the quantum foam became ever more turbulent. When one added the contributions of an infinite range of ever more turbulent quantum mechanical energies, one found an infinite sum. If superstring theory is to have any success, it must successfully address this recurring problem.

In 1988, David Gross and his student Paul Mende, both then at Princeton University, were able to show that as one increases the energy of the string, it becomes more affected by the quantum foam. However, this trend reverses as one crosses the Planck energy. Above the Planck energy, the size of a string increases, thereby reducing its ability to probe the "sub-Planckian" quantum foam. Unlike a point particle (like those used in the Standard Model), strings have a finite size. A point particle, having no size of its own, can in principle probe sizes down to zero, where the quantum foam is most turbulent. With its minimum size, there is a level of quantum foam that cannot affect a string. Thus the infinities that plague theories of quantum gravity involving point particles do not appear in superstring theory. This, more than anything else, is the crucial success of superstring theory.

There is an aspect of superstring theory that will sound familiar. The original string theory, when used to calculate probabilities (which is what one calculates using quantum mechanics), would occasionally come up with a negative number, which doesn't make sense, as probabilities must be between zero and one. Practitioners of the time were experimenting with the shape of the strings. Were they one-dimensional constructs that could vibrate only in the left-right direction? Were they two-dimensional and able to vibrate as depicted in Figure 8.22? What about full three-dimensional vibrations? As physicists experimented with the number of dimensions in which a string could vibrate, they realized that the "degree of nastiness" of the negative probabilities was worse for a small number of dimensions. Inspired, they considered the possibility of extra dimensions. Early calculations suggested that the negative probability problem could be resolved if there were ten dimensions ... nine spatial and one time-like. Improved calculations showed that what was needed was eleven dimensions ... ten spatial and one temporal. Because we experience only three space dimensions, it is necessary that the remaining spatial dimensions are "curled up" into a very small size. In contrast to our large extra dimensions discussion, in this instance the seven spatial dimensions are curled up into the very tiny Planck size. While this idea may be true, the smallness of these extra dimensions ensures the fact that they will not be experimentally observable for the foreseeable future.

We've spoken briefly about the nature of superstrings (the interested reader should peruse the suggested reading at the end of the book), but we've not discussed string theory in the same way we approached our discussion in Chapters 3 and 4. Let's now imagine how today's modern scattering experiments might be understood using the superstring worldview. Each pointlike particle, say an electron or photon, is replaced by a small string, which we draw for conceptual simplicity as a small loop. Let's consider the simple case of an electron and positron annihilating into a photon, before re-emerging into another electron-positron pair ($e^+e^- \rightarrow \gamma \rightarrow e^+e^-$). Ignoring, for clarity, the details of the vibration of the loops for the respective particles, we see in Figure 8.23 how a collision between two strings might appear.

We see that the "electron and positron" strings merge to form a single "photon" string before splitting back into two. All of the various Feynman diagrams of Chapter 4 can be drawn similarly, although for completeness they should include each particle's unique vibration pattern. Given the difficulty involved in drawing eleven-dimensional vibrations, I'm sure you'll forgive me if I don't attempt it.

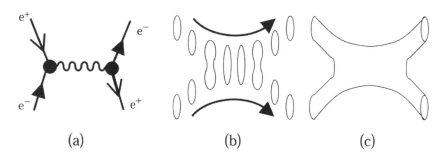

(a) (b) (c)

Figure 8.23 The annihilation of an electron/positron pair into a photon and re-emission of an electron/positron pair as understood in string theory. Two strings come together, merge and then split apart again.

Returning to the experimental flavor of this book, my (and others!) first reaction to hearing about superstring theory is "Great!! Tell us where to look for it!" Unfortunately, when confronted by this question, superstring theorists often hang their head and stub their toe into the dirt. The unfortunate fact is that the phenomena predicted by superstring theory occur at energies not accessible by any realistic future accelerator or cosmic ray experiments. Nonetheless, failure to make experimentally verifiable predictions does not make the theory wrong (as long as it doesn't make *wrong* predictions)... the world might indeed be correctly described by superstrings.

Superstring theorists can pursue explaining *why* the quarks and leptons have the mass that they do. Also, remember the "super" in superstrings. This explicitly assumes that supersymmetric particles (strings) should be discoverable. Correctly calculating the relationship between the known particles and their supersymmetric counterparts would be a real coup. Nonetheless, the failure of superstring theory to make experimentally verifiable predictions relegates it to the "interesting idea" bin for the foreseeable future.

In our discussion of superstring theory, we have raised a number of issues, all related to its possible candidacy for an "ultimate" theory. It passes the criteria of unifying the four forces, as the force-carrying bosons are all different vibrational modes of a single type of string. The myriad of quarks and leptons can be replaced in largely the same way. In fact, all phenomena can be explained by identical types of strings vibrating, merging and splitting in endless ways. The idea of generations can be explained and the number of observable generations has a plausible cause. The theory spans a great range of energies and does seem to satisfy the criteria set forth for "ultimate" status. The astute reader might muse as to what gives the strings their tension. Only further work will reveal as to whether this tension will be derivable from within the theory or will have to be supplied by experiment. Superstring theory is interesting and I expect that it will continue to fire the imagination of theorists for years to come.

In this chapter we have discussed three ideas that have no experimental support, yet all hope to extend our understanding of the universe in which we live. Yet, even Martinus Veltman, a theorist of such note that he shared the 1999 Nobel Prize in Physics, said in his book *Facts and Mysteries in Elementary Particle Physics*

> The reader may ask why in this book [i.e. his book] string theory and supersymmetry have not been discussed. String theory speculates that elementary particles are very small strings, and supersymmetry refers to the idea that corresponding to any particle there is another particle whose spin differs by 1/2, at the same time invoking a large symmetry between the two types.
>
> The fact is that this book is about physics, and this implies that the theoretical ideas discussed must be supported by experimental facts. Neither supersymmetry nor string theory satisfy this criterion. They are figments of the theoretical mind. To quote Pauli: they are not even wrong. They have no place here.

I'm not sure I completely agree with Veltman, as speculation often leads to new ideas, but I can't help nod a little as I read the above passage. In addition to my respect for his intellect (which was never in doubt), I must also add a respect for his wisdom. So you, gentle reader, should view all of these ideas with considerable skepticism. All of these ideas may prove to be utterly hogwash. Alternatively, some or all of them may contain a fragment of an even larger truth. My particle physics colleagues, experimentalists and theorists alike, will continue to explore these and many other ideas in the hope of gaining insight into the ultimate nature of reality; for as we see in the next chapter, the ideas discussed in this and earlier chapters concern perhaps the greatest question that can be asked ... "How did we get here at all?"

chapter 9

Recreating the Universe 10,000,000 Times a Second

I want to know how God created this world. I am not interested in this or that phenomenon, in the spectrum of this or that element. I want to know His thoughts; the rest are details.

— Albert Einstein

There are many marvelous books that are simply brimming with discussions of the newest ideas and discoveries pertaining to the cosmos. This is not one of those books. This book is fundamentally about particle physics, yet the two fields are inextricably linked. Cosmology, the field that studies the entire cosmos, across billions of light years and the 10–15 billion years since the creation of the universe, stands hand in hand with particle physics, which is concerned with the behavior of unstable particles with the most fleeting of lifetimes, many of which have not been generally present in the universe since the first instants following the Big Bang.

Given that these fields are seemingly so dissimilar, how is it that the study of particle physics can reveal so much about the birth and the ultimate fate of the universe? First, one must recall that in the tiny fractions of a second after the Big Bang, the universe was unimaginably hot. When matter (e.g. particles) is so hot, it is moving extremely quickly; that is to say, the matter (the particles) has (have) a lot of energy. And the study of highly energetic subatomic particles is exactly the topic that elementary particle physicists pursue. In the huge leviathan experiments with which you are now quite familiar, physicists collide particles together millions of times a second, routinely recreating the conditions of the early universe. Cosmology is fundamentally an observational science—in that we can only look out and see the universe—but we can't really do experiments (after all, creating and destroying universes is pretty exhausting work…conventional wisdom is that each one takes a week). We have but one universe and we learn about it by staring at it with ever more sophisticated instruments, trying to winnow out its secrets. In contrast, in particle physics we do *experiments*. We can change the energy of the particles. We collide baryons, mesons and leptons. We have control over the experimental conditions and directly observe the behavior of our experiments. Cosmologists can only infer the initial conditions of the universe by observation literally billions of years after the fact. Particle physics experiments can directly observe the behavior of matter under the conditions of the primordial inferno, thus the knowledge obtained from particle physics experiments is directly applicable to the study of cosmology.

In addition to the creation of the universe, cosmologists use the known laws of physics to describe the behavior of heavenly bodies. In general, they are very successful, yet they do occasionally experience failure. The rotation rates of the outer arms of galaxies are much too rapid to be explained by the matter that we can see (stars, planets, gas, etc.) So either the laws of gravity that we use to describe the world are wrong, or there are new phenomena to be discovered. We will discuss why cosmologists postulate the so-called "Dark Matter" (i.e. matter that makes its presence known solely through its gravitational effects and is somehow not observable in the traditional meaning of the word). Particle physicists potentially have something to say about this as well. How is it that particle physics can contribute to the discussion of the rotation of galaxies? This is because it is possible that we may discover massive particles that interact, not through the strong or electromagnetic force, but through only the weak force and perhaps not even that. Recall that after the primordial Big Bang was complete (a whole second after it began), the laws of physics and the populations of subatomic particles were frozen. As discussed in Chapter 7, by that time, there were essentially no antimatter particles and for every matter quark or lepton, there were about one billion (10^9) neutrinos and photons. If each neutrino had a small mass, this would contribute to the mass of the universe and perhaps explain the mystery. The discovery of neutrino oscillations, also discussed in Chapter 7, shows that neutrinos do have a mass and so perhaps the conundrum is solved. We'll talk more about this soon, but we believe that neutrinos cannot solve the galactic rotation problem by themselves. So again, we turn to particle physics, this time for more speculative theories. For instance, if supersymmetry turns out to be true, then there exists a lightest supersymmetric particle (or LSP). As we learned in Chapter 8, the LSP is thought to be massive, stable and does not interact with matter via any of the known forces except, conveniently, gravity. So the discovery of supersymmetry could directly contribute to studies of the large structures of the universe … galaxies, galaxy clusters and even larger structures.

In a single chapter, we cannot possibly describe all of the exciting developments and avenues of research followed by modern cosmologists. There are entire books, many listed in the bibliography, which do just that. Instead, we will follow the arrow of time backwards, discussing the various observations that are relevant to particle physics, pushing through the observation of the universe to the experiments performed in particle physics laboratories, past even that field's frontier and on to some of the ideas discussed in the previous chapter. By the end, I hope to have convinced you that the study of the very small and the highly energetic will supplement much of the beautiful vistas seen by the Hubble Telescope and other equally impressive astronomical observational instruments.

While in order to fully understand the universe you need to understand the particles and forces described in earlier chapters, to understand the universe in its cosmological or astronomical sense, it is gravity that reigns supreme. Even though in the particle physics realm gravity is the mysterious weak cousin of the better understood other forces, in the realm of the heavens, gravity's infinite range and solely attractive nature gives it the edge it needs to be the dominant force. The strong and weak force, both much larger than gravity at the size of the proton or smaller, disappear entirely when two particles are separated by as small a range as the size of an atom. Even the electromagnetic force, with its own infinite range, has both attractive and repulsive aspects. Averaged over the large number of subatomic particles that comprise a star, planet or asteroid, the attractive and repulsive contributions cancel out, yielding no net electromagnetic force at all. So gravity finally gets the attention that our senses suggest that it should.

For centuries, Newton's universal law of gravity was used to describe the motion of the heavens. It was only unseated in 1916 by the ideas of another great man, Albert Einstein. Einstein postulated his law of general relativity, which described gravity as a warping of space itself. Regardless of the theory used, we must focus on the fact that gravity is an attractive force. An attractive force makes objects tend to come closer together. Thus, after a long time, one would expect the various bits of matter that comprise the universe (i.e. the galaxies) would have all come together in a single lump. Given that we observe this not to be true, if we know the mindset of the astronomers of the early 1920s (during which time this debate raged), we can come to only one conclusion. While there certainly was discussion on the issue, the prevailing opinion was that the universe was neither expanding nor contracting, rather it was in a "steady state." Accordingly, Einstein modified his equations to include what he called a "cosmological constant."

The Shape of the Universe

The cosmological constant was designed with a single purpose ... to counteract gravity's pull and keep the universe in the static, unchanging state that was the consensus view at the time. Basically, the cosmological constant was Einstein's name for a hypothetical energy field that had a repulsive character. Because of its repulsive nature, it spreads out across the universe, filling it completely. (If you think about it, if every object repels every other object, the only way they can have the maximum distance between each other (in a universe of finite size) is to spread uniformly across the cosmos.) Essentially, the cosmological constant can be thought of as a uniform field, consisting of energy that is "self-repulsive." In a steady state universe, the strength of the repulsive cosmological constant is carefully tuned to counteract the tendency of gravity to collapse the universe, a point illustrated in Figure 9.1.

In 1929, Edwin Hubble presented initial evidence, followed by an improved result in 1931, which suggested that the universe was not static, but rather was expanding very rapidly. After much debate, an explanation emerged. In a cataclysmic explosion, termed the Big Bang, the universe was created at a single point and at a single time.

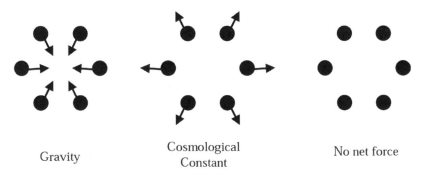

Gravity   Cosmological   No net force
                           Constant

Figure 9.1 Gravity is an attractive force in the universe. The cosmological constant provides an outwards pressure. In Einstein's early vision of the universe, the two forces were balanced, providing a static and non-changing universe.

Starting from a single spot in a place that can't even properly be called space, the matter that constitutes the universe was flung by the Big Bang outwards at great velocities. In an explosion of a house, like you might see in a war movie, the roof is blown off and ejected upwards very rapidly. As the explosive fireball expands, it cools off and it no longer forces the roof upwards. Eventually, the effects due to the force of gravity become dominant and the bits of the roof fall back to the ground. Similarly, the effect of the Big Bang is to fling the matter that makes up the beautiful stars and galaxies you see under a clear midnight sky across the universe. (In fact, the reality is more complicated, as the expansion of matter actually creates the universe as it goes. In addition, strictly speaking the Big Bang is still ongoing, as the universe continues to expand ... essentially we are in the later stages of the explosion. We'll gloss over these points right now and instead use the word "Big Bang" in a sloppy way that signifies the original explosion only.) Since the Big Bang is long over, one expects that the gravitational force between the constituents of the universe would cause the initial expansion to slow down and possibly even stop and crash the matter of the universe back together, like the bits of the roof crashing back to Earth. The fact that the universe was not static caused Einstein to remove from his equations the cosmological constant, calling it "the greatest blunder in his life." Ironically, nearly 80 years later the cosmological constant is making a comeback. More on this later.

As astronomers understood the phenomenon of the Big Bang and the slowing effect of the universe's self gravity, naturally a question arose. What happens to the matter in the universe after the initial explosion? Does the universe expand forever, slowing while it goes? Does it expand and eventually stop? Does the force of gravity cause it to eventually contract, making the matter of the universe race together in a "Big Crunch?" How can we resolve these questions?

Before we talk about these questions within the context of the fate of the universe itself, let's discuss a somewhat simpler example. Suppose you have a giant slingshot and you want to launch an object into a specific orbit around the Earth. This is a very high-tech sling-shot and can launch your object at any speed you want. As you choose your launch speed, you realize that three things can happen. Launch your object too slowly and it will crash back to Earth. Launch the object with too much energy and you'll fling it off into the dark depths of space. However, at a single particular velocity, which we call the "critical velocity," we are able to attain the desired orbit. One velocity among all possibilities is special.

In determining the fate of the expansion of the universe, whether it will expand forever or not, the critical parameter is the density of matter in the universe. Too much matter and the universe will eventually collapse, not enough and it will expand forever, never stopping. If the amount of matter is "just right," the universe will expand forever, moving ever slower until the expansion eventually stops in the infinite future. The whole thing has a very "Goldilocks" quality to it … too much, too little or just right.

We call the "magic" amount of mass needed to just stop the expansion of the universe in the far future the "critical density." Density in this context has the usual meaning, so one takes the ratio of the mass (or equivalently energy) of the universe to its volume. In order to easily communicate about this whole question, cosmologists have defined a quantity called Ω (omega), which is simply the ratio of the mass density of our universe (denoted ρ) to the critical mass density of the universe (denoted ρ_c). Mathematically, we say $\Omega = \rho/\rho_c$. If the mass density of our universe is equal to the critical density, then $\Omega = 1$. If our density is greater, then omega is greater than 1 ($\Omega > 1$), while obviously too low a mass density will make omega less than one ($\Omega < 1$). Thus, the determination of Ω will reveal the ultimate fate of the universe.

While our discussion thus far has been relatively intuitive, when the whole question is cast in Einstein's theoretical framework, the discussion becomes a bit murkier. Since many of the accounts you will read in newspapers and other sources are explained in Einstein's language, we'll talk a little about it here. Recall that Einstein's theory of general relativity cast gravity in a geometrical framework, describing gravity as a curvature of space itself. It should not surprise you that the question of the critical mass of the universe has a geometric analog. Since we are discussing the mass that permeates the universe, this mass gives the universe its shape. The concept of curved space is a pretty tricky one, requiring that one understands the distortion of our familiar three-dimensional space. As usual, intuition (and artistic talent) can fail us in this endeavor, so let's instead talk in two dimensions. If $\Omega = 1$, we can say that on average, the universe is "flat" like a plane. If the mass density of the universe is too high ($\Omega > 1$), the universe has a spherical shape, while if the density is too low ($\Omega < 1$), the universe has a "saddle" or "hyperbolic" shape.

In Figure 9.2, we see the three shapes that space can take. Think of two ants walking along two perpendicular lines in the grids of each type of space. In all cases each ant moves at a constant "local" speed. Local speed means how fast he is moving with respect to the place that his feet are touching. The counterintuitive thing one must realize is that due to the curvature of space, the ants in the three differently shaped spaces will separate at different speeds. Fundamentally, it is this aspect of space that will govern the fate of the universe.

Since you, gentle reader, have made it this far in this book, you are a curious person, with a deep-seated interest in the structure of the universe. I expect that you are becoming impatient. I can imagine what's going on in your mind. The burning question must be "Well? What is it? Is space curved or flat?" Cosmologists have finally been

Figure 9.2 Three types of space, flat, spherical and hyperbolic, or "saddle-shaped." Until recently the exact type of space that makes up the universe was not known. Recent work suggests that our universe is flat.

able to make the relevant measurements and they find … a drum roll please … space is flat. We know this because of subtle variations in the radio waves emitted by space itself. This measurement is somewhat beyond the scope of this book, although it is described in some of the suggested reading. We will revisit the radio waves from space in a little bit, albeit not at quite so technical a level as would be needed to fully convince you of the flatness of space. You'll have to trust me.

The Dark Side of the Universe

With the knowledge that space is flat (and $\Omega = 1$), we know what the mass (technically energy) density of the universe must be … it must be equal to the critical density discussed earlier. As a crosscheck, astronomers can look out at the universe and catalog the matter that they observe. They do this by looking out at the cosmos and cataloging stars. From what is known of stellar evolution, they can convert the brightness and color of each star they observe into a stellar mass. The visible mass of galaxies can be determined by similar studies and through the application of statistical techniques. What they find is that the amount of luminous matter in the universe is only about 0.5% of that needed to make space flat. So where is the missing matter?

This question is not a new one. Astronomers have long realized that the combination of the observed distribution of luminous matter and Einstein's law of gravity could not explain the rotation rates of galaxies. The rates at which a star orbits the center of an extended object like a galaxy is determined by two things. The first is the amount of matter (other stars and gas and such) contained within the spherical volume circumscribed by a star's orbit. The second parameter is the distance the star is from the galaxy's center. In a galaxy such as ours, with a large central bulge and long graceful and relatively sparse arms, these two effects compete. For stars at a radius greater than the extent of the central bulge, it's the size of the orbit that dominates.

When astronomers measured the speed of stars at various orbital radii in our own Milky Way galaxy (and other nearby galaxies), they found that the galaxies rotate differently than Einstein's theory would predict. The Milky Way rotates in a complex way, but essentially one expects the stars in the spiral arms to revolve more slowly as the radius of the orbit increases (much in the same way that Pluto moves much more slowly than Mercury). However, as shown in Figure 9.3, what one finds instead is that the rotational velocity of stars in the arms is independent of radius.

The favored (although not unique) explanation for this discrepancy is the idea that perhaps there exists matter throughout the galaxy that is not luminous. Luminous, in this context, means giving off electromagnetic energy. An object that we can detect, whether it emits visible light, infrared, ultraviolet, microwaves, radio, x-rays or other electromagnetic energy, is luminous.

Such a hypothesis is fairly arresting, if not exactly new. In the mid 1930s, Caltech astronomer Fritz Zwicky proposed non-luminous or dark matter to explain the motion of galaxies within galactic clusters. However, if dark matter exists, what is its nature and how could we

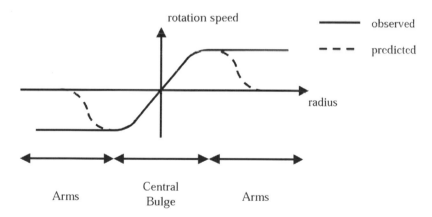

Figure 9.3 The rotation rates of galaxies are quite different than predicted from conventional gravitational theory and the observed distribution of matter in a galaxy. In contrast to predictions, in which the outer stars of the galaxy are expected to revolve more slowly, the revolution speed of the galaxy appears to be independent of radius. Observation of this fact has led to the idea of dark matter.

find it? Many options have been proposed which we introduce in increasing degrees of exoticness. Hydrogen gas within the galaxy, but not tied up in a star, can be excluded as it emits radio waves and is therefore luminous. The next most plausible explanation is the so-called "brown dwarfs." Brown dwarfs are essentially stars too small to ignite and burn. Somewhat larger than our own Jupiter, they can't quite make up their mind whether they are large planets or small, failed stars. There's nothing that forbids such objects from forming, indeed recent attempts to find planets around nearby stars have revealed objects that would qualify as being small brown dwarfs. However, since they are so small on the stellar scale, in order to make up the invisible mass that seems to permeate our galaxy, there needs to be a lot of them.

So how would you find invisible brown dwarfs? Essentially, you see them by the shadow they create. If brown dwarfs are so ubiquitous, you should be able to look at a distant star and eventually a brown dwarf would wander across of the line of sight between you and the relevant star, and you would see a dimming of the star's light. Space is large and stars are small, so any individual star is unlikely to be eclipsed in any reasonable amount of time, consequently astronomers simultaneously observe many stars. The usual approach is to look towards the center of our galaxy, which has the greatest concentration of stars and see if any of these are ever eclipsed. Long studies have seen very few such events, conclusively proving that a preponderance of brown dwarfs is not the explanation for dark matter, although the amount of matter tied up in brown dwarfs and related objects exceeds the mass tied up in luminous matter.

Another possible astronomical explanation of the dark matter question is black holes. We can rule out black holes as an explanation fairly easily. While black holes are, by definition, black (i.e. non-luminous), they play havoc with the matter that surrounds them. As matter encounters a black hole, it accelerates inwards. Accelerating matter usually radiates electromagnetic energy. Thus while the black holes are invisible, the lack of disturbances in the interstellar medium rules out the existence of so many black holes.

You may have heard of a super massive black hole at the center of our galaxy. While a consensus seems to have arisen that a black hole with the mass of millions of times greater than our Sun probably governs the galaxy's overall rotational dynamics, to explain the uniform rotational speed observed in the galactic arms requires a spherical and extended distribution of dark matter. Thus the central black hole, interesting though it may be, does not provide the explanation.

Collectively, these relatively mundane candidates for dark matter are called MACHO's (for MAssive Compact Halo Objects). The name stems from the fact that these objects have significant mass, are compact (like brown dwarfs, rather than gas clouds) and make their presence felt most strongly in the galactic halo (i.e. periphery) of the galaxy. All of the matter mentioned thus far is called baryonic, as it is made of common baryons (protons and neutrons). As the leptonic electrons contribute little to an atom's mass, their presence is ignored in the name.

Leaving the traditional explanations for dark matter, we now turn to our particle physics knowledge for options. If the Big Bang idea is true, neutrinos were produced copiously in the primordial inferno. One can calculate the number of neutrinos that should be present in the universe. It turns out that for every stable baryon (i.e. protons or neutrons) in the universe, there should be about 10^9 (one billion) neutrinos. While we don't know the mass of neutrinos, Chapter 7 suggests that we have enough information to make a reasonable guess as to their mass. If one combines the current best guess of the mass of the various flavors of neutrinos with the number of neutrinos inferred from the Big Bang model, one finds that the neutrinos can account for only about 1–4% of the mass needed to make the universe flat and only about 10% of that needed to explain the observed rotation rate of galaxies.

So, of the particles and objects that we know exist, we have only about 5% of the matter necessary to make space flat and about 15% needed to solve the galactic rotation problem. So now what? We need to find enough matter, first to explain the rotation of galaxies (which would be about 5 times the total potentially visible matter, i.e. all baryonic matter, even the dark stuff as it would emit light if it were heated enough) and then another source of mass to explain the flatness of the universe itself.

Let's first start with the matter needed to explain the rotation of galaxies. We need to find matter which is not affected by the electromagnetic force (or we could see it) or the strong force (or we could see it interact with ordinary matter). This form of matter may feel the weak force and by definition, it must feel the force due to gravity. While we have no real experimental evidence as to what sort of matter would make up this dark matter, we have found in Chapter 8 a hypothetical particle that might fit the bill. When we were discussing supersymmetry, we talked about the lightest supersymmetric particle or LSP. Because the LSP is the lightest of its brethren, there are no lighter supersymmetric particles into which it could decay. In addition, because supersymmetry is "conserved," these particles cannot decay into ordinary (and luminous) matter and therefore are stable. Further, since we have not detected the particle yet, if it exists, it must be electrically neutral, impervious to the strong force and massive. The LSP, while wholly theoretical, would prove to be an attractive candidate for the dark matter that governs galactic rotation.

Of course, since the LSP may not exist, there have been other particles proposed that might also prove to be the culprit. All of these particles are exotic, completely theoretical and quite possibly non-existent. However, the upcoming generation of particle physics experiments will be looking for heavy stable particles. Cosmologists will keep a close eye on these experiments, in the event that they provide cosmologically relevant information.

However, all the matter discussed thus far; baryonic luminous and "normal" dark matter (baryonic brown dwarfs and such) and non-baryonic "exotic" dark matter (LSPs or equivalent), while necessary to successfully describe galactic rotation measurements, can only account for about 30% of the matter needed to make space flat (or equivalently $\Omega = 1$). So now what? Cosmologists have now proposed another idea … dark energy. The dark energy can take several forms, one provided by Einstein's resurgent cosmological constant and another related idea called quintessence. The important point is that the dark energy provides a repulsive force. The cosmological constant is sort of a vacuum energy … something analogous to the Higgs field of Chapter 5. Basically, the vacuum itself is permeated by an energy field of a repulsive nature. Quintessence is somewhat more analogous to the discredited idea of the aether, the non-existent material which physicists once thought was needed to allow light to propagate. Quintessence, if true, would also be an energy field that permeates the universe. It can be disturbed and interact with itself. It is this aspect that distinguishes it from the cosmological constant, which is … well … constant. These two ideas make similar predictions and will require fairly precise measurements to say which of the two is correct, if either. If this all sounds rather fuzzy, this is because it is. This is research in progress. In research, confusion is good. It means that something doesn't hang together and you're about to learn something new. This is an exciting time in cosmology.

The whole idea of what constitutes the matter of the universe is a rather complicated one. The matter that makes up the beautiful and sparkling night sky is only responsible for 0.5–1.0% of the energy of the universe. The luminous matter is a very thin icing on a very dark cake. Table 9.1 shows the contribution of the various components to the makeup of the universe.

A skeptical reader might find this whole discussion to be suspect, as it seems very complicated. It very well might be that all of these different types of matter and energy are needed to fully describe the universe. On the other hand, it may be that there is a much simpler explanation … one we have not yet formulated. The possible solutions proposed here are not unique. For instance about 20 years ago, Mordehai Milgrom at the Weizmann Institute in Israel proposed a solution to the galactic rotation problem that did not invoke dark matter. He proposed a rather small modification to the laws of physics. His proposal would have no effect except in situations in

Table 9.1 Various components of the universe.

Material	Likely Composition	Source of Information	Percentage of the Mass of the Universe
Visible baryonic	Luminous matter, stars gas, etc.	Telescopes, etc.	~0.5%
Dark baryonic	Normal, but dark matter (brown dwarfs, planets, etc.)	Hydrogen and helium abundances in the universe, "eclipse" experiments	~5%
Exotic dark matter	Non-baryonic matter (no protons and neutrons) LSPs and other unusual matter to be discovered by particle physicists	Rotation speeds of galaxies, motion of galaxies within galactic clusters	~25%
Dark energy	Cosmological constant, quintessence, etc.	Observed flatness of space	~70%

which the acceleration was very small, as is the case in the outer arms of the galaxy. So, which of the two explanations is right? I don't know. Nobody does. These debates are what make the whole question so much fun. Luckily, experiments are now possible which may resolve the whole question. A full discussion of these ideas is outside the scope of this book, but the interested reader can peruse the suggested reading where these aspects of cosmological research are discussed in greater detail.

While the question of exactly what constitutes the universe is a burning one, there is another interesting question. Intricately interwoven with the question of what makes up the universe is the story of its birth and evolution. In this, there is one clearly favored explanation.

The Big Bang cosmology was originally suggested in 1922 by Aleksandr Friedmann at the University of Petrograd and developed independently about 5 years later by Georges Henri LeMaître, a Catholic priest turned astronomer. LeMaître later said that he had an advantage over Einstein, as his priestly training made him look favorably upon the idea that the universe had a distinct beginning. LeMaître called his progenitor of the universe the "primeval atom." Among other evidence, this cosmology was designed to explain the observed expansion of the universe, first discovered by Edwin Hubble in 1929. The term "Big Bang" was not offered by the proponents of the theory, but was intended to be a denigrating term, first suggested by a key opponent. Fred Hoyle was an architect of a competing theory, the so-called Steady State hypothesis (initially so beloved of Einstein). The Steady State theory postulated that the universe was in a … well … steady state, that is to say that matter was being created and consumed in equal quantities and thus on average nothing was changing. Hoyle, in a criticism of the competition, was unimpressed with the need for a unique event and thus offered the disparaging term "Big Bang" as a way to show how silly the theory was. Much to his chagrin, proponents loved the term and the Big Bang cosmology was named.

The Big Bang cosmology is the only one that clearly agrees with the observational evidence, as we will discuss in the following pages. Competing scientific theories and all ancient myths, including biblical ones, have been discredited. This is not to say that the Big Bang cosmology is without its mysteries. Details of what the universe looked like earlier, and how it got to be so smooth and homogeneous, are still topics of research and debate. Unfortunately, the popular press sometimes uses that debate to report sensational stories, "The Big Bang is Dead" being my particular favorite. Adherents of competing theories use these reports to try to convince others that the scientific community is in a much greater turmoil than it is. Biblical literalists insist that at best, the Big Bang cosmology be taught as a theory, on par with, but less true than their own Genesis-based ideas. Such an approach is nonsense. While the Big Bang cosmology is not without its own internal debates, no reputable scientist can dispute the evidence that the universe was once much smaller and hotter and that it is now expanding at great speed. The evidence for this is simply overwhelming. Theology based counterarguments must now join the same debate as scientists; Why are the laws of physics what they are? While an unlikely explanation, a deist answer to this question remains tenable.

When one considers how one might ascertain the nature of the universe immediately after the Big Bang, one is struck by the magnitude of the task. The Big Bang occurred between 10–15 billion years ago at an unknown point probably many billions of light years away (and quite possibly in a now-inaccessible dimension). Given that the primordial explosion was such a long time ago, it is difficult to infer any details. One might as well take air pressure measurements today and infer from them the details of that first nuclear detonation at Alamagordo, New Mexico.

Where Are the Galaxies?

As hard as the task may seem, astronomers actually have had an impressive success rate. Edwin Hubble found that other galaxies tend to be moving away from us. Even more interesting was the observation that the greater the distance to the galaxy, the faster it was moving away. Subsequent studies have verified Hubble's initial result and greatly improved the accuracy of his measurement. Scientists can use exquisitely precise telescopes, the Hubble Space Telescope being the most famous of them, and measure the speed of a galaxy. By knowing the relationship between speed and distance, they can determine the galaxy's distance. While the precise number assigned to the distance still has some experimental and theoretical uncertainties, at present we can see galaxies over 10 billion light years away.

A light year is the distance that light, that fleetest of messengers, can travel in a year. Light travels at 186,000 miles per second. In a year, it can travel 6×10^{12} (six quadrillion) miles and, in 10 billion years travel a whopping 6×10^{22} miles. These distances, while impressive, are not the most useful fact. The important thing to remember is that as fast as light travels, the size of space is incomparably greater. The Earth orbits the Sun at a distance of 93 million miles. It takes light a little over 8 minutes to travel from the Sun to our eyes. So the light you see from the Sun shows us not the Sun as it is right now, but rather as it was 8 minutes ago. The nearest star, Proxima Centauri, is 4.3 light years away. If you were to see it go nova the night you read this book (a highly unlikely prospect), any hypothetical people living there would be already dead for 4.3 years, as that's the amount of time it takes for the news to get here. The most important consequence of this observation is the following. The farther away an object is from Earth, the longer it takes for light to get here. When it does get here, you see the object not as it appears now, but as it appeared in the past. If you rigged three cameras on Earth to simultaneously record the Sun, Proxima Centauri and the nearest "real" galaxy to our own (M31, also called Andromeda), you'd be taking pictures of objects 8 minutes, 4.3 years and 2.2 million years in the past.

Once we realize this fact, it becomes obvious how to study the evolution of the universe. Take your most powerful telescopes and train them outwards, looking at ever more distant objects. The farther away you look, the farther back in time you see. If you're interested in how galaxies have changed over the years, look at our nearby galaxies and study their properties. To see a galaxy 2.2 million years ago (a cosmological blink of an eye), you merely need to look at our neighboring galaxy in Andromeda. As you look at galaxies at an ever-increasing distance, it is like looking at older and older snapshots. Each photo reveals something of an earlier era. Using such instruments as the Hubble Space Telescope (HST) and the Sloan Digital Sky Survey (SDSS), scientists have been able to image galaxies a mere billion or so years after the Big Bang. This was not long after the first stars formed and began to burn with their bright nuclear fire. As one looks back in time, galaxies begin to take on a different shape ... one more primitive. In this way, cosmologists interested in the physics of galaxy formation can view examples at all stages of development. In this, they are luckier than paleontologists. Cosmologists can see earlier "living and breathing" galaxies, while their dinosaur-hunting friends must content themselves studying dry bones.

While the study of the evolution of galaxies is interesting and a crucial effort for one wanting to understand the fate of the universe, in some sense, it doesn't address the question of why the universe is the way it is. A billion years after the Big Bang, the laws of physics had long since been determined. Well-understood nuclear and gravitational processes were shaping the stars and galaxies, but the question of why the nuclear fires burn as they do was still a mystery. To answer that question will require a journey further back in time. We'll continue that journey in a moment.

However, before we do, I'd like to take a moment to address a question raised by the observations by both the HST and the SDSS. This question involves the distribution of matter across the universe. One could imagine that matter was all lumped together, surrounded by an unimaginably vast void. Alternatively, matter could be distributed throughout the universe or spaced periodically like a giant honeycomb. So, what is the truth?

For nearly 100 years, astronomers have been doing three-dimensional maps of the universe. Even early astronomers could map the positions of objects on the surface of the sphere that is the heavens. With Hubble's insight, astronomers could determine an object's distance as well, thereby locating the object uniquely in space. On a purely stellar level, clearly matter isn't distributed uniformly. Each star contains a great concentration of matter, surrounded by the vastness of nearly empty interstellar space. One can expand the question further and ask if the stars are spread uniformly throughout space. On a distance scale of some few hundreds or thousands of light years, one finds that the stars are spread relatively uniformly. The situation changes when the entire galaxy is considered. Our own Milky Way galaxy is a spiral or barred spiral galaxy, with stars concentrated in long graceful arms that spiral out from a dense core. Other galaxies reveal different structures.

If one simply thinks of galaxies as clumps of matter, without too much thought going into the details of their structure, one can begin to ask questions that are more relevant to the structure of the universe. How are the galaxies arranged in the universe? It turns out that galaxies cluster together on the size scale of a few million light years. While such a distance is truly vast, it's one ten thousandth of the size of the visible universe as a whole. In 1989, Margaret Geller and John Huchra published a study in which they revealed a most marvelous map of the sky. Locating the galaxies out to a distance of 500 million light-years, they found the most delicate structure. This map of the universe showed galaxies arranged in long filaments across the sky, surrounding vast voids in which very little matter was found. Their data is shown in Figure 9.4. On the distance scales that they explored, the universe looked like soap bubbles with the galaxies arranged along the soap's film.

By the mid 1990s, several experiments redid Geller and Huchra's measurements, this time extending the distance investigated by a factor of ten. On this much larger distance scale, the bubbles look very small and the universe is much more uniform. Careful perusal of the images in Figure 9.4 indicates that the size of the voids in Geller and Huchra's measurements is the largest that the voids get. There do not appear to be even larger structures. The conclusion one must draw from this is the following. On the largest distance scales, roughly the size of the visible universe itself, matter is distributed uniformly throughout the cosmos. At the smaller scales of ribbons and bubbles of galaxy clusters, down through galaxies and a more stellar environment, gravitational interactions have made the universe more clumpy. The clumpiness, although interesting and incidentally crucial to life, does not reflect the beginnings of the universe. For that, the uniform distribution of matter is what must be explained. A newer idea called

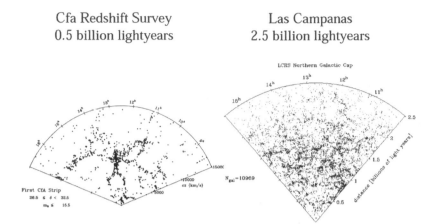

Figure 9.4 Experimental data from the Cfa and Las Campanas experiments. In both pictures, each dot represents an entire galaxy. The Cfa experiment looked out to a distance of 500 million light-years, looking for structure. The Las Campanas experiment greatly expanded that range. There appears to be clusters of galaxies, as well as spots where no galaxies exist. The largest structures in the universe seem to be about 100 million light-years in size. (Figure courtesy of John Huchra, for the Cfa Collaboration and Doug Tucker, for the Las Campanas Collaboration.)

cosmological inflation has been suggested to explain how the universe could be so uniform on such a large scale. Inflation suggests that a tiny fraction of a second after the Big Bang, the universe expanded extremely rapidly. We'll revisit this idea later when we talk about the conditions of the universe just fractions of a second after the Big Bang.

The Big Whisper

While observational astronomy using the electromagnetic spectrum (light, infrared, ultraviolet, x-rays, radio waves, etc.) to view heavenly objects has impressively contributed to our understanding of the universe at an earlier time, so far it has only been able to contribute for times more than a billion years after the Big Bang. To push our understanding even earlier requires a different approach. In 1945, Ukrainian émigré George Gamow took on a student, Ralph Alpher, who was to attempt to quantify the conditions immediately following the Big Bang. Joined shortly thereafter by another student named Robert Herman, Alpher set out to calculate the relative ratios of the elements that would be produced early in the universe. Like typical students, they followed their mentor's lead. Gamow had realized that in order for nuclear fusion to be able to produce elements other than hydrogen, the early universe had to be hot. What Gamow missed, but his students realized, was that if the universe was once a hot and dense fireball, in the intervening years it should have cooled considerably and it should be possible to view remnants of the original energy by looking out at the cosmos. While there was some question as to what the signature might be, something of a consensus arose that perhaps one would see a uniform radio or microwave background.

In 1964, Arno Penzias and Robert Wilson (this is a different Robert Wilson than Fermilab's first director) were working at Bell Laboratory in New Jersey. They were trying to make an absolute measurement of the radio emission of a supernova remnant called Cas A. Cas A is located in the constellation Cassiopeia and is, mostly due to its relative proximity, the brightest radio source in the sky. Making an absolute measurement is just about the hardest thing one can do. Making a relative measurement is much easier. In a relative measurement, one tries to compare two things. For instance, if one looks at two light bulbs, a 40-watt one and a 150-watt one, it's pretty easy to say that the 150-watt bulb is brighter. But to say exactly how many lumens the light is emitting (lumens are a unit of light like pounds are a unit of weight) is much harder.

Other people had measured the various radio sources in the sky and concluded that Cas A was the brightest source in the heavens. They also were able to even say how much brighter it was than its nearest competitor. However, in order to be able to compare their measurement to a calculation dealing with supernova, they needed an absolute number. They needed to be able to say unequivocally that Cas A put out so many units of radio energy. So the idea seems easy. One simply points an antenna at Cas A and records the radio energy received. There's only one problem. The fact is that everything emits radio waves. In the case of Penzias and Wilson, they were receiving radio waves from not only Cas A, but also from the antenna itself, the atmosphere, stray sources from those secret government labs in Area 51 that cause my Uncle Eddy to put tinfoil in his baseball cap, etc. Penzias and Wilson had a tough job ahead of them. They were able to calculate the amount of radio waves from all known sources and they subtracted out these effects, again aiming their telescope at the sky. This time they looked not at Cas A, but at empty space. They expected to see nothing, yet an unwanted radio hiss remained. In order to make their measurement, they needed to understand the source of this mystery. They calibrated and recalibrated their equipment. They climbed up into their antenna, evicted two pigeons, and cleaned up piles of bird poop. (Which goes to show you that the life of a research physicist is even more exciting than you think. Not only do we get the fast cars and beautiful women (or gray-eyed Counts for my more feminine colleagues), but sometimes we get the lucky bit of bird poop thrown in too.) Penzias and Wilson's efforts were appreciated by the custodial staff, but they didn't get rid of their mysterious hiss.

Penzias and Wilson were a bit depressed, as this unexplained radio noise would make their measurement a failure. As is usual at this point in an experiment, they started asking people for ideas. What did they miss? In January of 1965, Penzias was talking to Bernard Burke, who was a radio astronomer in his own right. Burke was aware of an effort by Jim Peebles at Princeton to find Gamow, Alpher and Herman's radio signal from the Big Bang. Finally, the pieces clicked into place. In 1965, Penzias and Wilson published an article in *Astrophysical Journal*, detailing their experimental results. This paper was accompanied by another paper, written by Peeble's Princeton group that interpreted their result. For the discovery of the radio signal remaining from the Big Bang, Penzias and Wilson received the 1978 Nobel Prize. Incidentally, they eventually published a measurement of the radio emissions of Cas A as well, although not to the same general acclaim as their serendipitous discovery.

It turns out that it is possible to convert the Big Bang's background signal into a temperature. The temperature of outer space is 2.7 degrees Kelvin or −455°F. An important question was "how uniform was this temperature?" Penzias and Wilson were able to scan the sky and they found that the radio emissions were remarkably uniform; any variation from perfectly uniform was less than 0.1%. This was the precision of their equipment, thus they couldn't say that this "background" radiation was nonuniform at the 0.01% level, but they could say that the uniformity was better than 99.9%. The temperature of the universe was everywhere 2.7 degrees Kelvin (K). Rounding the temperature upwards, we call this remnant radio radiation "the 3 K background." As is usually the case, earlier scientists had studied the cyanogen molecule in the interstellar environment and noted that it appeared to be surrounded by a bath of radiation between 2 and 3 degrees Kelvin. They missed the significance and yet more physicists joined the "If only…" club.

So why is this measurement interesting? The theory of the Big Bang suggests that at one time, the universe was much hotter and highly energetic photons were ubiquitous. At about 300,000 years after the Big Bang, the universe was a relatively cool 3,000 degrees Kelvin (about 5,000°F). All vestiges of quarks were gone and the universe was composed of the non-interacting neutrinos and the much more interesting protons, electrons, photons and the rare alpha particle (helium nuclei). Protons and electrons have the opposite electrical charge and thus they feel an attractive force. Get a proton and electron together and they really want to combine and become a hydrogen atom. Similarly, an alpha particle wants to grab two electrons and become a helium atom. However, highly energetic photons can knock the electrons away from the proton and thus electrically neutral atoms don't form. The photons jump from electrons to other electrons and back again like a hyperactive seven year old interfering with his older sister's date.

However, as the temperature drops below the 3,000 degrees Kelvin temperature, suddenly everything changes. The energy carried by the photons is no longer enough to separate the electrons and the protons. Now instead of a universe of separated electrical charge, the universe is full of neutral hydrogen and helium atoms. Since photons only interact with charged objects, the photons stop interacting and march undisturbed across the cosmos, much as their distant cousins, the neutrinos, were already doing. Thus, and this is the important part, these photons last interacted with matter 300,000 years after the cosmos came into being. These photons then are a snapshot of the universe only 300,000 years after the Big Bang. This pushes our understanding of the origins of the universe quite a bit closer to the beginning as compared to the studies of galaxies discussed earlier.

In a sense, the view that the 3 K background radiation is a highly uniform bath, recording the conditions of a much earlier epoch, has not changed much in the intervening years, although this is not to say that other measurements have not been made. In fact, in 1990 the COsmic Background Explorer (or COBE) satellite re-measured the 3 K background with exquisite precision. The full story of the significance of their results is beyond the scope of this book, but they are clearly described in George Smoot and Keay Davidson's book *Wrinkles in Time*. Smoot was a leading member of the group that measured the 3 K background radiation, while Davidson is a talented science writer, and this book is well worth your time. In the simplest terms, the COBE collaboration determined that the 3 K background did have a slight non-uniformity at the 0.001% level. To give you a sense of the magnitude of the accomplishment, they needed to measure the temperature with a precision of one part in 100,000. To give a concrete example, it's as if they accurately measured the length of a football field and found that it was off by one millimeter. These small variations in temperature reflect early variations in the density of the universe. These little variations in density have been amplified in the ensuing years to become the distribution of matter; the galaxies, galaxy clusters and so forth that we see now. In 2003, the WMAP experiment weighed in with even more precise measurements that confirmed COBE's results and improved our understanding of this early era. We'll return to these small density variations later.

At Three Minutes

Using the 3 K background radiation, we've pushed our understanding of the origins of the universe backwards in time, but we still have 300,000 years for which we haven't accounted. Observational cosmology has one more trick in its bag. As the universe cools, it passes through several phases, from the time when quark and lepton physics dominates, through the cooler phase of protons and electrons, to the time when hydrogen and helium atoms form. As the universe cooled, the quarks coalesced into protons or neutrons. Nuclear fusion would combine protons and neutrons to form alpha particles (the nuclei of helium atoms), as well as a few light elements and isotopes (deuterium, helium, lithium, etc.). This is called big bang nucleosynthesis (from the synthesis of nuclei). The Big Bang theory predicts 76% hydrogen and 24% helium and trace amounts of everything else. When we look out at the evening sky, we measure slightly different ratios (73% hydrogen, 26% helium, 1% everything else). The discrepancy is explained because in the ensuing years, nuclear fusion in the stars has converted some of the primordial mixture into heavier elements. In fact, if not for the stars that were born in earlier epochs, we would not exist. Carbon, oxygen and nitrogen all make up living tissues, while silicon, iron and other metals make up the planet on which we live. It is in the early stellar kilns that the elements so necessary for our existence were forged. One sees why Carl Sagan liked to say that we were all starstuff.

The small discrepancy between the hydrogen to helium ratio predicted by the Big Bang theory and what we observe might be a little troubling, leaving one with the sense that the stellar nuclear fusion explanation was just sweeping our ignorance under the rug. However, we recall the Hubble Space Telescope's ability to see galaxies at great distances and consequently in the distant past. By carefully studying the ratio of helium to hydrogen as we look farther back in time shows that at earlier times, this ratio was more like that predicted by the Big Bang theory. Of course, we are relieved.

Uniformity and Inflation

Before we begin our final journey from the Big Bang to the present, we need to make one last detour. The uniformity of the universe is rather remarkable. Look in any direction into deep space and the distribution of matter looks pretty much the same. Even more remarkable is the incredible uniformity of the radio background radiation (about 99.999%). Because (a) the universe is about 15 billion years old and (b) the radio background radiation is a snapshot of the universe a scant 300,000 years after the Big Bang, to see the radio background radiation is to look far back in time. If we look out in the deep void, we are looking nearly 15 billion years back in time. If we then turn 180°, we can also see 15 billion years in the past. However, since these two points are separated by 30 billion light years, light from one side of the universe cannot have reached the other side yet. Figure 9.5 illustrates how the two opposite edges of the universe are isolated from one another.

So how is it that the universe looks so uniform in every direction if light from one side hasn't reached the other side yet? Ordinarily, in order for two things to be the same temperature (recall 3 degrees Kelvin?), they need to touch in order to make the energy equal everywhere. Since the two sides of the universe have never touched, what we have is a mystery.

Beginning in 1979 and continuing on into the early 1980s, Alan Guth, then of Stanford Linear Accelerator Center, A.D. Linde of the Lebedev Physical Institute in Moscow and Andreas Albrecht and Paul Steinhardt, then at the University of Pennsylvania, published a series of papers in which they worked out the details of a new idea which could explain the uniformity of the universe. Taking its name from

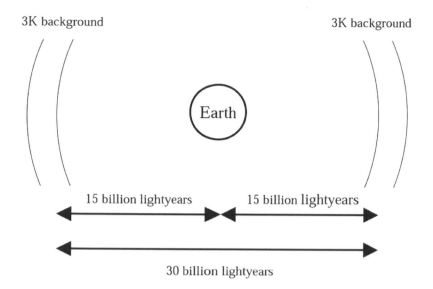

Figure 9.5 A cartoon showing that if the universe is 15 billion years old, then light from one side that we observe has not had time to get to the other side. How then is it possible that the 3 K background radiation is so uniform?

the economic situation in the United States in the late 1970s, this new idea was called inflation.

Basically, inflation is the idea that due to a phenomenon that we will discuss in a moment, early on the universe experienced a period of explosive growth. At a time of about 10^{-34} seconds, the universe began to expand rapidly, doubling every 10^{-34} seconds. So, at 2×10^{-34} seconds, the universe had doubled in size. By 3×10^{-34} seconds, it had quadrupled. By 10^{-33} seconds, the universe had expanded by a factor of $2^{10} = 1024$ times. 10^{-33} seconds later, the universe had expanded by 2^{20} or about a factor of a million. Thus if inflation turns out to be true, in a tiny fraction of a second, the universe would expand enormously. While nobody knows for sure how long the inflationary period lasted, informed speculation suggests that perhaps the inflationary period might have persisted for several hundred doubling periods.

Let's be more specific. If the universe began initially at a single point and exploded as the Big Bang theory suggests, by 10^{-34} seconds, the universe was a whopping 6×10^{-26} meters in diameter, or approximately 3×10^{-11} the size of a proton (although this number is entirely speculative). At this time, all pieces of the universe were in good thermal contact with one another. By about 10^{-32} seconds after the Big Bang, the universe exits the inflation period after several hundred doubling periods. For illustration purposes, let's choose 200 doubling periods. During this time, the universe expanded by a factor of $2^{200} = 2 \times 10^{60}$, thus the size of the universe would be about 10^{35} meters, enormously larger than the size of the observable universe.

The basic idea of inflation is shown in Figure 9.6. Initially, the universe is small enough that all parts are in good thermal contact. Inflation causes these points to be flung far apart. Since inflation causes the universe to expand faster than light can travel within the universe, these once-connected bits of the universe are now separated

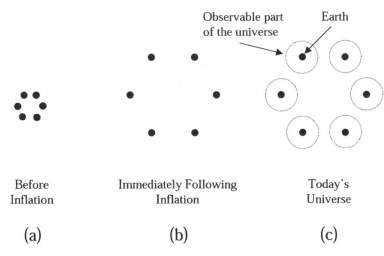

| Before Inflation | Immediately Following Inflation | Today's Universe |
| (a) | (b) | (c) |

Figure 9.6 A basic idea of how inflation might have occurred. (a) Early in the history of the universe, the points that make up the universe were in close proximity. (b) The universe then expanded at great speeds, separating points that were once in contact. (c) In today's universe, light from the respective points, once connected, has not yet traveled to the other spots. Thus while we can only see 15 billion light-years away (and thus 15 billion years into the past), the universe itself might very well be much larger.

by great distances, so great in fact that light, that fleetest of messengers, emitted by one point immediately after inflation, has yet to arrive at the other points. This idea shows that while we often say that the radius of the visible universe is 15 billion light-years, this is because that's the maximum distance we can see. Light from a star 20 billion light-years away will not arrive at our planet for another 5 billion years. In fact, this reminds us that the full universe could well be much larger than the paltry part we see.

Thus, we see how inflation can take a highly uniform and homogeneous universe of subatomic dimension and increase it to a much larger size, essentially instantaneously. Most importantly, the initial subatomic uniformity is maintained throughout the expansion. A fascinating speculation is that the slight non-uniformity seen in the 3 K radio background radiation might be the remnants of the quantum foam discussed in Chapter 8, expanded to cosmological size by the phenomenon of inflation.

While there have been other proposals to explain the uniformity of the universe, inflation explains another mystery. Earlier, we showed that measurements indicated that the overall geometry of the universe was flat (or equivalently $\Omega = 1$). We did not show why. Inflation provides an answer. Figure 9.7 shows that for example even a spherical universe, when expanded enormously, will appear flat.

One objection made by many people to the idea of inflation relates to something that physicists have been emphasizing in books like this, written for non-experts. One of the strongest and most

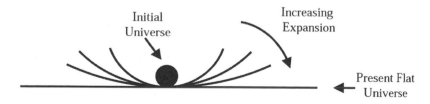

Figure 9.7 A demonstration of how the inflationary expansion of the universe might turn even a spherical universe into one that appears to be quite flat.

fundamental tenets of modern physics is "Nothing can go faster than light." Light would have taken at least 30 billion years to span the visible universe. Yet here I am claiming that the universe might have expanded from a subatomic size to much larger than the visible universe in a tiny fraction of a second. What gives? The answer is the following. Einstein's theory of relativity shows that nothing traveling within the universe can move faster than light traveling within the same universe. However, Einstein's theory does not place limits on the rate at which the universe itself can expand. So, it turns out that what would appear to have been a fatal flaw in inflation theory is actually not a problem at all.

Before we leave the topic of inflation, we should answer the question which I'm sure is bothering you. While it's all well and good to say that the universe suddenly expanded rapidly for just long enough to solve the uniformity problem and then the rapid expansion stopped, the fact is that this sounds a little bit like magic. Though it might be a bit much to expect that the inflation theory proponents have a detailed explanation as to the cause of the expansion, as the theory is relatively new and further progress requires experimental input, but they should at least be able to offer a plausibility argument.

In order for inflation to work, what is needed is a source of energy that becomes available and then disappears. This energy is what drives inflation. Luckily, we know of a physics mechanism that just fits the bill. It's called a phase transition. The most familiar phase transitions are when steam turns to water or when water turns to ice. For our discussion, let's concentrate on the second. As water cools, it loses energy and the temperature drops a corresponding amount. However, when water reaches 0°C, the rules change. Water must turn into ice. It takes a lot of energy to convert water from a solid to liquid and before ice can be formed, that energy must be released. Once the energy is released and the ice is frozen, the ice can also release energy with a corresponding drop in temperature, just as water did. But at the freezing point, lots of energy is released and the temperature is unchanged. A phase transition of this sort could have provided the energy that powered inflation.

It's all very well and good to talk about water and ice and somehow try to relate this to an almost unfathomable expansion of the universe, but in order for the idea to be truly plausible, one needs to think of phase transitions in particle physics. Luckily, there are several of varying levels of certainty, some indisputable, others strongly suspected, while others are purely theoretical. One familiar one (well moderately so) occurred 300,000 years after the Big Bang. Before that, protons, electrons and alpha particles (helium nuclei) roamed freely, buffeted by a constant barrage of photons. Once the temperature cooled enough so the electron energies were low enough that they could be captured by hydrogen and helium nuclei, atoms formed. Now that the electrons were bound closely into neutral atoms, suddenly the photons could move for long distances. Because photons interact with charged particles, when the electrons, protons and alpha particles made the transition from freely moving charged particles into neutral atoms, the photons could now travel freely. Essentially, the universe made the transition from being opaque to being transparent. Eventually, these photons became the radio background radiation discovered by Penzias and Wilson.

Another, more speculative phase transition is governed by the still undiscovered Higgs mechanism discussed in Chapter 5. Above a critical temperature (or equivalently energy), the force carrying bosons of the electroweak force were all massless. Once the universe cooled enough, the situation changed, leaving us with both the massless photon and the massive W and Z bosons and what appear to be two distinct forces. It is a phase transition like this that probably drove the inflation of the universe (if, indeed, inflation theory turns out to be true). The details of precisely which phase transition caused inflation are still unknown, although for purposes of discussion we will later make the plausible assumption that it was the breaking of the unity of the strong and the electroweak force that was the culprit. The best current thinking suggests that inflation was driven by the phase transition of a "scalar field," of which the Higgs field is the best understood. Supersymmetry provides many hypothetical scalar fields that could be the inflation-causing culprit.

We have now followed the path backwards in time, inspecting all of the major types of data available to observational cosmologists. From being able to view the universe at an age of one billion years, using the extraordinary instrument that is the Hubble Space Telescope, to the information of a much earlier age carried by the 3 K background radiation, one can understand an enormous stretch of time. From that time, 300,000 years after the Big Bang, when atoms were finally formed, we've pushed back further to the time where the atomic nuclei themselves coalesced a scant three minutes after the Big Bang. It is with well deserved pride that observational cosmologists can point back to the dawn of creation and say that they understand the physics of that early time and can use it to explain how we got where we are. The three minutes of mystery remaining is a tiny portion of the span of the lifetime of the universe. Nonetheless, when cosmologists mention their accomplishment to their particle physics brethren, the particle physicists can have only one response.

"That's cute."

Anticipating the crestfallen look we might see on our colleague's faces and not wanting to seem somehow crass and insensitive, particle physicists usually mull things over for a bit and choose their words a bit more carefully. They then follow with:

"No, no really … that's cute."

Realizing that this still sounds harsh, we explain. For all of the extraordinary successes of cosmological theory and observation, by the very late period of three minutes, the laws of physics had been cast in stone and the subsequent evolution of the universe is simply a manifestation of gravity (to pull matter together), the strong force (to hold the nuclei together), the electromagnetic force (to keep atoms and molecules together and to make chemistry work) and the weak force (to make the stars burn). Thus to fully understand the birth and evolution of the universe requires looking even further back in time to ever-increasing concentrations of energy. The questions we discussed in the last chapter: the question of why we have the forces we do; why are there three generations; and are the myriad phenomena we observe just different facets of some deeper truth, are all extremely relevant. To answer these questions, we need to better explore the microcosm to understand the cosmos. In a very real sense, modern particle physics explores the universe at time immediately following the Big Bang. Through the cousin disciplines of particle physics and cosmology, we have a real chance of someday understanding it all ... where do we come from and where are we going?

Back to the Beginning

As we discuss what particle physics can add to the debate, we take a different approach. Rather than starting now and looking ever further back in time, we instead start at the moment of the Big Bang and go forward in time. Since we don't know it all, we will necessarily begin with conjecture and by now, dear reader, you will begin to recognize the signposts as we travel. The language we have learned, of quarks and leptons, neutrinos and Higgs, supersymmetry and the fundamental forces, plays a crucial role in the beginnings of the universe.

The Big Bang theory really deals with the expansion of the universe from a smaller, hotter and unimaginably denser state to the extended form we see now. It doesn't really address the precise instant of creation, the moment when the universe passed from non-being into being. Nonetheless, with our vast knowledge of the nature of space and time, of quantum mechanics and general relativity, and of the manner in which matter behaves under hot and dense conditions, physicists can speculate as to the nature of the universe at the moment of creation. All of the matter and energy you see around you, everything in the vastness of space, was concentrated into a single point. Mind you, this was not just any sort of point; it was a quantum singularity. This had no size at all. Not only was all the matter and energy of the universe packed into a single point, space itself was packed into the same point. When the Big Bang occurred, the hot and dense matter did not expand into space the way hot gases from a firecracker do ... rather the expansion of the matter of the universe actually created space as it went (or maybe vice versa … it's a chicken or egg kind of thing). This all sounds some-what murky and perhaps it is, but one thing seems clear. At the moment of creation, there was no space and no time and all of the mat-ter of the universe was concentrated at a single point. You might ask yourself what were the conditions of the universe prior to the Big Bang. You might also ask yourself another question. If there was no space, then where *was* the quantum singularity? Was there some sort of "other" space of which we know nothing? My answer to these ques-tions is simple. I haven't a clue. There are physicists who ponder such questions, but they are generally regarded, as a Scottish colleague of mine has said, as being "dashing, but sketchy" (although you really need the brogue for effect). While it makes sense to ponder such ques-tions, there is sufficient mystery surrounding the conditions of the uni-verse in the tiniest moments following the Big Bang, that I believe that trying to quantify the nature of the pre-expansion universe to be essen-tially pointless. As our understanding of the early moments of the Big Bang improves, it will make an increasing amount of sense to ask the question of what happened before.

What triggered the event that caused the universe to expand is shrouded in mystery. Even some of my scientifically literate colleagues choose to invoke some form of God at this point. While they may be right, taking a somewhat more scientific tack, more physicists suggest that it was probably an effect of quantum mechanics that was the cul-prit. Surrounding the quantum singularity that was the universe was a quantum foam, with objects flickering into and out of existence in a mad frenzy. Quantum mechanics dictates only the probability that certain fluctuations of the foam will occur. Eventually, an extremely rare fluctuation will occur and perhaps such a fluctuation set in motion the course of events that led to the expanding universe in which we now live. Again, I regard this as a "how many angels can dance on the head of a pin" sort of thing (although I admit to lean-ing more towards the quantum mechanics answer). The fact is, that at such a concentration of energy, the laws of physics that govern the behavior of matter and energy may be quite alien from anything we can now imagine. I prefer to let the question wait until there is more experimental input. If you need a stronger statement, I recommend that you speak to your rabbi, priest, friendly neighborhood cosmologist or favorite bartender for additional input.

While the exact details of what happened before the Big Bang and even what provided the final trigger to initiate the cataclysmic explosion are not understood at all, once we have made a transition into the state where matter and energy (at unimaginable temperatures and densities, it's true) rule the universe, we can be more comfortable. We begin our story still guided by the speculation of Chapter 8, but as the temperature of the universe cools, we will begin to see phenomena about which we now know quite a bit. By the advanced age of a second or so, all of the interesting stuff will be over, with the universe left to evolve into the dizzyingly complex cosmos we now observe.

Our story of informed speculation begins, as you'd imagine, at the beginning. The size and the nature of the universe at the moment of creation are unknown. Possibly all of the matter of the universe was packed into a single mathematical point. If the string theory is correct, then perhaps the universe was simply very small, with the size of the string and some unknown interaction determining its size. We speculate that there was a single type of particle and a single force. If the string idea is correct, then all that existed was a single kind of string, vibrating in unimaginable ways.

At a time of 10^{-43} seconds, we have reached the Planck scale. The size of the universe could be any of a range of small numbers. I've seen estimates for the size of the universe from the Planck length to as large as $1/100$ of a centimeter. As this period of the universe is understood only through speculation, you can believe what you want within a range of reasonable values. The temperature of the universe was about 10^{32}°C with a density of 10^{90} kilograms in each cubic centimeter (note that lead has a density of 0.01 kilograms in each cubic centimeter). The pivotal thing that occurred at about 10^{-43} seconds is that the gravitational force somehow, by an unknown mechanism, became different (and weaker) from the other three forces. The split from the primordial uniformity to today's breathtaking diversity had begun.

The universe now existed in a way that was beginning to look mildly familiar. Particles and antiparticles existed (or strings vibrating in equivalent ways). As we don't understand why quarks and leptons are different, we don't really know when that separation occurred, but some theories suggest that it might have occurred in the time following the Planck time. Nonetheless, as best we know, the universe was able to expand and cool without any really dramatic things happening.

Even if the hypothesis of strings turns out to be correct, by 10^{-34} seconds the universe had expanded enough that this stringy nature was no longer evident, thus we revert to talking exclusively about particles. Our next pivotal moment occurred about this time. While at earlier times, the strong, electromagnetic and weak forces were all unified into a single force, at 10^{-34} seconds the strong force broke off into a separate force. Because the strong force is … well … strong, its disassociation was a bit more disruptive than gravity's earlier exodus from the unity of forces. Some (although by no means all) inflation proponents suggest that this phase transition (i.e. the moment when the laws of physics change) released the energy that drove the inflation of the universe. It is also here when the asymmetry between the matter and antimatter was solidified. For each billion antiparticles, there existed a billion and one particles. Both matter and antimatter particles still existed, just in ever so slightly different numbers.

With the energy made available from (perhaps!) the disassociation of the strong force, inflation commenced. The initial size of the universe is a matter of debate with estimates ranging from subatomic to the size of a basketball. Recall that inflation is supposed to double the size of the universe every 10^{-34} seconds. After a couple of hundred doubling periods, the universe had expanded to a size vastly larger than the visible universe. In the expansion, the density variation of the quantum foam would have been expanded to cosmologic proportions and would have eventually provided the seeds for the formation of galaxies (and also revealed itself to us in COBE and WMAP's extraordinary measurements).

From the end of the inflationary period at about 10^{-32} seconds to just below 10^{-10} seconds, there is no significant change. In fact, this domain of energy and temperature is sometimes called the desert. As we shall see in a moment, we may discover new phenomenon in this energy range very soon. But to the extent we now understand, the universe was expanding under the initial push of the Big Bang, with gravity trying to slow the process (although negligibly in such a small time). If supersymmetry is real, towards the end of this period the supersymmetric particles will no longer be created. The quarks and leptons existed, as did their antimatter analogs. The annihilation of matter and antimatter was ongoing and nearly complete at 10^{-10} seconds. The three generations of particles existed with equal probability, as the temperature (or equivalently energy) was above the threshold where the Higgs mechanism generates mass, thus all quarks and leptons were massless. Below this energy, our knowledge of the physics of the universe is pretty solid. The highest energy collision possible using the Fermilab Tevatron probes a time of about 4×10^{-12} seconds. (The Large Hadron Collider, set to commence operations in about 2007, could in principle probe back as far as 10^{-13} seconds.) But we should recall that protons and antiprotons are extended objects and the interesting collision occurs between the quarks and gluons carried inside, each carrying less energy than their parent proton or antiproton. Therefore, the bulk of Fermilab and CERN collisions probe the period of 10^{-10} seconds.

By 10^{-10} seconds, much had happened. Matter and antimatter particles had annihilated, leaving the one extra matter particle for every billion matter/antimatter pairs to go on to form everything. The universe had cooled below the Higgs transition energy, so the quarks and leptons had their respective masses. The particle generations were firmly in place. This milestone in the history of the universe marked the time when quarks had cooled enough to combine into baryons and mesons. While the mesons and most of the baryons would eventually decay, protons and neutrons would remain, although with too high an energy to combine to form elements. The universe consisted of protons, neutrons and neutrinos. Electrons and positrons still existed, not yet having annihilated to form the excess of electrons we now observe.

This all changed by the very late time of one second. Finally, the universe had cooled enough for electrons and positrons to annihilate. This created many residual photons (about one billion for each proton in the universe), which would eventually evolve to become the 3 K background radiation first observed by Penzias and Wilson. The density of the universe had dropped enough that neutrinos stopped interacting. In fact, it's generally true that most of the neutrinos created in the Big Bang last interacted a mere second after the beginning. Physicists ponder ways to measure these relic neutrinos, as they have remained essentially unchanged since the Big Bang. Unfortunately, one well understood phenomenon has affected them (i.e. the expansion of the universe). Like the 3 K background radiation in which the original, highly energetic photons of the early universe were lowered in energy into the radio waves observed by Penzias and Wilson, the expansion of the universe has lowered the energy of these neutrinos to a value that makes them difficult to detect. However, if somebody figures out a way to detect these neutrinos, we'll be able to view the universe a mere second after the Big Bang.

The age of the universe now advanced in human time scales. From one second until three minutes, the universe had cooled enough to allow protons and neutrons to bind together. When two protons and two neutrons bind together, they make the nucleus of a helium atom (also called an alpha particle). By one second, the history of the universe had passed out of the purview of particle physicists and moved on to that of nuclear physicists, who study the dynamics of protons and neutrons and how they combine. Physicists calculate that 76% of the baryons would be tied up in the protons of hydrogen nuclei and 24% in the protons and neutrons of helium nuclei. All the other elements made up just a tiny trace of the matter of the universe. Since protons and neutrons are made in equal quantities, the excess neutrons were left to decay (with a lifetime of 15 minutes) and form protons and eventually hydrogen. This is taken into account in the 76/24 ratio.

After three minutes, the universe had cooled enough so that no further nuclear fusion into helium and heavier elements would happen. The torch of understanding had passed to the atomic physicists. From the period of 3 minutes to about 300,000 years, the universe consisted of hydrogen nuclei (protons), helium nuclei (alpha particles) and electrons, all at a very high energy. The photons left over from the Big Bang would batter the particles, keeping them from combining. The neutrinos, as usual, passed through the universe oblivious to it all. By the late date of 300,000 years, the universe had been a hot plasma for all of its history. However at that epoch, an important thing happened. The universe had cooled enough so that the photons no longer had enough energy to knock an electron from a proton that had temporarily captured it into orbit. Try as the photons might, each time an electron passed near a hydrogen or helium nucleus, it would be captured, finally forming atomic helium or hydrogen. Since the atoms were now electrically neutral, they were invisible to photons. The photons finally stopped interacting and traveled unmolested across the cosmos to end up as an undistinguished, if remarkable, hiss in a radio antenna in Holmden, New Jersey.

After 300,000 years, astronomers take over. The hydrogen and helium gas that was the universe was free to travel, guided by gravity's implacable grip into huge clouds out of which condensed stars and galaxies, evolving finally into our beautiful cosmos and indeed us. This entire process is illustrated in Figure 9.8.

We've not spoken of the exotic particles that might make up the non-baryonic dark matter. If supersymmetry is the explanation, the lightest supersymmetric particles were probably formed until around 10^{-12} seconds after the Big Bang. This matter exists, dispersed throughout the cosmos, governed by gravity's pull. Since we don't know what the nature of the dark matter is, I just want to remind you that it is a mystery, not to be forgotten. I hope some young reader

Figure 9.8 An abbreviated history of the universe, emphasizing the first few moments after the Big Bang. The epochs mentioned in the text are shown.

will realize that there exists many such mysteries and be inspired to help solve them. Of course, they'll have to hustle, as I'd like to get there first!

As we close this chapter, we need to remember a few things. The universe is a marvelous place, the study of which has entranced generations of inquiring minds. How the cosmos came into existence remains an enduring question, but we are now beginning to unlock the mysteries that have puzzled those seekers of a bygone era. Astrophysicists now train sophisticated equipment to the heavens to fill in the later evolution of the universe. We now can do intricate experiments in particle physics laboratories that can reproduce the conditions of the early universe, a scant quadrillionth of a second after the beginning. Physicists are always thinking, trying to devise new techniques to get at even earlier conditions. One of the noblest endeavors ever undertaken by mankind is slowly succeeding. Perhaps some day we will understand it all.

In this book, we've discussed phenomena unimagined by our ancestors. While earlier cultures have often had various ideas on how the universe came into being, our own view is different, not only in detail, but also in principle. Our own view is governed by the observation of data. In addition, if our own cosmological ideas were a person, they would be nervous. Rather than being confident that this worldview is a revealed truth, never to be questioned, our anthropomorphized theory would realize that a single new observation (properly confirmed and cross-checked) could topple the entire edifice. A physics theory is like a baseball pitcher; it's only as good as its last observation. (Well, with pitchers it's their last game, but I trust that you'll pardon the mangling of the metaphor.) It is this constant vigilance that separates modern and earlier cosmologies. Scientists are actively trying to acquire new data to see how it fits into the picture. Theories that fail to describe the data are either revised or abandoned altogether. A modern example of our changing our most dearly held ideas includes the realization that the universe is not static and unchanging, but rather dynamic and growing (causing no less a great thinker than Albert Einstein to revise his theory). Another paradigm shift involved the experimental determination that space is flat, causing Alan Guth to add inflation to the traditional Big Bang cosmology. Each idea, we hope, brings us incrementally closer to the truth.

For all the discussion about the willingness of modern physicists to change their ideas, the fact is that the Big Bang theory has withstood an unparalleled effort to bring it down. Modern chemical and nuclear physics experimental data are incorporated, as well as the ever more precise astronomical measurements and the seemingly unrelated particle experiments towards which I've devoted my life. The Big Bang theory is assailed from all sides, not only by its fundamentalist detractors, but also even by its staunchest supporters. Yet, it survives. Like a Gibraltar in a sea of criticism, it stands.

This is not to say that all questions have been answered. The identity, indeed even the very existence, of dark matter is unresolved. The concept of inflation needs more study. The idea of superstrings or supersymmetry has not been established or refuted. The chronology of the universe before about 10^{-11} seconds is less solid than we'd like. Yet these all represent not flaws, but rather opportunities for study. If one of these efforts uncovers data that kills the Big Bang theory, so be it. Whatever would replace it would be incrementally closer to the truth, the discovery of which is every thinking person's goal.

I hope I've convinced you of the fundamental linkages between particle physics and cosmology. While I've emphasized the important role particle physics plays in understanding the early universe, it was cosmologists and astrophysicists who first realized the need for dark energy, dark matter, cosmic uniformity and inflation. Physicists of both kinds are needed to really get a handle on the ultimate questions that have puzzled curious minds for millennia. Only together will they reveal the truth.

c h a p t e r   1 0

Epilogue: Why Do We Do It?

Understanding is a lot like sex. It's got a practical purpose, but that's not why people do it normally.

— Frank Oppenheimer

Just why do we do it?

As we close this book, we should look back along the path we've traveled. Starting in the deep and misty past, our ancestors looked to the sky and pondered "Why?" In this, we have evolved little from those early seekers of the truth. In following our path, first started over 2500 years ago, we've explored the question, not only improving our understanding of the universe, but also the way in which we ask and answer questions. Our modern scientific method has proven to be the most powerful method thus far developed to attain the truth. In this, we *have* come a long way.

Yet much uncertainty remains. It is only through continued effort that we can continue to push back the frontiers of ignorance. This is not only true for the fields of particle physics and cosmology (although, as I said in the preface, I would argue they are the most interesting sciences), but all fields of scientific endeavor.

When I give public lectures, I occasionally come across a hostile heckler who remains unconvinced. He (somehow it's always a he) wants to use the opportunity to debate the merits of public funding of science research, which he regards as a colossal waste of money. While making the case for more research requires much more space than is available here, I'd like to sketch some main ideas.

I maintain that research is important, indeed necessary for the advancement of humankind. Efforts in medicine can improve health and extend life. Genetics studies can improve the yield of crops, as well as reduce the need for pesticides. There are many examples of directed research, in which the benefits are clear.

However, not all research has as obvious an outcome. Research into the physics of sparks seemed pretty fruitless at first. But as subsequent development through wireless telegraph, radio, television, cell phones and our modern connected world has shown, it was a worthwhile effort (although MTV and cell phones at concerts do somewhat belie the point). When physicists studied the electrical properties of semiconductors, nobody could foresee the first transistors and modern computers. When Alexander Fleming was studying moldy bread, the fact that the mold could kill bacteria was unknown. Yet with that information, we have been able to create antibiotics, thereby saving countless lives.

This is not to say that all research is successful. We've all seen old movies showing numerous outlandish attempts to fly, many of them comical and spectacular in their failure. Yet even with all of the failures, we fly. In a scant eight hours, we can travel from London to New York. And while the Wright brothers get all of the credit, those who failed were also important, as they found what didn't work.

Ecologically minded people decry the loss of biodiversity in our rain forests. While they are properly worried about the loss of species as a tragedy in its own right, they often use the fact that many of the extinct species could have provided new medicines to bolster their point. While not every new species will provide the substance that will cure cancer or AIDS, one of them might and to find it we have to look at them all. Failure is as crucial as success in scientific research, if somewhat less satisfying.

In my own field of particle and nuclear physics, we have also contributed to the greater good. The splitting of the atom has provided significant benefits to mankind. While the utopian vision of the 1950s of free electricity has not been realized, as fossil fuels are depleted nuclear power will inevitably provide a greater fraction of the energy budget. The problems with nuclear power are more psychological than technical. Those early experiments in which the uranium atom was split will keep you and your grandchildren warm. And this doesn't even mention the people saved by radiation treatments.

Particle physics has had at least two unintended spin-offs that could not have been anticipated. Superconductivity had been discovered by Heike Kamerlingh-Onnes in the early 1900s, and the opportunity to use it to make strong magnetic fields was clear. Relatively small regions with intense magnetic fields were available early on. However, when Fermilab decided to add a four-mile long ring of superconducting magnets in order to improve our research, such an effort was unprecedented. Nonetheless, the research commenced and, in 1987, the ring began data-taking operations. While Fermilab had a focused purpose, the technology then developed could be applied in several ways. Engineers with a medical interest were able to recraft the technology and make the large Magnetic Resonance Imaging (MRI) magnets so prevalent in today's hospitals.

Another enormously successful spin-off from particle physics research stems not so much from the characterization of the behavior of matter at high temperatures and densities, but rather through the need to efficiently communicate information. Modern particle physics experiments include five hundred or more physicists, spread across the world. Scientific collaboration is just that … collaboration. Physicists live by communication, bouncing ideas off one another, shooting down the bad ones and keeping the good. Because they do not live in a single geographic location, a method for inexpensive worldwide communication was needed.

This method needed to be able to exchange charts and graphs, text and large data files. Scientists at CERN realized the need and had enough technical acumen and funding to be able to solve the problem. In the end, their solution has grown into the World Wide Web. This is an example of a completely unexpected benefit. Next time you type www.anything.blah, remember that you are using a highly developed spin-off of particle physics research. Incidentally, the money made from the Web has more than paid for every particle physics experiment ever performed.

Not all publicly funded research reaps such a fruitful bounty. Some research efforts fail outright, while others succeed with no significant technical returns. Only a few hit the jackpot. A few, like Columbus' exploration, find something very useful, just not what they expected. The most conservative estimate I've seen suggests that the ratio of economic return on publicly funded research to dollars invested as ten to one. We in the technologically oriented world have a duty to pursue knowledge for the betterment of mankind.

But for all of research's undisputed benefits, the above discussion entirely misses the point. We do research for the same reason that we write beautiful poetry, create great art, build enormous monuments and put men on the Moon. We do this because we are human. It is our nature to explore, to create, to discover. We do these things, not because we can, but because we must.

Bob Wilson, Fermilab's first director was once called to testify before Congress to support additional funding to build a new accelerator. As is proper, he was being asked to justify the expenditure of public funds. During an interchange with Senator John Pastore, Bob was asked, "Is there anything connected with the hopes of this accelerator that in any way involves the security of this country?" Bob, being an honest guy, replied that he couldn't think of any. When Senator Pastore pressed him further, trying to clarify the answer by asking "It has no value at all in that respect?" Bob answered in a way that showed that he truly understood what curiosity-based research was really all about. Bob's classic response was

> It has only to do with the respect with which we regard one another, the dignity of men, our love of culture. It has to do with, are we good painters, good sculptors, great poets? I mean all the things we really venerate and honor in our country and are patriotic about. It has nothing to do directly with defending our country except to make it worth defending.

Bob always was one with a quick wit and an eloquent tongue.

The study of particle physics and cosmology embody one of the noblest efforts of mankind (well, OK, working for world peace is up there too). No other scientific endeavor, including the origin of life, tackles such grand questions. What is the nature of space and time? Where have we come from and where is it that we go? Why is it that we can exist at all? These questions vex the mind and border on the spiritual. The beauty of science is that these questions can not only be pondered, but also answered.

Henri Poincaré once said

> The scientist does not study nature because it is useful; he studies it because he delights in it, and he delights in it because it is beautiful. If nature were not beautiful, it would not be worth knowing, and if nature were not worth knowing, life would not be worth living.

Poincaré and Wilson express themselves more eloquently than I am able. My meager attempt to add to their inspiring words will be to leave you with the same words with which we began this book.

I hope that you have had as much fun reading this book as I had writing it. Science is a passion. Indulge it. Always study. Always learn. Always question. To do otherwise is to die a little inside.

But in the meantime, there's a great deal of work to do. If you'll excuse me, I have to get back to the lab. It's far too much fun to be away for long....

appendix A

Greek Symbols

Table A.1 Pronunciations of Greek letters.

Upper Case	Lower Case	Name	Pronunciation
A	α	alpha	al-fuh
B	β	beta	bay-tuh
Γ	γ	gamma	gam-uh
Δ	δ	delta	del-tuh
E	ε	epsilon	ep-si-lon
Z	ζ	zeta	zay-tuh
H	η	eta	ay-tuh
Θ	θ	theta	thay-tuh
I	ι	iota	eye-oh-tuh
K	κ	kappa	kap-uh
Λ	λ	lambda	lam-duh
M	μ	mu	myoo
N	ν	nu	noo
Ξ	ξ	xi	ks-ee
O	o	omicron	om-eye-kron
Π	π	pi	pie
P	ρ	rho	roe
Σ	σ	sigma	sig-muh
T	τ	tau	tau
Υ	υ	upsilon	oop-si-lon
Φ	ϕ	phi	f-eye
X	χ	chi	k-eye
Ψ	ψ	psi	sigh
Ω	ω	omega	oh-may-guh

a p p e n d i x   B

Scientific Jargon

Particle physics has its own language, as do most scientific fields. However, there are a few common topics, not specific to a particular field of study. One of them is scientific notation, which is a compact way to express numbers that are far from unity. In this book, we discussed the size of the observable universe, as well as the size of the proton. Given such a huge disparity in sizes, as well as our own typical size here on Earth, it is clear that we need to be able to express these completely different sizes in a succinct way. We do this via scientific notation. Basically, scientific notation is a quick way to "compact" the zeros. For instance, one can write a million as 1,000,000. But we can also see that another way one can get a million is by multiplying the number 10 together precisely six times ($10 \times 10 \times 10 \times 10 \times 10 \times 10$), which we can write as ten to the sixth power or 10^6. A billion (or thousand million, to my British colleagues) can be written as 10^9. The size of the observable universe, expressed in meters can be written as 10^{24} as opposed to the more cumbersome 1,000,000,000,000,000,000,000,000. A number that doesn't start with the digit "1" can also be expressed in scientific notation. For instance, 3,200 can be written as $3.2 \times 1,000$ or 3.2×10^3.

Small numbers can be written in a similar way. A number like 0.0001 can be written as $1 \div 10 \div 10 \div 10 \div 10$ or $1 \div 10^4$. Rather than having to carry around the "$\div$" sign, we can write 0.0001 instead as 10^{-4}. Similarly, a number like 0.045 can be written as 4.5×10^{-2}. While scientific notation can require more writing for numbers near unity, for very large or very small numbers, scientific notation is more efficient.

Scientific notation can be combined with the metric system to provide a very powerful way to communicate numbers. Rather than carrying around all the powers of ten, we give every factor of thousand a particular name (and for numbers near unity, every factor of 10 gets its own name). For instance, if one wanted to say that something had a length of one one-hundredth of a meter, they could say it was one centimeter long, as "centi" means "one one-hundredth." A kilometer is 1,000 meters, as "kilo" means "1,000." In particle physics, the most used units are meter (for length), second (for time) and electron volt (eV) (for energy). In order to combine the metric system numbering scheme with any unit, one puts a prefix, which denotes how many of something we have, with a base unit (like meter). An example of this is kilo-meter, or kilometer. In order to make a shorthand way of writing, we substitute "m" for meter, "s" for second and "eV" for electron volt. Table B.1 gives the list of usual prefixes, but for kilo, the prefix is "k." Thus a kilometer can be written as 1 km. This table shows how one can write a vast range of energies in a very compact way. As way of example, I use energy. But the discussion is valid also for meters and seconds as well.

Note in particle physics, energy usually exceeds kilo-electron volts (or keV). (An electron volt is defined to be the energy an electron would gain, having been accelerated by a one-volt battery.) However, sizes can be quite small, as the approximate 10^{-15} meter size of the proton is called one femtometer (or 1 fermi, in honor of Enrico Fermi).

Table B.1 Important methods to write big numbers. While the large numbers at the left are acceptable, the scientific notation is more compact. Even more compact is the use of prefixes to denote a particular large value. Remember that for electron volts, one pronounces all three letters (e.g. K-E-V for keV). It is permissible in the case of GeV to pronounce it as a single word "jev" and also a TeV can be pronounced "tev". Other single-word pronunciations are rare.

Voltage (Volts)	Scientific Notation	Word	Prefix	Symbol	Energy
0.000000000000000001	10^{-18}	Quintillionth	atto	a	1 aeV
0.000000000000001	10^{-15}	Quadrillionth	femto	f	1 feV
0.000000000001	10^{-12}	Trillionth	pico	p	1 peV
0.000000001	10^{-9}	Billionth	nano	n	1 neV
0.000001	10^{-6}	Millionth	micro	μ	1 μeV
0.001	10^{-3}	Thousandth	milli	m	1 meV
0.01	10^{-2}	Hundredth	centi	c	1 ceV
0.1	10^{-1}	Tenth	deci	d	1 deV
1	10^{0}	One	—	—	1 eV
10	10^{1}	Ten	deka	da	1 DeV
100	10^{2}	Hundred	hecto	h	1 heV
1,000	10^{3}	Thousand	kilo	k	1 keV
1,000,000	10^{6}	Million	Mega	M	1 MeV
1,000,000,000	10^{9}	Billion	Giga	G	1 GeV
1,000,000,000,000	10^{12}	Trillion	Tera	T	1 TeV
1,000,000,000,000,000	10^{15}	Quadrillion	Peta	P	1 PeV

a p p e n d i x C

❖

Particle-Naming Rules

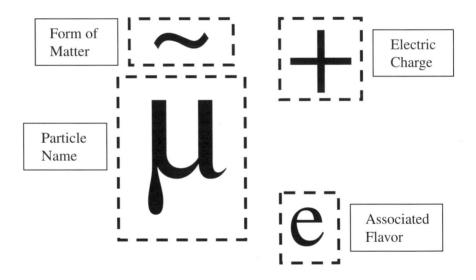

Not all particles will use all of the four fields shown above. For instance, the specific example given above is nonsense. This is because it is actually unusual for any particular particle to use all of the fields. Representative particle names will be given at the end of the Appendix.

Particle Name

Particle names are much like a language. They follow a certain set of rules, but not strictly. Typically particle names are either Roman or Greek symbols. In the example above, the Greek letter μ (mu) is used. This denotes a muon.

Electric Charge

Particles can have positive, negative or neutral electric charge. Positive particles are denoted "+," negative ones denoted "−," while neutral particles are written "0." The way one says something indicating electrical charge is "plus," "minus" and either "zero" or "naught." Often, if a particle is electrically neutral, the "0" is omitted entirely. It is possible to have double the electric charge and one would indicate this via a "++," which is pronounced "double plus."

If one simply writes "+" or "−," this implies an electric charge equivalent to that carried by a proton or electron respectively. The quarks carry fractional electric charge, thus this must be explicitly stated, for instance the up quark would be written "+2/3" and the down quark would be written "−1/3." A μ^+ is pronounced "mu plus," a π^- is called a "pi minus," a π^0 is a "pi zero or pi naught" and a Δ^{++} is a "delta double plus."

Form of Matter

There are three forms of matter that one can indicate in this way. If there is no symbol in that space, this indicates ordinary matter. An overline indicates antimatter and is pronounced "bar," for instance an antimatter charm quark is denoted $\bar{c}$ and pronounced "c bar." A tilde (~) is used to indicate a supersymmetric particle. Thus a $\tilde{W}$ is the supersymmetric analog of the W boson. However, one makes the names of supersymmetric analog of bosons by removing some letters at the end of the name of matter particle bosons and adding the phrase "ino" (i.e. graviton→ gravitino, photon→ photino, W → wino, Z → zino, gluon → gluino, and higgs → higgsino). For the supersymmetric analog of fermions, one simply adds an "s" to the beginning of the name (e.g. electron → selectron, muon → smuon, quark → squark, etc.)

There is a special case for the forms of matter. For the charged leptons (electrons, muons and tau leptons), one usually does not indicate the antimatter version via an overline. While an overline is acceptable in this case and physicists would understand you, it is traditional to indicate the matter and antimatter nature of charged leptons only via their electric charge. Matter charged leptons are negatively charged, while antimatter ones are positive.

Associated Flavor

Associated flavor is used mostly for the neutrinos, but not exclusively. For neutrinos, one usually denotes which flavor neutrino it is by putting the symbol of the associated charged lepton (electron, muon, tau lepton). Thus, an electron neutrino is ν_e, a muon neutrino is denoted ν_μ and a tau neutrino is ν_τ.

Associated flavor is also used for mesons and baryons to indicate an unusual quark content. The rules here are a bit arcane. For instance a meson containing a bottom quark and a down antiquark is called a "B meson." However, the electrically identical meson which contains a bottom quark and a strange antiquark is denoted B_s and pronounced "B sub S." In this case, the bottom quark nature of the meson is reflected in the particle name, while the down or strange quark content is listed in the associated flavor column. This use of associated flavor is better left for experts.

Fundamental Particles

Table C.1 Known and suspected fundamental particles, including normal matter, antimatter and supersymmetric particles.

Charged Leptons					
Matter		**Antimatter**		**Supersymmetric**	
e^-	electron	e^+	positron	$\tilde{e}$	selectron
μ^-	muon	μ^+	antimuon	$\tilde{\mu}$	smuon
τ^-	tau	τ^+	antitau	$\tilde{\tau}$	stau

Neutrinos					
Matter		**Antimatter**		**Supersymmetric**	
ν_e	electron neutrino	$\bar{\nu}_e$	electron antineutrino	$\tilde{\nu}_e$	electron sneutrino
ν_μ	muon neutrino	$\bar{\nu}_\mu$	muon antineutrino	$\tilde{\nu}_\mu$	muon sneutrino
ν_τ	tau neutrino	$\bar{\nu}_\tau$	tau antineutrino	$\tilde{\nu}_\tau$	tau sneutrino

Quarks					
Matter		**Antimatter**		**Supersymmetric**	
u	up	$\bar{u}$	anti-up, u-bar	$\tilde{u}$	sup
d	down	$\bar{d}$	anti-down, d-bar	$\tilde{d}$	sdown
c	charm	$\bar{c}$	anti-charm, c-bar	$\tilde{c}$	scharm
s	strange	$\bar{s}$	anti-strange, s-bar	$\tilde{s}$	sstrange
t	top	$\bar{t}$	anti-top, t-bar	$\tilde{t}$	stop
b	bottom	$\bar{b}$	anti-bottom, b-bar	$\tilde{b}$	sbottom

Gauge Bosons			
Matter		**Supersymmetric**	
γ	photon	$\tilde{\gamma}$	photino
g	gluon	$\tilde{g}$	gluino
W	W boson	$\tilde{W}$	wino
Z	Z boson	$\tilde{Z}$	zino
G	graviton	$\tilde{G}$	gravitino
h	higgs	$\tilde{h}$	higgsino

Non-Fundamental Particles (Baryons and Mesons)

The baryons and mesons are not fundamental particles, thus they are somewhat less interesting than the fundamental particles listed on the previous page. Nonetheless, there are a few which recur frequently. You should recall from our discussion of Chapters 2 and 3 that there are literally hundreds of possible baryons and mesons. We only list a few common ones here.

Table C.2 A handful of typical baryons and mesons. This is not an exhaustive list.

Baryons		Mesons	
p	proton	π	pion
n	neutron	K	kaon
Δ	delta	ρ	rho meson
Λ	lambda	J/ψ	"J-Sigh"
Ξ	cascade	Υ	upsilon
Ω	omega	B	B meson

appendix D

❖

Essential Relativity and Quantum Mechanics

Once you strip away the mathematical veneer and the need to calculate physical quantities, physics is fundamentally an easy science. In colleges and universities around the country, classes with names like "Conceptual Physics" (but more commonly known as "Physics for Poets") are continuously taught. In these classes, the ideas of physics are taught, devoid of the mathematics that mystifies an unsettlingly large fraction of today's college students. Newton's laws, the nature of the electric force, how friction works, how planes fly and why boats float are all explained using the relatively clumsy language of words, rather than the more concise and elegant language of mathematics. Nonetheless, even presented in an unnatural language, Physics for Poets classes are successful. The students are able to understand the ideas that govern the world, even if they don't know how to calculate precise physical quantities.

Conceptual Physics is relatively easy until one reaches quantum mechanics and Einstein's theory of relativity. These two fields are usually grouped together and called "Modern Physics." (This always bugged me, as Einstein's two theories of relativity were published in 1905 and 1916 and quantum mechanic's heyday was the 1920s ... hardly modern. Modern Physics should properly be reserved for the subjects discussed in this book, but some things are just too entrenched to change.) Both relativity and quantum mechanics are more difficult to teach, as they describe phenomena not only which most people never observe, but they tell us that the world acts in ways that are in direct conflict with our common experiences. Relativity states that how fast your clock ticks depends on how fast you're going. It also says that two runners, running the same race at the same time over the same path, will say that they ran a different length, depending on how fast they ran.

Similarly, quantum mechanics states that one cannot predict the outcome of any specific atomic experiment and further that you can't know the location of any subatomic particle. Even if you figured out exactly where an electron was, you couldn't figure out how fast it was going. Such mind-bending counterintuitive behavior has captured the imagination of generations of students (including mine).

With such seemingly bizarre behavior, you might expect that it would take a fair bit of effort to understand the two ideas and you'd be right. We don't have the time nor space to cover these ideas in detail here. Thus, what we do instead is completely ignore the question of *why* and simply tell *what* sorts of behaviors are relevant to understand particle physics. Even then, we will be selective in our choices.

In order to understand relativity, you need to know what it means to talk about a particular reference frame. Let's leave particle physics aside and talk about baseball and bugs. Let's say a major league pitcher has thrown a ball and it is traveling at 90 miles per hour. What does this mean? It means that a bug, sitting on home plate and holding a radar gun (stay with me here) will measure 90 mph. On the other hand, another bug, sitting on the baseball, pointing a similar radar gun at the baseball, will measure zero miles per hour, as the bug is moving as fast as the baseball. Thus, we see that the speed of the baseball is not uniquely determined. Depending on how fast each bug is moving, he'll get a different answer. This is where the word "relativity" in Einstein's theory of relativity comes from; the speed of the baseball as measured by each bug is always relative to the bug's motion.

We could generalize the discussion to other bugs, including ones traveling in cars or fighter jets and each would measure a different speed. How can we say anything definitive about the speed of the baseball, if what we measure is relative to what we're doing? Simple... we just choose a particular "special" situation (or reference frame), the one in which we are traveling at exactly the same speed as the baseball. If we do, we measure that the baseball is not moving (i.e. at rest). We call this the rest frame.

One thing that occurs in relativity that is entirely counter to everyday experience is the fact that in addition to velocity measurements being dependent on the details of how you are moving, so do time measurements. Two people, observing the same event (like watching a ball drop), will say the fall took a different amount of time. This is true if (and this is the crucial part) they are moving at different speeds. The bottom line is that a person's perception of time depends on how fast they are going. Why this should be true requires some explanation, but for the purposes of this book, we simply take it as an experimentally-verified given.

For elementary particles, there is only one time that really matters and this is the particle's lifetime. We have quoted lifetimes for various particles (e.g. the lifetime of a pion is typically 2.6×10^{-8} seconds). However, if perceived time depends on how fast something is moving, it stands to reason that different people, moving at different speeds, would measure a different amount of time between when the particle was created and when it decayed. So, what do the numbers quoted throughout the text mean? The numbers quoted are the so-called lifetime in the rest frame. This is the shortest time one can measure. All other people who have a non-zero velocity with respect to the particle will measure a longer lifetime.

This effect has practical consequences. For instance, while a pion thinks that its lifetime is typically 2.6×10^{-8} seconds (26 ns), even if it travels at the speed of light, it will only travel 8 meters (25 feet) before decaying. At large accelerators where pion beams are desired, one realizes that the pion thinks that the lab is racing past it. Therefore, a person on the ground will measure a longer lifetime than the canonical 26 ns. If a particle lives longer, it can travel further and this greatly reduces the technical challenges of having to pack everything into a short space. For the daring, one can calculate just how much longer the particle will live due to this effect. Einstein defined a quantity γ (which, confusingly enough, has nothing to do with γ radiation) which is the ratio of how long a Fermilab guy thinks a pion lives, compared to what the pion itself thinks (A pion thinks? You know what I mean...) γ is defined to be the energy of the particle beam, divided by the rest mass (more on this later) of the particle. The rest mass of a pion is 0.140 GeV, so a beam of 140 GeV (entirely reasonable) pions will live 140/0.140 = 1000 times longer. In the case of such a pion, it can travel about five miles, which is very big compared to the size of a typical particle physics laboratory.

This effect also is exploited in DØ and CDF, as well as nearly all other collider experiments. One thing that both groups want to identify with good efficiency is particles containing a bottom quark. Since both top quarks and Higgs bosons decay into bottom quarks, finding bottom quarks is a prerequisite for identifying these interesting collisions. A hadron carrying a bottom quark will live about 1.5×10^{-12} seconds. Traveling at the speed of light, such a particle will travel an average of a mere 0.5 millimeters before decaying. Given that a hadron carrying a bottom quark will decay into a mere handful of particles, while the collision itself generates typically on the order of a hundred, it's hard to identify the relevant few from the many. However, because of the fact that the interesting hadron carrying a bottom quark can have an energy of 10–40 GeV and the hadron has a mass of about 5 GeV, the effect we are discussing here buys us a factor of 2–8. Thus we experimenters in the laboratory see the bottom-quark-carrying hadron travels typically 0.9–3.6 millimeters, just enough to do the job. Even so, we need to build the highly complex silicon vertex detectors mentioned in Chapter 6.

Another useful aspect of relativity is $E = mc^2$ which, I'm afraid I have to tell you, is wrong. Actually, it's more correct to say that it's a special case. In words, the equation says that energy equals mass (times a multiplicative factor) and vice versa. This is true. However, there are more types of energy than simply mass energy. As discussed in the text, there is also motion energy. A form of "motion energy" (in loose terms) familiar to those who have taken high school physics is momentum. (Yes, yes fellow physicists. I know that momentum is not energy. Shush...) Since energy must include both mass and motion forms of energy, the equation must reflect this fact. Thus Einstein's *real* equation is

$$E^2 = [mc^2]^2 + [pc]^2$$

Where E is the total energy, p is the momentum of the particle, c is the speed of light and m is the rest mass of the particle. We see that in the case of no momentum (i.e. $p = 0$), our familiar $E = mc^2$ returns.

Particle physicists are lazy (or clever, or perhaps they are the same thing). In order to make our calculations easier, we carefully choose the units in which we do our calculations. (This is entirely equivalent to choosing ounces, pounds or tons to calculate a weight.) Thus, we choose to express all velocities as a fraction of the speed of light. Therefore, by definition the speed of light (c) must be 1. This simplifies the above equation considerably, as it then reduces to $E^2 = m^2 + p^2$.

Now comes the fun part. Like time, both energy and momentum are affected by the speed of the observer. However, m (the mass) is not. This is a specific and unchanging number. You probably have heard or read that the mass of a particle varies with velocity. THIS IS NOT TRUE. Such an assertion comes mostly from a pedagogical approach in how the idea of relativity is introduced. When professors introduce relativity to their classes, they want to maintain as great a contact with earlier Newtonian physics as possible. Relativity is mind-blowing enough that all extraneous additional sources of confusion are assiduously avoided. It turns out that one can (incorrectly) invent a new term called "relativistic mass," which changes with the observer's relative velocity. The benefit of this approach is that you can keep some of Newton's equations, as long as everywhere there used to be mass, one replaces it with the new relativistic mass. But that's just for the student's sake. Actually Einstein's equations are both correct and different in detail from Newton's. It is very important to realize that the mass of a particle does not change with velocity.

Many people with whom I have spoken seem to have a resistance to the idea that mass doesn't change with velocity. Indeed, the greatest resistance comes from those laymen with a most sophisticated appreciation of modern physics. Because of this, let's divert for just a moment to offer an analogous (at least in potential for confusion) question, for which the answer changes as your appreciation of the question increases: What is the value of 0/0?

When you first encounter this question, you are very young ... say second or third grade. You are just learning your simple division facts. Thus, the teacher might tell you that $0/0 = 0$. This statement, while wrong, allows the teacher to gloss over the point and concentrate on the task at hand, which is understanding the basic concept of division.

A few years later, when the concept of division is no longer new and mastery of its more subtle points is desired, your teacher probably told you that $0/0 = 1$, as did $1/1$, $2/2$, $3/3$ and so on. Again, the question of the right answer to 0/0 is subsumed by the greater lesson.

Soon after you learn $0/0 = 1$, you are told that this is wrong, as it is impossible to divide anything by zero. Thus 0/0 is impossible. If you had a more sophisticated algebra teacher, they might have told you that $0/0 = x$ can be written as $0 = 0 \cdot x$. Since anything times zero is zero, this means that x can be any number and thus "x is undefined." This is clearly a long ways from $0/0 = 1$. Finally, as you enter calculus with its limits, you find out that any particular instance of 0/0 can have a correct answer that is determined by precisely how the problem is posed.

So we see that at each step of the way, the answer to the question "What is the value of 0/0?" changes. This is because for a young child trying to master the idea of division, all of the other ideas (although closer to the truth) are a distraction.

So it is with relativity and the concept of a variable mass. For the person being introduced to relativity, the first idea to master is the fact that there is an ultimate speed, above which one cannot go faster. Phrasing this idea in terms of a velocity dependent mass aids the student with this counterintuitive idea. Once this idea seems natural (or perhaps only somewhat unnatural), then the fact that it is inertia and not mass that really increases is introduced. Because mass and inertia are the same at low velocities, the earlier approximation (variable mass) seems natural. But you, gentle reader, are now a sophisticated student of science. It's time to face the facts…mass *does not* increase as velocity increases. For those who persist in using the relativistic mass idea, they would say that the rest mass does not change with velocity.

Given this crucial fact, we can proceed with an important question. How do you measure the mass of a particle that lives for such a short amount of time that you never see it, but instead only its decay products? This is really slick. You need two things. The first is Einstein's equation relating energy, momentum and mass and the second is the recollection that energy and momentum are conserved. Let's illustrate using an example. Suppose that you have a particle that decays into two daughters (e.g. a Higgs boson decaying into a bottom quark and a bottom antiquark, $H \rightarrow b\bar{b}$). To keep things general, let's call the parent particle "A" and the two daughters "1" and "2." The essential points of the decay are illustrated in Figure D.1.

Before the decay, there is only particle A, so we have its energy E_A, its momentum p_A and its mass M_A. After the decay, we have only particles 1 and 2, so we can write their energy as $E_1 + E_2$, their momentum as $p_1 + p_2$ and their mass as m_1 and m_2 respectively. (Note we don't sum the masses, as that's not interesting.) Since energy and

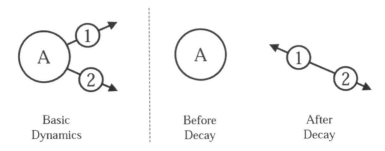

Figure D.1 The simple idea of a single particle decaying into two daughter particles. Thus, particle A disappears and reappears as particles 1 and 2.

momentum are conserved (which means that they are the same before and after the decay), one can write $E_A = E_1 + E_2$ and $p_A = p_1 + p_2$. Now we're ready to go. Before the decay, we have

$$\underbrace{M^2_A}_{\text{velocity independent}} = \underbrace{\underbrace{E^2_A}_{\text{velocity dependent}} - \underbrace{p^2_A}_{\text{velocity dependent}}}_{\text{velocity independent}}$$

This is the first cool thing. While E and p both depend on the speed of the observer, how they depend compensates, so the mass is unchanged. Now we can take the next step, which is to bridge the information after the decay with the information from before the decay. Substitute in for E_A and p_A

$$\underbrace{M^2_A}_{\text{Parent mass only}} = \underbrace{(E_1 + E_2)^2 - (p_1 + p_2)^2}_{\text{daughter variables only}}$$

Thus, we see if we carefully measure the energy and momentum of the decay products, we can get the right mass of the parent, every time, even though both the energy and momentum of both the parents and the daughters vary with velocity. The mass never does.

While there is much more to learn about relativity, it is only these two phenomena that are mentioned in the text and so we move on to quantum mechanics. Quantum mechanics is even more bizarre than relativity. Niels Bohr said "Anyone who is not shocked by quantum mechanics simply doesn't understand it." While mastery of quantum mechanics takes considerable study, we discuss only two aspects here, without too much worry about *why*. Again, we concentrate on *what*.

Probably the most important thing about quantum mechanics is the fact that even practicing physicists don't really understand it. Engineered in the 1920s, the basic formulation was constructed by several physicists, with Erwin Schroedinger providing the central equation. Paul Dirac added special relativity (and incidentally predicted antimatter) and Richard Feynman, Sin-Itiro Tomonaga and Julian Schwinger fixed up some loose ends. Quantum mechanics had evolved into modern quantum field theory, yet one thing remained unchanged. None of the equations could predict the outcome of any individual experiment. While that does sound like a fatal flaw in a theory, the reality isn't quite as bad. It's true that if I give you two particles and tell you absolutely everything that can be known about them, even modern quantum field theory cannot tell you in detail how a particular collision will evolve. Quantum field theory predicts only probabilities. It will predict only what is the relative likelihood of the various collision possibilities. It does *not* tell you what will occur in any particular collision. Thus, the only way one can verify any prediction from modern theories is to take many measurements (i.e. many collisions) and view what sorts of behaviors occur. From the relative frequencies of occurrences, one constructs measured probabilities and compares those measurements to the theoretical predictions.

While the idea that our beautiful theories fail to predict the behavior of particles in a single particular collision is unsettling, dealing only with probabilities isn't so bad, as long as we can take many measurements. It is another aspect of quantum mechanics that provides for a much livelier time than one would expect from our large-scale intuition.

The law of conservation of energy is one of the most fundamental tenets of physics. In a system where energy is not allowed to be added or removed, the energy of the system is unchanged (or conserved to use physicist's lingo). However, at the quantum mechanical level, one finds a tiny loophole in this core principle of physics. This loophole is provided by Heisenberg's Uncertainty Principle. Werner Heisenberg's principle is elegantly expressed in mathematics as

$$\Delta E \, \Delta t \geqslant \hbar/2$$

where ΔE is the amount of energy that isn't conserved, Δt is the length of time for which energy isn't conserved and $\hbar/2$ is just a small number, specifically 3.3×10^{-22} MeV•s. Since this is a pretty counterintuitive idea, let's talk about it using the clumsy language that is English and intersperse our discussion with more natural examples. Basically, what the uncertainty principle says is that it is possible for energy to not be conserved, as long as the non-conservation doesn't persist for too long. Heisenberg's equation has the basic form $(xy = 1)$, which can be written as $(y = 1/x)$. As x increases, y decreases. Energy non-conservation is a big deal, but in certain contexts it is a *really* big deal.

Consider a bit of empty space, containing no energy. With no energy, nothing can change, as energy is the catalyst of change. However, Heisenberg's equation suggests that energy can fluctuate for a short while. If so, perhaps there might be enough energy to create a particle; $E = mc^2$ after all. In order to balance the energy books, the particle will have to disappear quickly so that the energy is released and that particular spot will return to zero average energy. We call these ephemeral particles which temporarily violate the law of conservation of energy "virtual particles." All of the usual rules must be maintained, so a particle can only be created in tandem with its associated antiparticle. So, we can calculate how long pairs of virtual particles are allowed to persist.

We use for example the light electron and the heavy top quark, illustrated in Table D.1. To provide a sense of scale, we remind

Table D.1 Time and distance a virtual electron/positron pair and top/antitop quark pair can exist from Heisenberg's Uncertainty Principle. The heavier top quarks can exist for a much shorter time and in a much smaller space than the lighter electrons.

Particle	Mass (m)	ΔE needed ($2 \times m$)	Time in existence (seconds)	Distance traveled (fm)
Electron	0.511 MeV	1.022 MeV	3.3×10^{-22}	100
Top quark	175 GeV	350 GeV	9.4×10^{-28}	0.0003

ourselves that 1 fm (or 1 femtometer or 1 fermi or 10^{-15} meters, all the same thing) is about the size of a proton. In contrast, a small atom is about 100,000 times larger in size. Thus, we see that the characteristic distance in which virtual electron/positron pairs can persist is small compared to an atom, although fairly large compared to a proton. In contrast, because of their much greater mass, top/antitop quark pairs can exist only for an even shorter time and therefore they must be situated much more closely together (about 1/3500 times the size of a proton).

The uncertainty principle helps us to understand the quantum foam idea discussed in Chapters 8 and 9. At the very small size scales, empty space isn't so empty. Virtual pairs of particles are created, persist for a short time and then annihilate each other in order to balance the energy books. The very face of space itself is constantly changing, much like a foam (and hence the name) in which bubbles are constantly coming into existence and then popping into oblivion. Further, as one looks at space with an ever more powerful microscope (i.e. as smaller things can be resolved), the energy available (and thus the mass of the particles created) grows. It is this aspect of quantum mechanics which is the basis for the assertion in the text that as we probe ever smaller sizes, the quantum foam becomes ever more turbulent. It also shows why probing smaller sizes means one must add the effects of ever larger ephemeral energy and also why the minimum size of superstrings is so attractive.

One of the most exciting possibilities of modern cosmology theory and measurements is the idea that the slight non-uniformity in the 3K background radiation observed by COBE and WMAP (and discussed in Chapter 9) might be a signature of the primordial quantum foam, locked into place for all eternity by Alan Guth's inflation.

I realize that this journey through relativity and quantum mechanics was very quick and left things out and finessed others. This is because this isn't a book about quantum mechanics or relativity. Nonetheless, certain ideas played a prominent enough role in the main text and warranted special mention. The interested reader should peruse the suggested reading for books that dwell on these fascinating topics in greater detail.

appendix E

Higgs Boson Production

In Chapter 5, I told you a small fib. It really wasn't so much of a fib as a decision to gloss over a technical point, the explanation of which would break up the explanatory flow. In this Appendix, we can spend some time to get into the deeper details.

The first omission concerns the statement that the Higgs boson is created in proton/antiproton collisions through the coalescence of two gluons... the Higgs boson then decays into a bottom quark/antiquark pair. Diagramatically, we say gg $\rightarrow$ H $\rightarrow$ b$\bar{\text{b}}$, or you can take another look at Figure 5.10 or E.1a. A reader who has been proceeding carefully through this book will have realized that such a statement is pure and utter hogwash. It can't possibly be true, at least in the simple way stated above. The two facts that prove that this can't happen have been mentioned in the text. The first fact is that Higgs bosons interact more with massive objects and less with less massive ones, interacting not at all with massless ones (Chapter 5). The second fact is that gluons are massless (Chapter 4). Taken together, these show that Higgs bosons do not directly interact with gluons and therefore cannot be directly created by them. So what gives? How can we say gg $\rightarrow$ H with a straight face? It's because it's true, albeit with a small sophistication.

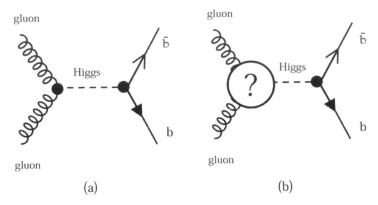

Figure E.1 (a) The naïve model of how the Higgs boson is created. (b) A model of Higgs boson creation that highlights the fact that there is a mystery where the gluons merge to form the Higgs boson.

In Chapter 5, we drew a Feynman diagram representation of the creation of a Higgs boson via gluon coalescence, which we reproduce in Figure E.1a. Two gluons fuse and produce a Higgs particle. While this is sort of true, there must be more to the story. We deal with this by realizing that what is really being said is that two gluons approach the interaction point and one Higgs boson exits. What *isn't* discussed is the details of what goes on at the moment of interaction. Thus I draw in Figure E.1b a circle over the actual point of creation of the Higgs boson. Inside the circle, many different sorts of interactions are possible. The only constraint is that all of the interaction be contained completely within the circle. This satisfies the "two gluons approach, while one Higgs boson leaves the interaction point" condition.

Since the Higgs boson interacts more with massive particles, it would prefer to interact with the massive top quark, the most massive of the known particles. Luckily, gluons can interact with quarks, even top quarks. Thus, a Feynman diagram in which gluons interact with top quarks, which in turn interact with a Higgs boson would be ideal and one fitting this set of criteria is given in Figure E.2.

In this interaction, the topmost gluon temporarily splits into a top-antitop quark pair. The bottommost gluon then interacts with the

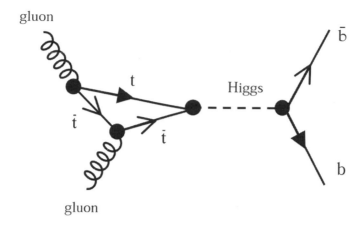

gluon

t̄

t

Higgs

t

b̄

b

gluon

Addition of top quarks
in Higgs boson creation

Figure E.2 More realistic version of Higgs boson creation, showing the intermediary stage whereby gluons first convert into top and antitop quark pairs.

antitop quark, deflecting back to the top quark with which it annihilates, creating a Higgs boson. Which gluon first converts into the tt pair is arbitrary, as is whether the second gluon interacts with either the top or antitop quark. Thus there are four possible cases, which are shown in Figure E.3. (Note E.3a is a repeat of E.2.)

The laws of quantum mechanics tell us we can't know, even in principle, which of the four diagrams of E.3 caused the interaction and to do the calculations, we must add in all four, but the final answer must be independent of which process actually occurred. Indeed, quantum mechanics states that for each creation of the Higgs boson, all four diagrams contributed.

Typically we draw the single Feynman diagram of Figure E.4 that covers all four cases. Because of the inherent ambiguity in whether or not the second gluon interacted with either the top quark or antiquark, we drop the little "⁻" which denotes antimatter and present the Figure E.4.

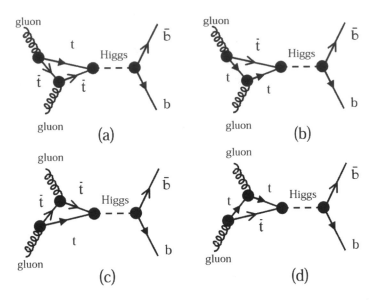

Figure E.3 More sophisticated version of Higgs boson production, showing the many ways in which gluons can create top/antitop quark pairs.

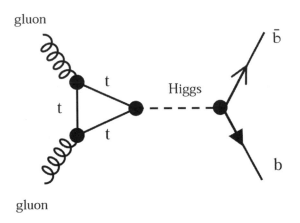

More Correct Higgs Production
Feynman Diagram

Figure E.4 Generic Feynman diagram for Higgs production. The top quark loop is meant to indicate all combinations of top/antitop quark loops.

Thus, we see the two gluons briefly convert into $t\bar{t}$ pairs, which create the Higgs boson, which in turn eventually decays into the bottom quark and antiquark pair. So $gg \rightarrow H \rightarrow b\bar{b}$ is OK, as long as we realize that it includes an intrinsically non-observable intermediate stage. As mentioned in Chapter 4, each Feynman diagram is a compact way to write an equivalent mathematical equation. I'll let you imagine this one although, as you might imagine, it's moderately difficult.

The critical reader might ask one additional question. Why is it that we have the Higgs boson interacting with top quarks on the left and bottom quarks on the right? We discuss in Chapter 5 the conditions which limit the Higgs boson's decay into primarily bottom-type quarks and antiquarks. While the Higgs boson will decay into the heaviest particle possible, it is thought to be unlikely that the Higgs boson is heavy enough to decay into a $t\bar{t}$ pair (although some of my colleagues will look for just that).

So why does the left side of Figure E.4 include top quarks? We know from Chapter 4 that the combined mass of a top/antitop quark pair is about 350 GeV, far higher than the 115–190 GeV anticipated mass of the Higgs boson. We also know that top quark pairs are very hard to make. So what's up?

The answer lies in a tricky bit of physics. In Appendix D, we introduced the Heisenberg Uncertainty Principle, which states that the energy in a system can spontaneously change, as long as the change persists for a short enough time. In Figure E.4, the "top quark loop" only exists for a fleeting moment, so Heisenberg's principle applies. Further, since mass and energy are equivalent, Heisenberg's principle allows the mass of the top quarks involved in the loop to be something other than the 175 GeV measured by the DØ and CDF experiments. These fleeting particles, which carry mass other than their "right" mass, are called virtual particles. As long as they live for only a short time, their existence does not violate any laws of physics.

Of course, top quarks are not the only particles that can be virtual. Bottom quarks can also be virtual, with a mass temporarily exceeding their measured value. However, when all factors are taken into account, it is the top quark loop that plays the dominant role in the creation of Higgs bosons.

One final point must be made. The top quark loop is thought to be the most important contributor to Higgs boson creation, when all known particles are considered. However, physicists hope to discover additional heavy particles. If one considers supersymmetry, introduced in Chapter 8, the top quark loop might be replaced by one including supersymmetric quarks or *squarks*. In fact, the discovery of the Higgs boson, with a measurement of a creation rate that is different than that predicted from known particles, may be the first experimental evidence for supersymmetry or some other unknown phenomenon.

appendix F

Neutrino Oscillations

Neutrino oscillations can be a mathematically tricky concept, so I have placed a few concepts in this Appendix. The first one is simply the full set of fusion processes that are going on in the Sun. I restrict my discussion to those processes involved in converting protons (i.e. hydrogen nuclei) into helium nuclei. These are given in Table F.1. It is clear from the table that the primary process for the creation of neutrinos is in the so-called "pp" process, in which two protons are fused to create a ^2H nucleus (also called a deuteron, which is simply a nucleus containing one proton and one neutron). However, the first process that was studied was the relatively rare process called ^8B, which produces very energetic neutrinos.

The second topic that is interesting (but technical) is the idea of neutrino oscillation. Neutrino oscillations are pretty cool, not only because of the interesting physics that they reveal, but also because they are mathematically pretty tractable. I won't do the derivation here; rather I will point you to the further reading (e.g. Perkin's book, listed in the reading for Chapters 3 and 4). The calculation is pretty simple if the following words don't strike terror in your heart. It uses quantum mechanical time-dependent wave function evolution. (Trust me, there are people who pee their pants in excitement over

Table F.1 Primary mechanisms whereby two protons are fused into helium nuclei. While not crucial for our discussion, the names listed (e.g. pep) denote the particles that go into the reaction, in this case, two protons and an electron.

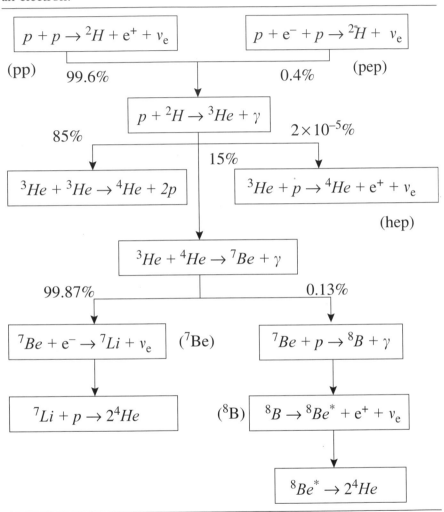

this stuff. It's especially interesting as this is one of the few cases in which you can do the entire calculation on a single piece of paper.) If that all seems like simply too much effort, we'll just take the final equation and explain what it means.

Neutrino oscillation theory deals with one phenomenon. Suppose that you have a beam consisting of only a particular flavor of neutrinos. These could be electron-, muon- or tau-type neutrinos. These neutrinos are in a beam of energy E and we direct the beam at a detector placed at a distance away from where they are created. We label this distance L, for length.

Let's consider the simplest case, the one in which a single flavor of neutrino is oscillating into a single different kind of neutrino. It could be $\nu_e \leftrightarrow \nu_\mu \leftrightarrow \nu_\mu \leftrightarrow \nu_\tau$ or $\nu_e \leftrightarrow \nu_\tau$, but we'll just call the other neutrino type 2, with mass m_2 and just say $\nu_1 \leftrightarrow \nu_2$. We start with a large number of neutrinos, all of type 1. As they travel on their way to the detector, they oscillate, with some fraction turning into type 2, before turning back into only type 1. After all the math is done, we find that the probability that a particular neutrino of type 1 has oscillated into type 2 is

$$Pr(\nu_1 \rightarrow \nu_2) = \sin^2 2\theta \sin\left[\frac{1.27\Delta m^2 L}{E}\right]$$

where L is the length between the neutrinos' creation and the detector in kilometers, E is the neutrino beam energy in GeV, $\Delta m^2 = m^2_2 - m^2_1$ in eV^2 and 1.27 comes from the calculation and has the right units to make everything work. $\sin^2(2\theta)$ is just a fancy way to say how fast the neutrinos oscillate into the other flavor. L and E are known. Δm^2 and θ are unknown and it is these two quantities that experiments measure. The Δm^2 term is the difference between the masses of the two types of neutrinos and reflects the slight difference in speed at which the two neutrino species move. The $\sin^2(2\theta)$ term, properly understood, reveals something about underlying physics. Technically θ is pretty tricky and beyond the scope of even this Appendix. Basically, it stems from the fact that the neutrinos of a particular flavor (e.g. ν_e, ν_μ and ν_τ) don't have a unique and well-defined mass, while neutrinos with a well measured mass do not have a well-defined flavor. If you want more details, please peruse the suggested reading.

Let's look quickly at the equation. If one wants to make measurements, you can only vary L and E. When an experiment is designed, one chooses L, for instance the Minos experiment shoots a beam of neutrinos from Fermilab to the Soudan mine in Minnesota, a distance of several hundred miles. But since neutrino experiments are so massive (recall Super Kamiokande's 50,000 tons of water?), you can't move them. The best you can do is spend the money to build multiple detectors at fixed locations. However, by varying E, the beam energy, you can probe different amounts of oscillation. Some experiments have been designed very carefully to be able to change the beam energy, although, as you'd imagine, neutrino beams are pretty tricky to build.

Because E and L are well known in accelerator experiments and known to a degree in cosmic ray experiments, what each group does is measure the ratios of the two different types of neutrinos in which they are interested and compare the measurements with the ratios of the same neutrinos at the source. With a single measurement, each experiment cannot uniquely identify Δm^2 and θ, but they can determine a set of numbers that work. In analogy, one might have measured a bunch of things and in the end, one can write the result as $(\text{unknown } \#1)^2 + (\text{unknown } \#2)^2 = (\text{measured})^2$. If measured $= 1$, then $(\text{unknown } \#1) = 1$ and $(\text{unknown } \#2) = 0$ works, as does $(\text{unknown } \#1) = 0$ and $(\text{unknown } \#2) = 1$. Many other combinations are also possible. This measurement doesn't tell you either unknown uniquely, but it does reveal the range of allowed values. In the end, you combine the results of many experiments to determine the correct values of your two unknowns.

In Figure F.2, three ellipses are drawn, each the result of a particular experiment. Each experiment says that they don't know the value of each unknown, but the truth is inside their respective ellipse. We see that the intersection of the three ellipses provides for a better estimate of the unknowns than any of the individual experiments.

In the case of neutrino oscillations, the curve describing the allowed combinations of Δm^2 and θ are much more complicated than

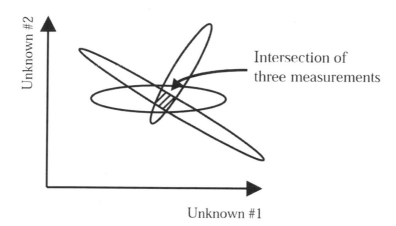

Figure F.2 If three different experiments cannot determine two unknowns uniquely, but rather only the range of allowed values (depicted by the respective ellipses), by taking three measurements, the space that is common to all three is much smaller than that observed by any one experiment. Thus, one can determine the "true" values of the two unknowns with improved precision.

simple ellipses. By the end of the decade, we should have enough independent measurements to close in on the right answer. In the meantime, experiments will continue to take data, each with a different E and L. In addition, the various experiments also have some control over Δm^2 and θ (by looking at the different possible oscillations, $\nu_e \leftrightarrow \nu_\mu$, $\nu_\mu \leftrightarrow \nu_\tau$ or $\nu_e \leftrightarrow \nu_\tau$). Eventually, the truth will be revealed.

❖

Further Reading

There are many marvelous other books out there that cover material similar to what I've discussed here. I list some of those books below. The organization is such that I will list books that are globally interesting first and then a few that are specific to a particular chapter, segregated by chapter. If a globally interesting book has a particularly good treatment of a specific chapter's topic, I mention it there too.

Generally good books include *The Particle Garden* by Gordon Kane (Perseus Books 1996). While somewhat less detailed than this book, it covers similar material. Leon Lederman and Dick Teresi wrote *The God Particle* (Houghton-Mifflin, 1983), nominally about the Higgs Boson, but the Higgs encompasses only a small fraction of the book. Their history of the early years of scientific investigation is rather good. Other books include *Q is for Quantum: An Encyclopedia of Particle Physics* by John Gribbin (Touchstone Books, 2000), *A Tour of the Subatomic Zoo: A Guide to Particle Physics* by Cindy Schwarz (Springer-Verlag, 1996), *Quarks and Gluons: A Century of Particle Charges* by H.Y. Han (World Scientific, 1999) and *The Charm of Strange Quark: Mysteries and Revolutions of Particle Physics* by R. Michael Barnett, Henry Muehry, Helen R. Quinn and Gordon Aubrecht (Springer Verlag, 2000). Note that the last two books focus more on the physics and less of the historical context.

Chapter 1: Early History

There are a vast number of books on the early history of physics prior to 1900. Some possibilities include *The Birth of New Physics* by I. Bernard Cohen (Norton, 1985), *Physics, the Human Adventure: From Copernicus to Einstein and Beyond* by Gerald Holton and Stephen Brush (Rutgers University Press, 2001) and *Before Big Science: The Pursuit of Modern Chemistry and Physics 1800–1940* by Mary Jo Nye (Twayne Publishers, 1996). You might find in your library the *Dictionary of Scientific Biography*, which is sixteen volumes long (Scribner, 1970–1980).

Chapter 2: The Path to Knowledge
(History of Particle Physics)

Probably one of the best books available detailing the history of particle physics in the 20th century is *The Second Creation* by Robert P. Crease and Charles C. Mann (Rutgers University Press, 1996). I cannot recommend this book highly enough. It is extremely well written and very interesting.

For some personal recollections of the fall of parity and the discovery of muon neutrinos, one should read Leon Lederman and Dick Teresi's *The God Particle*.

For information on J.J. Thomson, I suggest *J.J. Thomson and the Discovery of the Electron*, by E.A. Davis and I.J. Falconer (Taylor and Francis, 1997). For the discovery of antimatter, Carl David Anderson's *The Discovery of Antimatter* (World Scientific, 1999) has many personal recollections.

Chapters 3 and 4: Quarks and Leptons and
Forces: What Holds It All Together

Books rarely cover only particles or forces, so I lump these two chapters together.

Gordon Kane's *The Particle Garden* is very good on this topic. All of the other books listed at the beginning all cover this material. For an account of the discovery of the *W* and *Z* bosons, you might want to try *Nobel Dreams* (Random House, 1986) by Gary Taubes. Taubes is a journalist and his book reveals some of the personalities and their interaction in a search for a Nobel Prize winning discovery.

More technical treatments include *Introduction to Elementary Particles* by David Griffiths (John Wiley & Sons, 1987), a textbook aimed at advanced undergraduate college students. For the extremely brave or foolish, there are two more books to consider. The first is *Introduction to High Energy Physics* by Donald Perkins (Addison-Wesley, 1987), which is a graduate level textbook for students wanting to become particle physicists. While difficult, it exhibits his experimentalist mindset, so it can be read, although there will be some mathematical parts which you would skip. The second book is *Quarks and Leptons: An Introductory Course in Modern Particle Physics* by Francis Halzen and Alan Martin (John Wiley & Sons, 1984). This book is also for aspiring particle physicists, but this time for ones with a more theoretical bent. This book is extremely difficult and occasionally looks like it is written in a different language, which it is. Nonetheless, it is *the* textbook used, so most particle physicists have used it.

Chapter 5: Hunting for the Higgs

There is not a tremendous amount written on the Higgs, available for a layman. Gordon Kane's *The Particle Garden* does a reasonable job, although with less detail than given here. In addition, Lederman's *The God Particle* is named after the Higgs boson, although it does not cover it in any technical detail.

Chapter 6: Accelerators and Detectors: Tools of the Trade

I am unaware of any books written at this level on this topic. Gordon Kane's *The Particle Garden* discusses a little of the detectors and sociology of experimental particle physics, but since he's a theorist, my experimentalist's pride says we should ignore it (although, in all honesty, it's pretty good). There are somewhat higher-level texts available. Good ones include *Techniques for Nuclear and Particle Physics Experiments: A How-To Approach* by William R. Leo (Springer-Verlag, 1994), *Data Analysis Techniques for High-Energy Physics* by R. Fruhwirth, M. Regler, R.K. Bock, H. Grote and D. Notz (Cambridge University Press, 2000) and *Introduction to Experimental Particle Physics* by Richard Fernow (Cambridge University Press, 1989). A fairly difficult book on accelerators is *An Introduction to the Physics of High Energy Accelerators*, by D.A. Edwards and M.J. Syphers (Wiley-Interscience, 1992). For a non-specific book, you might peruse any of the large number of high school and first-year college textbooks. The July 2000 Scientific American article on the topic is *The Large Hadron Collider* by Chris Lewellyn Smith discusses accelerators, but not in very much detail.

Sharon Traweek is a sociologist who wrote a book *Beamtimes and Lifetimes: The World of High Energy Physicists* (Harvard University Press, 1988, paperback 1992). Using the research techniques and language associated with scholarly sociological research, she talks about the culture of particle physicists. Not all physicists agree with her perception of our culture, but the book is an interesting read.

Chapter 7: Near Term Mysteries

There are very few books for the public that discuss these topics. One slightly dated book on neutrinos is *The Elusive Neutrino: A Subatomic Detective Story* by Nickolas Solomey (W.H. Freeman, 1997). I am unaware of a modern treatment of CP violation (including the most recent research), but Leon Lederman's *The God Particle* covers the topic, but without the experimental results of the last several decades. In addition, one can find some information in the following Scientific American articles. *The Asymmetry between Matter and Antimatter* by Helen Quinn and Michael Witherell (October 1998) discusses recent knowledge in (surprise!) the asymmetry between matter and antimatter. *Detecting Massive Neutrinos* by

Edward Kearns, Takaaki Kajita and Yoji Totsuka (August 1999) covers the first solid evidence for atmospheric neutrino oscillations, while *Solving the Solar Neutrino Problem* by Arthur B. McDonald, Joshua R. Klein and David L. Wark (April 2003) covers SNO's recent results.

Chapter 8: Exotic Physics (The Next Frontier)

In this chapter, I discuss three topics. Each topic has few sources specifically addressed to that particular branch of physics, while remaining accessible to the layman. *Supersymmetry* by Gordon Kane (Perseus Press, 2000) is, I believe, unique in that it is a single topic book discussing supersymmetry. It is not geared to the beginner, but anyone who has read this book will find that one tractable. In addition, Kane has recently published an article in the June 2003 article of Scientific American. Brian Green's *The Elegant Universe* is a very nice, focused description of superstrings. The early chapters are exceptional, while the later ones get increasingly technical. While many details are revealed, he never loses his audience. Green also has an article *The Future of String Theory* in the November 2003 issue of *Scientific American*.

The topic of large extra dimensions has no popular book of which I am aware devoted to it. There was a *Scientific American* article written by the theory's architects, *The Universe's Unseen Dimensions* by Nima Arkani-Hamed, Savas Dimopoulos and Georgi Dvali, in the August 2000 issue. This article is repeated in the Fall 2002 special issue of *Scientific American* that concentrates on cosmology. For the simpler idea of higher dimensions, Edwin A. Abbott's *Flatland: A Romance of Many Dimensions* (Dover, reprint 1992) is very nice, even though it was written a century ago.

Chapter 9: Recreating the Universe 10,000,000 Times a Second

Because cosmology is a large field and I could only devote a short amount of space to it, I should like to say that there is an enormous amount of brilliant writing on the topic. A good general account of the Big Bang is *The Big Bang Theory* by Karen C. Fox (John & Sons, Inc., 2002). This book includes an account of the ekpyrotic universe theory, which is a competitor idea to inflation. A somewhat dated book, but still encyclopedic, is *The Big Bang* by Joseph Silk (W.H. Freeman, 1980). It covers many topics although not in great detail. It describes some of the more reputable alternatives to the Big Bang. Another similar book is *The First Three Minutes* by Steven Weinberg (Basic Books, 1977).

For those interested in some of the personalities involved in a historical account of the various aspects of cosmology, I recommend *Blind Watchers of the Sky* by Rocky Kolb (Addison-Wesley, 1996). Rocky is a particularly gifted communicator of science and his book conveys that fact. Another book of interest is *Genesis of the Big Bang* by Ralph Alpher and Robert Herman (Oxford University Press, 2001) a personal account of the two physicists who predicted the 3K background radiation.

One book that I think is extremely good, and is not often mentioned in similar reading lists is *Einstein's Greatest Blunder?* by Donald Goldsmith (Harvard University Press, 1995). *Wrinkles in Time* by George Smoot and Keay Davidson (William Morrow, 1993) is also worthwhile. Smoot was a leader in the COBE experiment's version of their search. It is well written and easy to understand. You might follow reading that book with *Afterglow of Creation* by Marcus Chown (University Science Books, 1996). It also details a similar account, but also reveals some of the objections of some of Smoot's collaborators with what they perceived as an overly one-sided account of COBE's successes. A more recent book that details some of the physics of the 3K background radiation (as opposed to emphasizing the experiment) is *How the Universe Got its Spots* by Janna Levin (Princeton University Press, 2002). Presented in a series of unsent letters to her mother, Levin's book is written in a way that one will either love or find somewhat distracting. Her explanations however, are first rate.

Martin Rees' book *Our Cosmic Habitat* (Princeton University Press, 2001) is a mixture of the cosmological as well as the astronomical. He gives a more detailed account for why we believe that space is flat than was possible, given the limited space available in this book. He also projects the future of the universe more explicitly than some other books.

A nice recent book, which has some discussion of CP violation in a cosmological setting, is *The Accelerating Universe* by Mario Livio (John Wiley & Sons, 2000). This book also focuses on inflation and the flatness of the universe. The architect of cosmic inflation, Alan Guth, wrote *The Inflationary Universe* (Addison-Wesley, 1997). Another book one might read is *Quintessence: The Mystery of the Missing Mass* by Lawrence Krause (Basic Books, 1999). The name is self-explanatory.

Leaving books, I offer some articles in *Scientific American*, which may be found at your public library. Specifically, there was a special edition *The Once and Future Cosmos*, published in the fall of 2002. As is usual in *Scientific American* articles, each consists of some few pages, which discuss many of the topics covered in this chapter. *The Search for Dark Matter* by David Cline (March 2003) covers experimental searches for dark matter not covered in this text. A viable, although highly controversial, alternative to the need to invoke dark matter to explain galactic rotation curves can be found in *Does Dark Matter Really Exist?*, by Mordehai Milgrom, in the August 2002 issue.

Finally, I offer a few books for the very brave. These are not written for the layman; in fact they are actual physics journal articles, combined into a book. Only the most motivated readers are encouraged to pursue these. The first is *Cosmological Constants: Papers in Modern Cosmology*, edited by Jeremy Bernstein and Gerald Feinberg (Columbia University Press, 1986) and the second is *Cosmology and Particle Physics*, edited by David Lindley, Edward (Rocky) Kolb and David Schramm (American Association of Physics Teachers, 1991).

Appendix D: Essential Relativity and Quantum Mechanics

There are many books on these two complex topics. I offer only a few here. George Gamow wrote two books as part of a series, which are now available in the combined book *Mr. Tompkins in Paperback* (Cambridge University Press, reissue 1993). Originally published in 1939, the language is a bit archaic (he refers to the "gay tribe of electrons" to indicate their continuous motion). Nonetheless, the descriptions are very accessible. Another good book is Robert Gilmore's *Alice in Quantumland: An Allegory of Quantum Physics* (Copernicus Books, 1995). As the title suggests, Alice visits a place where things are odd. It's quite understandable. Lewis Carroll Epstein's *Relativity Visualized* (Insight Press, 1985) is a nice introduction to special relativity. If you insist on learning from the master, you might peruse Albert Einstein's *Principle of Relativity* (Dover Publications, 1924) or his *Relativity: The Special and the General Theory* (Crown Publishing, reprint 1995), although for my own taste, ol' Al writes at a fairly high level.

Richard P. Feynman is an entertaining guy and I offer three of his books as useful further reading. They are *Six Easy Pieces* (Perseus Books, 1995), *Six Not-So-Easy Pieces* (Perseus Books, 1997) and *QED: The Strange Theory of Light and Matter* (Princeton University Press, 1985). The two "Six" books are extracts from a larger trio of books called *The Feynman Lecture in Physics* and are excerpts from lectures he gave at Caltech in the early 1960s. They have math in them, but they are entertaining and lucid even so. His *QED* is perhaps the only popular discussion on the subject of Quantum ElectroDynamics. Told by an architect of the theory, it is extraordinarily illuminating.

Moving towards more textbook-like suggestions, unfortunately each containing moderately sophisticated mathematics, William H. Cropper's *The Quantum Physicists and an Introduction to Their Physics* (Oxford University Press, 1970) includes a liberal mix of history and physics. Robert Resnick and David Halliday have written *Basic*

Concepts in Relativity and Early Quantum Theory (John Wiley & Sons, 1985). This is probably one of the most tractable of the textbooks. Two others of similar level and quality are Kenneth Krane's *Modern Physics* (John Wiley & Sons, 2nd edition, 1995) and *Quantum Physics of Atoms, Molecules, Solids, Nuclei and Particles* (John Wiley & Sons, 2nd edition, 1985) by Robert Eisberg and Robert Resnick.

Glossary

accelerator   a large device which can be used to increase the velocity (and energy) of subatomic particles.

antilepton   an antimatter lepton.

antimatter   a substance that can annihilate matter, forming pure energy.

antiquark   an antimatter quark.

atmospheric neutrino problem   the observation that the numbers of neutrinos generated in the atmosphere are not distributed as expected. Strong calculations indicate that there should be two muon-type neutrinos for each electron-type neutrino. Experiments show a one-to-one ratio. This is taken as strong evidence for neutrino oscillations.

atom   the smallest unit of matter that retains its chemical identity. Once the constituents of an atom are removed from the atom, it no longer acts the same way chemically.

BABAR   a detector based at SLAC, attempting to measure CP violation in B mesons.

background   something that looks like what you're looking for, but really isn't.

baryon   any particle containing three quarks. The proton and neutron are the most common baryons.

beamline   a series of magnets used to guide a beam of high energy particles to a destination. In many ways, this is analogous to a series of lenses and prisms that guide light.

BELLE   a detector based at KEK, attempting to measure CP violation in B mesons.

beta radiation   a type of radiation discovered in nuclear decay. Beta particles are electrons, created in a force mediated by the weak force. Beta particles were known before the atomic electron was discovered.

Big Bang   a theory that suggests that the universe began in a massive, primordial explosion. Significant observational evidence supports this theory.

blath   a hypothetical direction in a fourth dimension. This term is unique to this book and is not standard. Just as right-left, front-back and up-down describe directions in our familiar three dimensions, blith-blath describes directions in the fourth dimension.

blith   see blath.

BNL   Brookhaven National Laboratory, located on Long Island, New York.

Booster   the third accelerator in the Fermilab accelerator chain. Accelerates a particle from 401 MeV to 8 GeV.

bosinos   the supersymmetric analog of bosons.

boson   any of a family of particles that have a quantum mechanical spin that is "integral" ($\ldots -2, -1, 0, +1, +2, \ldots$).

bottom   a moderately heavy quark with a mass of about 4.5 times that of a proton, carrying an electrical charge of $-1/3$. From generation III.

broken symmetry   a phenomenon whereby something that was once uniform becomes no longer uniform. An example might be the humidity in the air on a summer day. During the day, the air and water are mixed uniformly. After a rain storm, the water is on the ground while the air is not... the symmetry previously enjoyed by air and water no longer exists.

bubble chamber an early detector technique whereby charged particles crossed a superheated liquid. In their passage, the particles left a track, much like a jet contrail, which can be photographed.

calorimeter a device for measuring the energy of high energy particles. Usually consists of a mix of dense material (like metal), interspersed with light material (gas, plastic or liquid).

CDF an acronym for the Collider Detector at Fermilab, the original of two large experiments currently operating at Fermilab. Co-discoverer of the top quark.

CEBAF the Continuous Electron Beam Accelerator Facility, an accelerator located at the Thomas Jefferson National Accelerator Facility in Newport News, VA.

Cerenkov radiation a phenomenon whereby a charged particle traverses a transparent medium at a speed faster than light passes through the same medium. When this occurs, a blue light is given off. Used often in large water-based detectors in order to detect neutrinos.

CERN Conseil Européen pour la Recherche Nucléaire, the European Council for Nuclear Research, located on the Swiss/French border. Long the home of the LEP accelerator, they are currently building the LHC, which will surpass the Fermilab Tevatron in energy in about 2007 or so. In 1954, the official name was changed to the European Organization for Nuclear Research, however the acronym did not change.

Cfa a group who surveyed the distances to surrounding galaxies. By plotting the position of the galaxies, they were able to get the first measurement of the structure of the universe out to a distance of 500 million light-years.

charm a moderately heavy quark with a mass of about 1.5 times that of a proton, carrying an electrical charge of $+2/3$. From generation II.

CMS the Compact Muon Solenoid. One of two large detectors that will operate at the LHC.

COBE *CO*smic *B*ackground *E*xplorer, an orbiting detector of the 3K background radiation. The experiment was able to show non-uniformities in the 3K background radiation at 0.001%.

Cockroft-Walton the first accelerator in the Fermilab accelerator chain. Accelerates an H^- particle from rest to 750 keV.

collider any accelerator in which counter-rotating beams of particles are collided.

collision when two particles collide and the ensuing debris. Also called an event.

color a property of quarks and gluons. Essentially it is the charge that causes the strong force. No observable particles can have color (similar to atoms having no net electric charge).

confinement the premise that quarks and gluons cannot be observed by themselves, but only contained within a meson or baryon.

conserved the property of being unchanged. There are many physical properties that exhibit this behavior, momentum, energy, angular momentum, electric charge and so on. This is a very powerful behavior, as if one measures one of these conserved quantities at any one time, that quantity is known forever.

cosmic rays originally thought to be the particles observed in various experiments, the meaning of the term "cosmic" rays has become more complex. Particle from space (protons or atomic nuclei) hit the atmosphere at high energy. They cause a shower consisting primarily of pions, which decay into photons and muons. The muons and photons are detected at ground level. We call all of the particles at all stages cosmic rays, where originally only the muons carried the title.

cosmological constant a concept, originally proposed by Albert Einstein, which he added to his equation of General Relativity. Basically, the cosmological constant provided a repulsive pressure that offset gravity's attractive nature. For a while, the cosmological constant was thought to be an error, although recent measurements have made it experience a renaissance.

cosmologists   physicists who study the beginning of the universe.

CP violation   charge conguation and parity violation, possibly a partial explanation for the matter dominance of the universe. Generally the equations describing subatomic behavior are unchanged if one first swaps all directions (left ↔ right, in ↔ out, up ↔ down) and then swaps matter with antimatter. In 1964, when a reaction that did not seem to respect this symmetry was discovered, physicists discovered an asymmetry between matter and antimatter.

cyclotron   an early form of particle accelerator which accelerated particles in a spiral path.

DØ   a large detector, currently operating at Fermilab. Co-discoverer of the top quark. Named for one of the six collision points on the accelerator ring (called AØ, BØ and so on). The author's experimental home for about 10 years.

dark energy   an energy that is thought to permeate the universe and provide for the universe's perceived flatness.

dark matter   hypothetical matter which, if it exists, accounts for the fact that the outer arms of galaxies seem to rotate more quickly than expected.

decay   the conversion of one particle into two or more different particles.

Delphi   one of four large experiments located at the LEP accelerator.

Delta   a particular baryon, with spin of 3/2. The quark content consists of all possible combinations of up and down quarks, thus there are four unique Delta particles ($\Delta^-, \Delta^0, \Delta^+, \Delta^{++}$). The Δ^{++} lead to the proposal of color as a quark property.

DESY   Deutsche Elektronen Synchrotron, the German Electron Synchrotron, in Hamburg, Germany.

detector   an apparatus that detects particle collisions. Specifically, one attempts to measure as much about the collision as possible, for instance the position, trajectory and energy of each of the hundred or more particles that are created in a typical collision.

deuteron   the nucleus of an atom which is an isotope of hydrogen. It is a single particle, which contains within it a single proton and a single neutron "stuck together" like they are attached by Velcro.

discrete   see quantized.

DoE   the Department of Energy. A United States governmental agency responsible for a large fraction of the particle physics research budget. Parent agency of Fermilab.

down   the second lightest of the quarks, carrying an electrical charge of $-1/3$. From generation I.

electromagnetic force   force felt between two electrically charged particles. Responsible for holding the atom together.

electron   a particle carrying negative electric charge usually found in a dispersed cloud surrounding the nucleus of an atom. Electricity is explained as the movement of electrons.

electron volt   a unit of energy, often abbreviated eV. A particle carrying the same charge as the proton, accelerated by an electric potential difference of one volt is said to have one electron volt of energy.

electroweak force   a combined force, which includes both the weak and the electromagnetic force. Understanding just how these two forces are unified into one electroweak force is currently a forefront topic of experimental research.

Electroweak Symmetry Breaking   the phenomenon whereby a single and higher energy theory becomes two distinct forces: electromagnetism and the weak force.

energy conservation   a crucial and fundamental property of the universe. Energy is always conserved (that is, unchanged). Energy can change forms, but the total energy in any system is always the same.

eV   see electron volt.

event   see collision.

exchange   a particle exchange is the process in which two incoming particles affect one another's trajectory through the emission and absorption of a third particle. Exchange in this context is more like an exchange of money (typically one-way) rather than an exchange of gifts.

experimentalists   physicists who make and interpret measurements.

Fermilab   Fermi National Accelerator Laboratory, located outside Chicago, Illinois. Currently the laboratory with the highest energy accelerator in the world.

fermion   any of a family of particles that have a quantum mechanical spin that is "half integer" (i.e. ... $-5/2$, $-3/2$, $-1/2$, $1/2$, $3/2$, $5/2$ ...).

Feynman diagrams   cartoons, first worked out by Richard Feynman, which had a one-to-one correspondence to complicated mathematical equations. These diagrams make setting up the initial calculation of a particular particle interaction very easy.

fixed target   an experiment in which a beam of highly energetic particles is made to hit a stationary particle target.

flavor   another word for "type". For instance, the electron and muon are different flavors of leptons.

fluorescence   a phenomenon whereby a material glows when illuminated by light. The color of the glow and the color of the illumination are often different.

flux   the amount of objects or a field that pass through an area. For instance, one might talk about the flux of water passing through a hula-hoop placed in a river or the flux of gravity through the same hula-hoop held horizontally above the surface of the Earth.

fragmentation   the process whereby a quark or gluon converts into a jet.

GALLEX   the *GALL*ium *EX*periment a neutrino detector in which gallium is converted into arsenic.

gamma radiation   a type of radiation from the nucleus of the atom. Gamma radiation is a very high energy photon.

generation   a term describing the fact that there appears to be two additional "carbon copies" of the particles that make up the common universe. Why there should be exactly three (which seems to be the case) is unknown.

GeV   a Giga electron Volt, i.e. one billion (10^9) electron Volts

gluino   a hypothetical fermion that is the supersymmetric analog of the gluon.

gluon   massless particle that mediates the strong force. Found inside the nuclei of atoms, as well as inside all hadrons.

graduate student   a person who has completed a bachelor's degree and is working on an advanced degree (either a Masters or a doctorate). In practice, this is an apprenticeship program, in which the student learns by doing.

Grand Unified Theory   a hypothetical idea that suggests that all of the physics that describes the universe can be stated as a single theory from which all phenomena can be understood.

gravitino   a hypothetical fermion that is the supersymmetric analog of the graviton.

graviton   hypothetical (i.e. not yet discovered) massless particle that is believed to mediate the gravitational force.

gravity   the weakest force known. While it is outside the current scope of particle physics research, it holds the universe together.

GUT   see Grand Unified Theory.

hadron   any particle that feels the strong force. Contains within it quarks, gluons and sometimes antiquarks.

Heisenberg Uncertainty Principle   an idea, first realized by Werner Heisenberg, that it was possible to have energy not be conserved as long as the non-conservation was for a short period of time.

HEP   High Energy Physics, the study of particle interactions at high collision energy.

HERA   the Hadron Electron Ring Accelerator, an accelerator in Hamburg, Germany which accelerates positrons and protons and collides them together. Primary purpose is to carefully study the structure of the proton.

Higgs boson a hypothetical particle that is thought to give other particle's their mass. The search for this particle is the primary reason for the current Fermilab data-taking effort, which commenced in March 2001.

higgsino a hypothetical fermion that is the supersymmetric analog of the Higgs boson.

Homestake detector a large tank of perchloroethylene (dry cleaning fluid), in which the first evidence for neutrino oscillations from the Sun was observed.

hyperon baryons that carry at least one strange quark.

inflation an idea that at an early point in the history of the universe it underwent a rapid expansion. This idea explains both the uniformity and observed flatness of space. Inflation has not been refuted but is difficult to definitively prove.

interaction also called a collision or an event. The situation in which two particles affect one another, as in a typical collision.

interaction see collision.

ionization the phenomenon whereby a particle carrying electric charge, in its passage through matter, knocks the electrons off the atoms contained within the matter. This causes the original particle to lose energy.

isotope a variant of a particular type of nucleus. The identity of an atom is defined by the number of protons contained within the nucleus. The number of neutrons is generally between 1–2 times as many as the number of protons, but is not any specific number. Atoms containing an unusual number of neutrons in the nucleus are called isotopes of the usual element.

ISR the Intersecting Storage Rings, accelerators at CERN which collided proton beams.

jet a series of subatomic particles, typically mesons, traveling in generally the same direction. Looks like a "shotgun" blast of particles and is the signature of a quark or gluon scatter.

kaon   a meson containing a strange quark or antiquark.

KEK   Kou Enerugi Kasokuki Kenkyu Kikou, the High Energy Accelerator Research Organization, in Tsukuba, Japan. The home of the Belle detector.

keV   a kilo electron Volt, i.e. one thousand (10^3) electron Volts.

L3   one of four large experiments located at the LEP accelerator.

lambda particle   a baryon containing two "light" quarks (i.e. some combinations of up and down quarks) and one strange quark.

LAMPF   the Los Alamos Meson Production Facility. An accelerator at the Los Alamos Laboratory used to study mesons. No longer operating.

large extra dimensions   the idea that there may be more dimensions in the universe than our familiar four (three space and one time). The additional dimensions will be much smaller than the ones we observe, but the "large" in large extra dimensions means large compared to the Planck scale. These dimensions could be as large as a millimeter, although it is likely that, if they exist, they will be much smaller.

Las Campanas   a group who surveyed the distances to surrounding galaxies. By plotting the position of the galaxies, they were able to make a measurement of the structure of the universe out to a distance of 5 billion light-years. Named for the place where their telescopes are located.

LBL   the Lawrence Berkeley Laboratory, a laboratory perched on the hill above the University of California, Berkeley. While the laboratory no longer has high-energy accelerators, physicists from the laboratory are involved in experiments at other laboratories. Site of the discovery of the antiproton.

LEP   the Large Electron Positron accelerator at CERN in Europe. As its name suggests, it accelerates electrons and positrons. So far, this is the largest (although not highest energy) accelerator in the world, with a circumference of about 18 miles. It was designed to make detailed measurements of the Z boson and it accomplished this task admirably. Later, its energy was raised to characterize the W boson. During its final months of operations, it might have observed events made by the Higgs boson, although this claim is weak. No longer operating.

lepton   an elementary particle (i.e. one with no known substructure) that does not feel the strong force. There are two types of leptons: electrically charged and electrically neutral. The most familiar type of electrically charged lepton is the electron. The neutral leptons are called neutrinos.

LHC   the Large Hadron Collider. This CERN accelerator will accelerate two counter-rotating beams of protons to extremely high energies, exceeding any currently possible. It is being built inside the same tunnel as the now-defunct LEP accelerator. It is scheduled to start operations in about 2007.

limit   one type of measurement possible by scientific experiments. In the event that something is not observed, one can at least exclude some of the event's possible characteristics. Usual limits are when one rules out a range of allowed masses for a hypothetical particle. The particle is then said to have a mass greater or less than the reported number, indeed if the particle exists at all.

LINAC   an accelerator that accelerates charged particles in a straight line, using electric fields. Also the second accelerator in the Fermilab accelerator chain. Accelerates a particle from 750 keV to 401 MeV

LSND   Liquid Scintillator Neutrino Detector, a detector at LAMPF that has published a controversial neutrino oscillation result.

Main Injector   the fourth accelerator in the Fermilab accelerator chain. Accelerates a particle from 8 GeV to 150 GeV. Began commissioning in 1999.

Main Ring   was the fourth accelerator in the Fermilab accelerator chain. Accelerates a particle from 8 GeV to 150 GeV. Originally the highest energy accelerator in the Fermilab chain (at a time when there were only four Fermilab accelerators), it is no longer used, having ceased operations for the final time in the very beginning of 2000.

MAP   the Microwave Anisotropy Probe, a follow on experiment to COBE, intended to measure the uniformity of the 3K background radiation to unprecedented precision. Has exquisitely succeeded. Renamed WMAP in 2002.

meson   any particle containing a quark and an antiquark.

mesotrons   particles carrying a mass between that of the electron and proton. The modern name is meson.

MeV   a Mega electron Volt, i.e. one million (10^6) electron Volts.

Mini-Boone   the *BOO*ster *N*eutrino *E*xperiment, a prototype of a possible future larger Boone experiment. The purpose of this experiment is to confirm or refute the LSND result.

MINOS   Main Injector Neutrino OScillation experiment. A beam of neutrinos from Fermilab's Main Injector is aimed at the Soudan 2 mine in Minnesota, which contains a huge detector.

MSSM   the minimal supersymmetric model, a particular way to meld supersymmetry into the Standard Model.

muon   the charged lepton of the second generation. Basically a heavy electron, although it is an unstable particle, decaying approximately in a millionth of a second.

neutrino   an electrically neutral particle that feels only the weak force and perhaps gravity. Because a neutrino interacts so weakly with ordinary matter, one generated by the Sun could pass through 5 light-years of solid lead.

neutrino oscillations   the idea that the different neutrino flavors can "morph" into one another. While phenomena that can be explained by neutrino oscillations has been observed for about 30 years, the first direct evidence was announced in 1998. Continues to be a topic of active research.

neutron   an electrically neutral particle found in the nucleus of an atom. Contains quarks within it.

NLC   the Next Linear Collider, a proposed new accelerator that will collide electrons and positrons at unprecedented energies. So far, only preliminary planning has been done. No site has been chosen, no significant money appropriated, etc.

NSF the National Science Foundation. A United States governmental agency responsible for a large fraction of the science research budget.

nucleus the small and dense core of an atom, consisting of protons and neutrons.

Opal one of four large experiments located at the LEP accelerator.

parity the idea that in some physics equations, if one swaps all directions (left ↔ right, in ↔ out, up ↔ down), the equations appear to be unchanged.

particle physicists physicists who study the behavior of subatomic particles at the highest energy currently achievable.

parton a particle found inside a proton, neutron or any baryon or meson. Quarks and gluons are partons.

phosphorescence a phenomenon whereby a material glows after being illuminated by light. After the illumination ceases, the material continues to glow for a while.

photino a hypothetical fermion that is the supersymmetric analog of the photon.

photomultiplier a piece of detector apparatus that can convert a single photon into millions or tens of millions of electrons.

photomultiplier tube a detector component that can convert a single photon into millions of electrons in a tiny fraction of a second.

photon massless particle that mediates the electromagnetic force. All electromagnetic phenomenon can be explained by the exchange of many photons.

phototube see photomultiplier tube.

Physical Review Letters the leading American physics journal, specializing in short and topical papers of general physics interest.

pion the lightest of the mesons. Originally thought to be the mediator of the strong nuclear force.

Planck Scale   the "natural" energy, size and time scales for the unification of forces. Studying this regime far exceeds current technology.

Plum Pudding   an obsolete model of the atom in which small and hard electrons carrying negative electric charge exist inside a goopy and positively charged fluid.

PMT   see photomultiplier tube.

positron   the antimatter analog of an electron.

postdoc   short for post-doctoral associate. This is a person who has completed their Ph.D. degree and is temporarily working for a senior physicist. They use this time to continue learning skills and try to find a permanent position. This is often considered to be the most fun time in a physicist's life, as the pay is decent and the responsibilities of seniority are held at bay.

propagator   the exchanged particle in a Feynman diagram.

proton   an electrically charged particle found in the nucleus of an atom. Contains quarks within it.

QCD   Quantum Chromodynamics, the theory of the strong force.

QED   Quantum Electrodynamics, the theory of the electromagnetic force, including both relativity and the idea that the force can be described as an exchange of photons.

quantized   the idea that things can come in unit quantities. An example might be water, which appears to be continuous, yet actually exists in individual water molecules.

quantum mechanics   a physics theory, originally worked out in the 1920s, which governs the behavior of subatomic particles at very small sizes. Predicts very counterintuitive behavior.

quantum number   a bit of jargon pertaining to quantum mechanics. Since small particles have integer units of their properties (say charge or spin), one can characterize their (say) charge by a single number, indicating how many units of charge they carry. Many of the properties of subatomic particles can be given as quantum numbers.

quark   one of the known elementary particles, containing no known substructure. Quarks are generally found in the nucleus of an atom, although exotic quarks can be manufactured in large particle accelerators.

quark-gluon plasma   the idea that at sufficiently large energy, the quarks and gluons will no longer be confined within a hadron and intermix freely.

quintessence   one of several possible explanations for an energy field that pervades the universe. This could be responsible for the cosmological constant.

radio-frequency   the mechanism whereby one actually makes electric fields inside a particle accelerator. Essentially the same technology that powers radio transmitters generates the accelerating electric field.

relativity   a theory, originally postulated by Albert Einstein, which governs the behavior of objects traveling at speeds that are a significant fraction of the speed of light. This theory is very well established, but predicts counterintuitive behavior.

RHIC   the Relativistic Heavy Ion Collider, an accelerator based at BNL. All manner of hadrons are accelerated from protons up to gold nuclei. Two counter-rotating beams are made to collide to study the possibility of creating a quark-gluon plasma.

RIP   Research In Progress. I stole the phrase from Gordy Kane's book "The Particle Garden."

SAGE   the Soviet-American Gallium Experiment. Confirmed the solar neutrino problem.

sbottom   a hypothetical boson that is the supersymmetric analog of the bottom quark. Often called the bottom squark.

scharm   a hypothetical boson that is the supersymmetric analog of the charm quark. Often called the charm squark.

scintillator   a material that emits a quick pulse of light when traversed by a charged particle. Often plastic with a slightly purple tinge.

sdown   a hypothetical boson that is the supersymmetric analog of the down quark. Often called the down squark.

selectron   a hypothetical boson that is the supersymmetric analog of the electron.

shower   a large number of particles, created by the impact of a single particle. Showers can be initiated by electrons, photons and hadrons. In an electron shower, an electron passes near an atom and emits a photon. The electron repeats this behavior, emitting many photons. Each of the photons can produce electron/ positron pairs, each of which can also cause photon emission. Each daughter particle has lower energy than its parent. The shower turns a single high energy particle into thousands of low energy ones.

signal   the type of physics events for which you're looking.

silicon detector   a detector consisting of small strips of silicon, often 0.02 millimeters wide and considerably longer. Used near the collision point in order to disentangle the trajectories of the hundred or so particles that exit the collision. Larger detectors will have many particles hit the same piece of the detector and will only be registered as a single particle. This will continually exert pressure to make silicon detectors with ever-smaller elements.

SLAC   the Stanford Linear Accelerator Center, a large linear accelerator laboratory located in Palo Alto, California. The home of the BABAR detector.

SLC   the Stanford Linear Collider. A particle accelerator at SLAC, specializing in electron-positron collisions. A competitor to the LEP accelerator. The home of the SLD detector.

SLD   the Stanford Linear Detector, an experiment at SLAC that competed with the various LEP experiments.

SNO   the Sudbury Neutrino Observatory. A laboratory in Sudbury, Canada, which is able to make definitive measurements of solar neutrinos.

solar neutrino problem   the observation that the number of neutrinos detected from the Sun seems to be much less than expected.

Soudan 2   originally a proton decay experiment, this detector now conducts neutrino research. The detector is located in the Soudan mine in Minnesota.

SPEAR the Stanford Positron Electron Accelerator Ring, an accelerator at SLAC. Two notable discoveries were the J/ψ meson (i.e. the charm quark) and the τ lepton. This accelerator was turned over to synchrotron light generation in 1990.

special relativity a theory which describes the behavior of objects at very high speeds.

spin quantum mechanical angular momentum. While wrong in detail, each particle can be thought of as spinning. In fact, this is not true, rather spin is a property of the particle, like its charge or mass.

spokesman the leader of an experiment. Many experiments have two spokesmen who serve at the same time. Typically elected by the physicists on the experiment. There has been some push to use the term spokesperson, which seems to me just a tad too PC. An experiment's spokesman may be any gender, nationality or ethnicity.

Spp̄S an upgrade of the SPS, which allowed counter-rotating beams of protons and antiprotons to collide.

SPS the Super Proton Synchrotron, an accelerator at CERN, which could accelerates protons to several hundred GeV.

squarks the supersymmetric analog of quarks.

SSC the Superconducting SuperCollider, America's next generation accelerator, intended to compete with the LHC. Cancelled by Congress in the fall of 1993.

sstrange a hypothetical boson that is the supersymmetric analog of the strange quark. Often called the strange squark.

Standard Model all of the physical principles we currently understand to explain the subatomic particle realm. Of all known phenomena, only gravity is not included.

stop a hypothetical boson that is the supersymmetric analog of the top quark. Sometimes called the top squark.

strange a moderately heavy quark with a mass of about 0.3 times that of a proton, carrying an electrical charge of $-1/3$. From generation II.

strangeness a property of baryons and mesons which cause the particles to be created in great profusion (indicating that their creation is mediated by the strong force) but which decay very slowly (indicating that their decay is governed by the weak force). Eventually it was shown that this was caused by the creation of a new quark type called strange.

strong force the force holding the nucleus of the atom together. Carried by gluons. Felt between quarks and gluons. The charge that causes the strong force is whimsically called color, although it has no relationship to the common meaning of the word color.

subatomic anything that is smaller than the size of an atom.

sup a hypothetical boson that is the supersymmetric analog of the up quark. Often called the up squark.

Super Kamiokande the Super Kamioka Nucleon Decay Experiment, originally a proton decay experiment, this detector now conducts neutrino research. Super-K, as it is known, announced the first direct evidence for atmospheric neutrino oscillations in 1998.

superstrings a theory which supposes that at a very small size, what appear to be pointlike particles are, in fact, small oscillating strings. A theory containing this idea can successfully include gravity into the pantheon of known forces. There is no experimental evidence to support or refute this idea.

supersymmetry a purely theoretical idea whereby all equations governing particle behavior do not change if all fermions are swapped with bosons. This idea, while theoretically appealing, predicts many additional particles, none of which have been observed even after extensive searches. Finding supersymmetry, if indeed the world exhibits this property, is a high priority of both the Tevatron and the LHC.

symmetry an important concept in modern physics theories. This is the idea that a particular equation will not change if a parameter is changed. For instance, if one decides that east is a positive direction or west is a positive direction will not affect the measurement of how long it takes a ball to drop.

synchrotron   a type of accelerator in which the particles are guided in a circle using magnetic fields. The acceleration region is a small region where an electric field speeds up the particles.

synchrotron radiation   electromagnetic radiation from charged particles when they are accelerated. Accelerators made to create this type of radiation have been used to study the atomic structure of various materials and biological samples.

TASSO   the Two Armed Spectrometer Solenoid, an experiment at the DESY laboratory. Most notable achievement was the first observation of the gluon.

Technicolor   a competitor theory to the Higgs boson.

TeV   a Tera electron Volt, i.e. one trillion (10^{12}) electron Volts.

Tevatron   an accelerator at Fermilab that can accelerate protons and antiprotons. Currently the highest energy accelerator in the world and the fifth accelerator in the Fermilab accelerator chain. Its name comes from the fact that it is designed to accelerate particles to an energy of one tera electron volt (1 TeV). Notable discoveries have been the discovery of the top quark, the bottom quark and the tau neutrino. Currently operating.

theorists   physicists specializing in coming up with new models and the related calculations.

thermal equilibrium   the idea that in a system there are no concentrations of energy. Therefore, any energy flow in one direction is exactly balanced by a counter-flow of energy.

thesis advisor   each graduate student chooses a faculty member whose job it is to mentor the student.

TJNAF   the Thomas Jefferson Nuclear Accelerator Facility, a laboratory outside Newport News, Virginia. Primary purpose is to study relatively low-energy QCD.

top   the heaviest known quark with a mass of about 175 times that of a proton, carrying an electrical charge of +2/3. From generation III.

tracker a device for measuring the trajectory of charged particles. The simplest tracker can be thought of as a plane of parallel wires (like a harp). When a particle crosses near a wire, an electrical signal is generated on the wire and detected. By a series of planes of wires, the trajectory can be reconstructed by observing which wires were hit. Technologies other than wires can also provide the information as to where the particle crossed the plane. A common new technology uses silicon strips of very tiny size. In addition, planes of plastic fiber optics made of scintillating plastic can be used.

trigger the act of deciding which of the many collisions that occur should be recorded. Since there are literally millions of collisions that occur for each one that can be recorded, the detector must decide which collisions are interesting and should be recorded.

UA1 one of two large experiments located at the SPS accelerator in CERN. Discoverer of the electroweak bosons.

UA2 one of two large experiments located at the SPS accelerator in CERN.

unification the process whereby two seemingly-dissimilar forces are shown to be two aspects of a single underlying and more fundamental force.

unitarity the principle whereby if one adds up all of the possible probabilities for all possible interactions, they must sum to 100%. This means that a particle must do *something* (although not interacting at all is one of the possibilities).

unity a fancy way to say "one."

up the lightest of the quarks, carrying an electrical charge of $+2/3$. From generation I.

U-Particle a particle, proposed by Hideki Yukawa, which was to mediate the nuclear force between protons and neutrons. The modern name for this particle is the pion.

vertex any spot in a Feynman diagram at which a particle is emitted or absorbed.

virtual particles   particles that have temporarily violated the laws of conservation of energy and momentum. This is possible due to the vagaries of the Heisenberg Uncertainty Principle.

V-particles   particles discovered in the 1950s which turned out to be the creation of particles carrying the strange quark.

W boson   one of two massive bosons that mediate the weak force. The *W* boson is electrically charged and can change the identity of the particles with which it interacts. It is this particle that can decay the heavier particle generations into the lighter ones.

weak boson   massive particles that mediate the weak force. Two types exist, the electrically charged *W* bosons as well as the electrically neutral *Z* boson.

weak force   the weakest of the forces studied by particle physicists. Carried by the weak bosons, the *W* and the *Z* particles. Can change the identity of particles involved in the weak force. Responsible for the burning of the Sun and partially for volcanoes, which are both driven by radioactive decay.

wino   a hypothetical fermion that is the supersymmetric analog of the *W* boson. Also a person who has studied too much particle physics, preferring to spend their time wandering city streets, consuming inexpensive beverages from brown paper bags.

WMAP   the Wilkinson Microwave Anisotropy Probe, a renamed version of MAP. See MAP.

x-rays   part of the electromagnetic spectrum carrying considerable energy.

Yukon   a particle predicted by Hideki Yukawa that mediated the strong force within the nucleus of an atom. The modern name for this particle is the pion.

Z boson   one of two massive bosons that mediate the weak force. The *Z* boson is electrically neutral and acts much like a massive photon.

zino   a hypothetical fermion that is the supersymmetric analog of the *Z* boson.

Index

1/05